*Understanding Delta-Sigma Data Converters*

IEEE Press
445 Hoes Lane
Piscataway, NJ 08854

**IEEE Press Editorial Board**
Stamatios V. Kartalopoulos, *Editor in Chief*

| M. Akay | M. E. El-Hawary | F. M. B. Periera |
| J. B. Anderson | R. Leonardi | C. Singh |
| R. J. Baker | M. Montrose | S. Tewksbury |
| J. E. Brewer | M. S. Newman | G. Zobrist |

Kenneth Moore, *Director of IEEE Book and Information Services (BIS)*
Catherine Faduska, *Senior Acquisitions Editor*
Anthony VenGraitis, *Project Editor*

IEEE Solid-State Circuits Society, *Sponsor*
SSC-S Liaison to IEEE Press, Stuart K. Tewksbury

# Understanding Delta-Sigma Data Converters

**Richard Schreier**
Analog Devices, Inc.

**Gabor C. Temes**
Oregon State University

IEEE Press

A JOHN WILEY & SONS, INC., PUBLICATION

Copyright © 2005 by the Institute of Electrical and Electronics Engineers, Inc. All rights reserved

Published by John Wiley & Sons, Inc., Hoboken, New Jersey.
Published simultaneously in Canada.

No part of this publication may be reproduced, stored in a retrieval system or transmitted in any form or by any means, electronic, mechanical, photocopying, recording, scanning or otherwise, except as permitted under Section 107 or 108 of the 1976 United States Copyright Act, without either the prior written permission of the Publisher, or authorization through payment of the appropriate per-copy fee to the Copyright Clearance Center, Inc., 222 Rosewood Drive, Danvers, MA 01923, (978) 750-8400, fax (978) 646-8600, or on the web at www.copyright.com. Requests to the Publisher for permission should be addressed to the Permissions Department, John Wiley & Sons, Inc., 111 River Street, Hoboken, NJ 07030, (201) 748-6011, fax (201) 748-6008.

Limit of Liability/Disclaimer of Warranty: While the publisher and author have used their best efforts in preparing this book, they make no representation or warranties with respect to the accuracy or completeness of the contents of this book and specifically disclaim any implied warranties of merchantability or fitness for a particular purpose. No warranty may be created or extended by sales representatives or written sales materials. The advice and strategies contained herein may not be suitable for your situation. You should consult with a professional where appropriate. Neither the publisher nor author shall be liable for any loss of profit or any other commercial damages, including but not limited to special, incidental, consequential, or other damages.

For general information on our other products and services please contact our Customer Care Department within the U.S. at 877-762-2974, outside the U.S at 317-572-3993 or fax 317-572-4002.

Wiley also publishes its books in a variety of electronic formats. Some content that appears in print, however, may not be available in electronic format.

*Library of Congress Cataloging-in-Publication Data is available.*

ISBN 0-471-46585-2

Printed in the United States of America.

10 9 8 7 6 5 4 3

# Contents

*Foreword* **xi**

References **xii**

**CHAPTER 1**    *Introduction* **1**

1.1 The Need for Oversampling Converters **1**
1.2 Delta and Delta-Sigma Modulation **4**
1.3 Higher-Order Single-Stage Noise-Shaping Modulators **9**
1.4 Multi-Stage (Cascade, MASH) Modulators **10**
1.5 Bandpass $\Delta\Sigma$ Modulators **13**
1.6 $\Delta\Sigma$ Modulators with Multi-Bit Quantizers **15**
1.7 Delta-Sigma Digital-to-Analog Converters **16**
1.8 History; Performance and Architecture Trends **17**

**CHAPTER 2**    *The First-Order Delta Sigma Modulator* **21**

2.1 Quantizers and Quantization Noise **21**
    *2.1.1 Binary Quantization* **28**

2.2 MOD1 as an ADC  **29**
2.3 MOD1 as a DAC  **34**
2.4 MOD1 Linear Model  **36**
2.5 Simulation of MOD1  **38**
2.6 MOD1 under DC Excitation  **41**
    *2.6.1 Idle Tone Generation  42*
    *2.6 2 Graphical Visualization  45*
2.7 Stability of MOD1  **49**
2.8 The Effects of Finite Op-Amp Gain  **50**
    *2.8.1 Linear Systems Perspective– Degraded Noise Shaping  50*
    *2.8.2 Nonlinear Systems Perspective– Dead Zones  51*
2.9 Decimation Filters for MOD1  **54**
    *2.9.1 The Sinc Filter [9]  55*
    *2.9.2 The $Sinc^2$ Filter  58*
2.10 Conclusions  **60**

**CHAPTER 3**  *The Second-Order Delta-Sigma Modulator*  **63**

3.1 The Second-Order Modulator: MOD2  **63**
3.2 Simulation of MOD2  **67**
3.3 Nonlinear Effects in MOD2  **71**
    *3.3 1 Signal-dependent quantizer gain  71*
    *3.3.2 Stability of MOD2  74*
    *3.3.3 Dead-band behavior  77*
3.4 Alternative Second-Order Modulator Structures  **79**
    *3.4.1 The Boser-Wooley Modulator  79*
    *3.4.2 The Silva-Steensgaard Structure  80*
    *3.4.3 The Error-Feedback Structure  81*
    *3.4.4 Generalized Second-Order Structures  82*
    *3.4.5 Optimal Second-Order Modulator  84*
3.5 Decimation Filtering for Second-Order $\Delta\Sigma$ Modulators  **86**
3.6 Conclusions  **89**

**CHAPTER 4**  *Higher-Order Delta-Sigma Modulation*  **91**

4.1 High-Order Single-Quantizer Modulators  **91**
4.2 Stability Considerations in High-Order Modulators  **97**

*4.2.1 Single-Bit Modulators 98*
*4.2.2 Multi-Bit Modulators [12] 104*
4.3 Optimization of the NTF Zeros and Poles 107
*4.3.1 NTF Zero Optimization 107*
*4.3.2 NTF Pole Optimization 111*
4.4 Loop Filter Architectures 115
*4.4.1 Loop Filters with Distributed Feedback and Input Coupling– The CIFB and CRFB Structures 115*
*4.4.2 Loop Filters with Distributed Feedforward and Input Coupling– The CIFF and CRFF Structures 121*
4.5 Multi-Stage Modulators 122
*4.5.1 The Leslie-Singh (L-0 Cascade) Structure [16] 123*
*4.5.2 Cascade (MASH) Modulators 127*
*4.5.3 Noise Leakage in Cascade Modulators 132*
4.6 Conclusions 136

**CHAPTER 5**  *Bandpass and Quadrature Delta-Sigma Modulation 139*

5.1 The Need for Bandpass and Quadrature Modulation 139
5.2 Bandpass NTF Selection 145
*5.2.1 Pseudo N-path transformation 149*
5.3 Architectures for Bandpass Delta-Sigma Modulators 151
*5.3.1 Topology Choices 151*
*5.3.2 Resonator Implementations 154*
5.4 Bandpass Modulator Example 161
5.5 Quadrature Signals 166
5.6 Quadrature Modulation 172
5.7 Conclusions 176

**CHAPTER 6**  *Implementation Considerations For $\Delta\Sigma$ ADCs 179*

6.1 Modulators with Multi-Bit Internal Quantizers 179
6.2 Dual-Quantizer Modulators 182
*6.2.1 Dual-Quantization MASH Structure 182*
*6.2.2 Dual-Quantization Single-Stage Structure 183*
6.3 Dynamic Element Randomization 184

6.4 Mismatch Error Shaping  **186**
    *6.4.1 Element Rotation or Data-Weighted Averaging*   *189*
    *6.4.2 Individual Level Averaging*   *191*
    *6.4.3 Vector-Based Mismatch Shaping*   *192*
    *6.4.4 Element Selection Using a Tree Structure*   *196*
6.5 Digital Correction of DAC Nonlinearity  **199**
    *6.5.1 Digitally-Corrected Multi-Bit ΔΣ Modulator with Power-Up Calibration*   *200*
    *6.5.2 Digitally-Corrected Multi-Bit ΔΣ ADC with Background Calibration*   *202*
6.6 Continuous-Time Implementations  **205**
    *6.6.1 A Continuous-Time Implementation of MOD2*   *207*
    *6.6.2 Inherent Anti-Aliasing in CT ΔΣ ADCs*   *212*
    *6.6.3 Design Issues for Continuous-Time Modulators*   *213*
6.7 Conclusions  **216**

## CHAPTER 7     *Delta-Sigma DACs*   *219*

7.1 System Architectures for ΔΣ DACs  **220**
7.2 Loop configurations for ΔΣ DACs  **222**
    *7.2.1 Single-Stage Delta-Sigma Loops*   *223*
    *7.2.2 The Error Feedback Structure*   *224*
    *7.2.3 Cascade (MASH) Structures*   *226*
7.3 ΔΣ DACs Using Multi-Bit Internal DACs  **229**
    *7.3.1 Dual-Truncation DAC Structures*   *230*
    *7.3.2 Multi-bit Delta-Sigma DACs with Mismatch Error Shaping*   *232*
    *7.3.3 Digital Correction of Multi-Bit Delta-Sigma DACs*   *236*
    *7.3.4 Comparison of Single-Bit and Multi-Bit ΔΣ DACs*   *238*
7.4 Interpolation Filtering for ΔΣ DACs  **239**
7.5 Analog Post-Filters for ΔΣ DACs  **243**
    *7.5.1 Analog Post-Filtering in Single-Bit ΔΣ DACs*   *244*
    *7.5.2 Analog Post-Filtering in Multi-Bit ΔΣ DACs*   *251*
7.6 Conclusions  **253**

## CHAPTER 8     *High-Level Design and Simulation*   *257*

8.1 NTF Synthesis  **257**
    *8.1.1 How* `synthesizeNTF` *works*   *260*

*8.1.2 Limitations of* `synthesizeNTF` *262*

8.2 NTF Simulation, SQNR Calculation and Spectral Estimation  263

8.3 NTF Realization and Dynamic Range Scaling  266

*8.3.1 The ABCD Matrix  271*

8.4 Creating a SPICE-Simulatable Schematic  273

*8.4.1 Voltage Scaling  273*

*8.4.2 Timing  274*

*8.4.3 kT/C Noise  280*

8.5 Conclusions  281

**CHAPTER 9**  *Example Modulator Systems*  283

9.1 SCMOD2: General-Purpose Second-Order Switched-Capacitor ADC  283

*9.1.1 System Design  284*

*9.1.2 Timing  286*

*9.1.3 Scaling  288*

*9.1.4 Verification  289*

*9.1.5 Capacitor Sizing  292*

*9.1.6 Circuit Design  294*

9.2 SCMOD5: A Fifth-Order Single-Bit Noise-Shaping Loop  298

*9.2.1 NTF and Architecture Selection  298*

*9.2.2 Implementation  302*

*9.2.3 Instability and Reset  311*

9.3 A Wideband 2-0 Cascade System  311

*9.3.1 Architecture  312*

*9.3.2 Implementation  315*

9.4 A Micropower Continuous-Time ADC  317

*9.4.1 High-Level Design  318*

*9.4.2 Circuit Design  322*

9.5 A Continuous-Time Bandpass ADC  326

*9.5.1 Architecture/Analysis  328*

*9.5.2 Subcircuits  333*

9.6 Audio DAC  337

*9.6.1 Modulator Design  338*

*9.6.2 Interpolation Filter Design  344*

*9.6.3 DAC and Reconstruction Filter Design  355*

9.7 Conclusions  357

*9.7.1 The ADC State-of-the-Art  357*

9.7.2 *FOM Justification*    *359*
9.7.2 References    362

**APPENDIX A**    *Spectral Estimation*    *365*

A.1 Windowing    366
A.2 Scaling and Noise Bandwidth    373
A.3 Averaging    377
A.4 An Example    379
A.5 Mathematical Background    383

**APPENDIX B**    *The Delta-Sigma Toolbox*    *389*

Demonstrations and Examples    390
Summary of Key Functions    391
    `synthesizeNTF`    *393*
    `predictSNR`    *395*
    `simulateDSM`    *396*
    `simulateSNR`    *398*
    `realizeNTF`    *400*
    `stuffABCD, mapABCD`    *401*
    `scaleABCD`    *402*
    `calculateTF`    *403*
    `simulateESL`    *404*
    `designHBF`    *405*
    `simulateHBF`    *408*
    `findPIS`    *409*
Modulator Model Details    410

**APPENDIX C**    *Noise in Switched-Capacitor Delta-Sigma Data Converters*    *417*

C.1 Noise Effects in CMOS Op Amps    419
C.2 Sampled Thermal Noise    423
C.3 Noise Effects in an SC Integrator    425
C.4 Integrator Noise Analysis Example    433
C.5 Noise Effects in Delta-Sigma ADC Loops    435

# Foreword

In the next few remarks, we shall explain our reasons for writing this book. Since its inception in the early 1960s, delta-sigma modulation has evolved through a number of generations and now stands as one of the more popular methods for constructing high-performance analog-to-digital and digital-to-analog converters. The concept of noise-shaping, which is central to delta-sigma modulation, continues to be refined and to be applied to new areas. The technical literature contains a wealth of information regarding various aspects of delta-sigma modulation, but the sheer volume of this material acts as a barrier to the designer who is embarking on his first trip into the land of delta-sigma.

Earlier works to which we contributed served the need for a guide by cataloging publications and republishing a subset of them [1], or by enlisting the help of a collection of authors [2]. This book, in contrast, provides a coherent and consistent entry-level reference into the world of delta-sigma. The introductory chapter is, in essence, a generalist's guide. That chapter is devoted to introducing main roads and to identifying major landmarks, and so should chart some territory which is unfamiliar to the reader. Do not despair if you feel a little lost, since if you understand that chapter completely you do not deserve to be called a beginner! The remainder of the text plots a more deliberate course through the countryside, one which allows time for exploring some of the side roads and for pondering a number of the more remarkable edifices which dot the landscape.

Unfortunately, the text is by no means all-inclusive. In order to have a non-zero probability of completing this work in a reasonable amount of time, we have had to make difficult decisions regarding what material would be appropriate for an introductory text and what material could be omitted. We hope that the reader will find the route we have chosen makes a pleasant and satisfying journey, and we trust that the many researchers and designers who cut roads through the wilderness will find their hard work adequately acknowledged. As always, the blame for any omissions or misdirections lies with us. Let us know if you encounter any.

Bon voyage!

## References

[1] *Oversampling Delta-Sigma Data Converters*, J. C. Candy and G. C. Temes, Eds. New York: *IEEE Press*, 1991.

[2] *Delta-Sigma Data Converters*, S. R. Norsworthy, R. Schreier and G. C. Temes, Eds., Piscataway NJ: IEEE Press, 1997.

# CHAPTER 1  *Introduction*

In this introductory chapter, the need for oversampling data converters will be discussed, and their performance contrasted with that of Nyquist-rate converters. Delta modulation and delta-sigma modulation will be described and compared. The basic architectures for delta-sigma ($\Delta\Sigma$) modulators will be presented, and newer trends identified.

## 1.1 The Need for Oversampling Converters

Computational and signal processing tasks are now performed predominantly by digital means, since digital circuits are robust and can be realized by extremely small and simple structures which can in turn be combined to obtain very complex, accurate and fast systems. Every year, the speed and density of digital integrated circuits (ICs) is increased, enhancing the dominance of digital methods in almost all areas of communications and consumer products. Since the physical world nevertheless remains stubbornly analog, data converters are needed to interface with the digital signal processing (DSP) core. As the speed and capability of DSP cores increases, so too must the speed and accuracy of the converters associated with them.

# 1 Introduction

Fig. 1.1 illustrates the block diagram of a signal processing system with analog input and output signals, plus a central digital engine. As shown, the analog input signal (usually after some amplification and filtering) enters an analog-to-digital converter (ADC) which transforms it into a digital data stream. This stream is processed by the DSP core, and the resulting digital output signal is reconverted into analog form by a digital-to-analog converter (DAC). The DAC output is usually also filtered and amplified to obtain the final analog output signal.

Data converters (both ADCs and DACs) can be classified into two main categories: *Nyquist-rate* and *oversampled* converters. In the former category, there exists a one-to-one correspondence between the input and output samples. Each input sample is separately processed, regardless of the earlier input samples; the converter has no memory. Thus, applying a digital input word containing bits $b_1, b_2, \ldots b_N$ to a Nyquist-rate DAC ideally results in an analog output ↑ LSB

$V_{LSB} = \dfrac{V_{REF}}{2^N}$

$$V_{out} = V_{ref}(b_1 2^{-1} + b_2 2^{-2} + \ldots + b_N 2^{-N}), \tag{1.1}$$

(where $V_{ref}$ is the reference voltage) regardless of any previous input word. The accuracy of conversion can be evaluated by comparing the actual value of $V_{out}$ with the ideal value given by (1.1).

≠ Nyquist frequency

As the name implies, the sampling rate $f_s$ of Nyquist-rate converters can be as low as Nyquist's criterion requires, i.e., twice the bandwidth $f_B$ of the input signal. (For practical reasons, the actual rate is usually somewhat higher than this minimum value.)

$f_s/2 = f_B$

In most cases, the linearity and accuracy of Nyquist-rate converters is determined by the matching accuracy of the analog components (resistors, current sources or

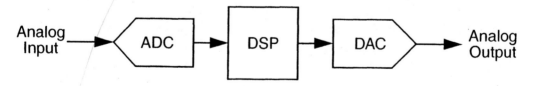

Figure 1.1: A modern signal processing system contains a DSP core bracketed by data converters.

DNL — each step size the same

INL — each transition is close to ideal

capacitors) used in the implementation. For example, in the $N$-bit resistor-string DAC shown in Fig. 1.2, the resistors must have a relative matching error less than $2^{-N}$ to guarantee an integral nonlinearity INL less than 0.5 LSB. Similar matching requirements prevail for ADCs and DACs constructed from current sources or switched-capacitor (SC) branches. Practical conditions restrict the matching accuracy to about 0.02%, and hence the *effective number of bits* (ENOB) to about 12, for such converters.

In many applications (such as digital audio), higher resolution and linearity is required, perhaps as much as 18 or even 20 bits. The only Nyquist-rate converters capable of such accuracy are the integrating or counting ones. These, however, require at least $2^N$ clock periods to convert a single sample, and hence are too slow for most signal-processing applications.

Oversampling data converters are able to achieve over 20 ENOB resolution at reasonably high conversion speeds by relying on a trade-off. They use sampling rates much higher than the Nyquist rate, typically higher by a factor between 8 and 512,

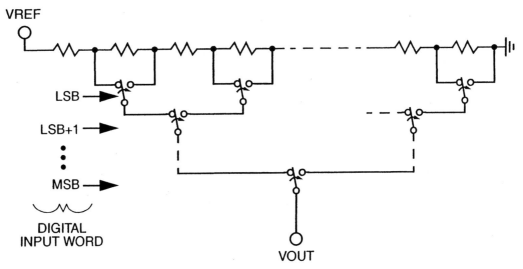

**Figure 1.2: A resistor-string DAC. LSB denotes the *least significant bit* and MSB the *most significant bit* of the digital input.**

and generate each output utilizing all preceding input values. Thus, the converter incorporates memory elements in its structure. This property destroys the one-to-one relation between input and output samples. Now only a comparison of the complete input and output waveforms can be used to evaluate the converter's accuracy, either in the time or in the frequency domain.

A common measure of a converter's accuracy is the signal-to-noise ratio (SNR) for a sine-wave input. The relationship between ENOB and SNR for an ideal Nyquist converter with sine-wave excitation is $SNR = 6.02 ENOB + 1.76$. The inverse relationship is often applied to oversampling converters to convert an SNR into an effective number of bits.

As will be shown in later chapters, the implementation of oversampling converters requires a considerable amount of digital circuitry, in addition to some analog stages. Both need to be operated faster than the Nyquist rate. However, the accuracy requirements on the analog components are relaxed compared to those associated with Nyquist-rate converters. The cost paid for high accuracy thus includes faster operation and added digital circuitry; both of these are getting cheaper as digital IC technology advances. Hence, oversampling converters are gradually taking over in many applications previously dominated by Nyquist-rate ones.

## 1.2 Delta and Delta-Sigma Modulation

Next, oversampling analog-to-digital converters processing *baseband signals* (i.e. signals with spectra centered around dc) will be discussed. Such data converters contain several stages. Analog and digital filter stages may be used before and after the stage (called the *modulator,* or *converter loop*) which performs the actual analog-to-digital conversion. The two main types of oversampling modulators are the *delta* modulator and the *delta-sigma* modulator. Fig. 1.3a shows a basic delta modulator used as an ADC. It is a feedback loop, containing an internal low-resolution ADC and DAC, as well as a loop filter (here, an integrator). It is a nonlinear system (due to the quantizing effect of the ADC) as well as a dynamic one (due to the memory in the integrator), and hence its analysis is a difficult mathematical task. Simple qualitative understanding of its operation can, however, be gained by using a linearized model of the internal ADC which consists of a unity-gain buffer and an additive quantization noise $e$. Assuming perfect operation of the DAC as well as a reference voltage $V_{ref} = 1$ V and a sampling rate $f_s = 1$ Hz, the discrete-time linear system of Fig. 1.3b results. Analyzing this, it can easily be shown that the (digital) output signal at time $n$ (i.e. $t = n/f_s$) is

$$v(n) = u(n) - u(n-1) + e(n) - e(n-1). \tag{1.2}$$

The name *delta modulator* is derived from the fact that the output is based on the difference (delta) between a sample of the input and a predicted value of that sample. In the general case, the loop filter may be a higher-order circuit, which generates a more accurate prediction of the input sample $u(n)$ than $u(n-1)$, to subtract from the actual $u(n)$. This type of modulator is sometimes called a *predictive encoder*.

The advantage of this structure is that for oversampled signals the difference $(u(n) - u(n-1))$ is much smaller than $u(n)$ itself, on average, and hence larger input signals can be allowed. There are, however, several disadvantages. The loop filter (integrator for the first-order loop shown) is in the feedback path, and hence its nonidealities limit the achievable linearity and accuracy. Also, in the demodulator, a DAC and a demodulation filter (for first-order modulators, an integrator) are

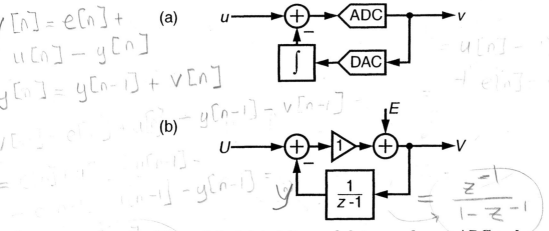

Figure 1.3: (a) A delta modulator used as an ADC and (b) its linear z-domain model.

needed. The filter has a high gain in the signal band, and hence will amplify the nonlinear distortion of the DAC as well as any noise picked up by the signal between the modulator and demodulator.

An alternative oversampling structure which avoids the shortcomings of the predictive modulator is shown in Fig. 1.4a. It is again a feedback loop, containing a loop filter as well as an internal low-resolution ADC and DAC, but the loop filter is now in the forward path of the loop. Replacing as before the quantizer (ADC) by its linear model, the linear sampled-data system of Fig. 1.4b results. Analysis gives

$$v(n) = u(n-1) + e(n) - e(n-1). \qquad (1.3)$$

Thus, the digital output contains a delayed, but otherwise unchanged replica of the analog input signal $u$, and a differentiated version of the quantization error $e$. Since the signal is not changed by the modulation process, the demodulation operation does not need an integrator as was the case for the delta modulator. Hence, the amplification of in-band noise and distortion at the receiver does not take place. Furthermore, the differentiation of the error $e$ suppresses it at frequencies which are small compared to the sampling rate $f_s$. In general, if the loop filter has a high gain in the signal band, the in-band quantization "noise" is strongly attenuated, a process now commonly called *noise shaping*.

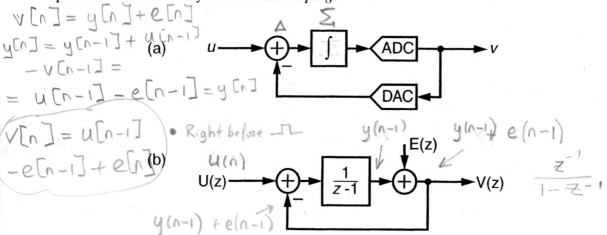

Figure 1.4: (a) A delta-sigma modulator used as an ADC and (b) its linear z-domain model.

Any nonlinearity of the ADC is simply combined with the quantization error $e$, and is thus suppressed in-band along with $e$. Nonlinear distortion in the DAC, however, affects the output signal without any shaping, and hence it represents a major limitation on the attainable performance. This effect can be handled in various ways. The simplest, and historically earliest, method is to use single-bit quantization. In this case, the input/output characteristic of the DAC consists of only two points, and hence the DAC's operation is inherently linear.[†] For multi-bit (typically, 2-5 bit) quantization, digital correction or dynamic matching techniques may be used. These will be discussed in Chapter 5.

It can be shown that the system of Fig. 1.4 can be obtained from that of Fig. 1.3 by cascading an integrator or summing block with the delta modulator. Hence, the structure of Fig. 1.4 came to be called a *sigma-delta* ($\Sigma\Delta$) *modulator*. Alternatively, one can observe the differencing at the input, followed by the summation in the loop filter, and hence call the structure a *delta-sigma* ($\Delta\Sigma$) *modulator*. Both terms have been used in the past to denote the first-order system of Fig. 1.4 with a single-bit quantizer. Other systems with higher-order loop filters, multi-bit quantizers, etc. are most properly called *noise-shaping modulators*, but it is common to extend the term $\Delta\Sigma$ modulator (or $\Sigma\Delta$ modulator) to these systems as well. This text follows the accepted usage. Examples of these general $\Delta\Sigma$ modulator systems will be briefly discussed in the next section.

The output noise due to the quantization error in the $\Delta\Sigma$ modulator is $q(n) = e(n) - e(n-1)$, as (1.3) shows. In the z-domain, this becomes $Q(z) = (1 - z^{-1})E(z)$, and in the frequency domain, after $z$ is replaced by $e^{j2\pi fT}$, the *power spectral density* (PSD) of the output noise is found to be

$$S_q(f) = (2\sin(\pi fT))^2 S_e(f). \quad (1.4)$$

[†]. More precisely, the DAC operation is *affine*, rather than linear. Since the input-output behavior of a memoryless binary DAC can be represented exactly by $w = kv + c$, where $k$ is the DAC gain and $c$ is the DAC offset, a binary DAC is linear (in the strict sense of the term) only if $c = 0$. However, since the dc offset of a converter is often immaterial, the distinction between an affine DAC and a truly linear DAC is usually unimportant.

Here, $T = 1/f_s$ is the sampling period, and $S_e(f)$ is the 1-sided PSD of the quantization error (noise) of the internal ADC. For "busy" (i.e., rapidly and randomly varying) input signals, one may approximate $e$ with white noise of mean-square value $e_{rms}^2 = \Delta^2/12$ where $\Delta$ is the step size of the quantizer, and thus

$$S_e(f) = \frac{\Delta^2}{6 f_s} \tag{1.5}$$

The filtering function $1 - z^{-1}$ is called the *noise transfer function* (NTF). The squared magnitude of the NTF as a function of frequency is illustrated in Fig. 1.5.

As Fig. 1.5 illustrates, the NTF of the $\Delta\Sigma$ modulator is a highpass filter function. It suppresses $e$ at frequencies around 0, but the NTF also enhances $e$ at higher frequencies around $f_s/2$.

We introduce next the *oversampling ratio*

$$OSR = \frac{f_s}{2 f_B} \tag{1.6}$$

where $f_B$ is the maximum signal frequency, i.e. the signal bandwidth. $OSR$ defines how much faster we sample in the oversampled modulator than in a Nyquist-rate converter.

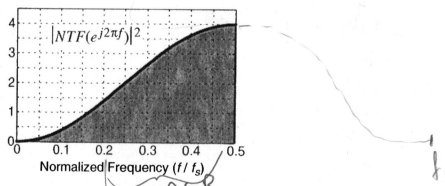

Figure 1.5: Noise-shaping function for the $\Delta\Sigma$ modulator shown in Fig. 1.4.

Integrating $S_q(f)$ between 0 and $f_B$ gives the in-band noise power. By (1.4)-(1.6), and assuming $OSR \gg 1$, to a good approximation

$$q_{rms}^2 = \frac{\pi^2 e_{rms}^2}{3(OSR)^3}. \tag{1.7}$$

As expected, the in-band noise decreases with increasing $OSR$. However, this decrease is relatively slow; doubling the $OSR$ reduces the noise only by 9 dB, and hence enhances the ENOB by only about 1.5 bits. Even for $OSR = 256$, ENOB < 13 bits results, assuming single-bit quantization is used.

## 1.3 Higher-Order Single-Stage Noise-Shaping Modulators

An obvious way to increase the resolution (i.e., the ENOB) of the $\Delta\Sigma$ modulator is to use a higher-order loop filter. By adding another integrator and feedback path to the circuit of Fig. 1.4, the structure of Fig. 1.6 results. Linearized analysis gives

$$V(z) = z^{-1}U(z) + (1-z^{-1})^2 E(z). \tag{1.8}$$

This indicates that the NTF is now $(1-z^{-1})^2$ in the z-domain, which applies a shaping function of $(2\sin(\pi fT))^4$ to the PSD of $e$. It follows that the in-band noise power is (to a good approximation for $OSR \gg 1$)

$$q_{rms}^2 = \frac{\pi^4 e_{rms}^2}{5(OSR)^5}. \tag{1.9}$$

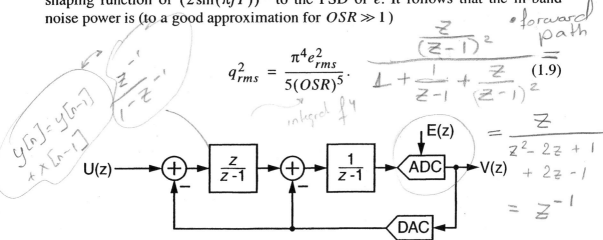

Figure 1.6: A second-order delta-sigma modulator.

Hence, doubling *OSR* results in about 2.5 bits of additional resolution. This is a much more favorable trade-off than that of the first-order modulator. For example, if we assume that single-bit quantization with $\Delta = 2$ results in $e_{rms}^2 = 1/3$, (1.9) indicates that ENOB is about 19 bits for $OSR = 256$, whereas a first-order modulator only achieves an ENOB of about 13 bits under the same assumptions. (Chapter 3 will show that this simple comparison is slightly flawed because a second-order single-bit modulator exhibits *quantizer overload* and thus has $e_{rms}^2 > 1/3$. A more accurate value is 17 ENOB for $OSR = 256$. Despite this 2-bit discrepancy, the ENOB of a second-order modulator does increase by 2.5 bits for each doubling of *OSR*.)

In principle, by adding more integrators and feedback branches to the loop, even higher-order NTFs can be obtained. For an $L^{th}$-order loop filter resulting in $NTF(z) = (1 - z^{-1})^L$, the in-band noise power is approximately

$$q_{rms}^2 = \frac{\pi^{2L} e_{rms}^2}{(2L + 1)(OSR)^{2L+1}} \quad (1.10)$$

and the number of bits added to the resolution by doubling the *OSR* is given by $L + 0.5$. The in-band noise as a function of *OSR* based on (1.10) is plotted in Fig. 1.7. Here 0 dB corresponds to a quantization noise power of $e_{rms}^2$.

For high-order loops, stability considerations, which have thus far been ignored, reduce the achievable resolution to a lower value than that given by the above equations and Fig. 1.7. For high-order, single-bit modulators the difference is substantial, amounting to more than 60 dB for a $5^{th}$-order modulator. This topic will be discussed in detail in Chapter 4.

## 1.4 Multi-Stage (Cascade, MASH) Modulators

An increasingly popular structure, which eases the stability problems associated with high-order modulators, is the *cascade* modulator, also called the *multi-stage* or *MASH* (for Multi-stAge noise-SHaping) modulator. The basic concept is illustrated in Fig. 1.8. The output signal of the first stage is given by

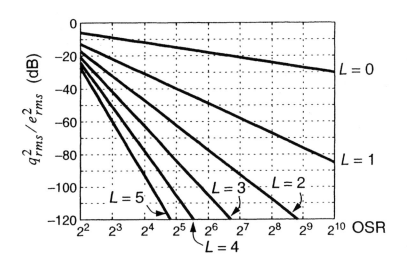

**Figure 1.7:** Theoretical in-band noise power for $L^{\text{th}}$-order $\Delta\Sigma$ modulators.

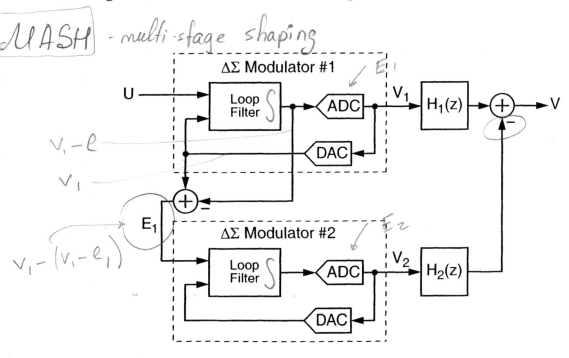

**Figure 1.8:** A multi-stage delta-sigma modulator.

# 1 Introduction

$$V_1(z) = STF_1(z)U(z) + NTF_1(z)E_1(z), \qquad (1.11)$$

where $STF_1$ and $NTF_1$ are the signal and noise transfer functions, respectively, of the first stage. The second stage is added to improve the SNR beyond what $NTF_1$ can provide.

As shown in Fig. 1.8, the quantization error $E_1$ of the input stage is found in analog form by subtracting the input to its internal quantizer from its output. $E_1$ is then fed to another $\Delta\Sigma$ loop forming the second stage of the modulator, and converted into digital form. Hence, the output signal of the second stage in the z-domain is given by

$$V_2(z) = STF_2(z)E_1(z) + NTF_2(z)E_2(z), \qquad (1.12)$$

where $STF_2$ and $NTF_2$ are the signal and noise transfer functions, respectively, of the second stage. The digital filter stages $H_1$ and $H_2$ at the outputs of the two modulator loops are designed such that in the overall output $V(z)$ of the system the first-stage error $E_1(z)$ is cancelled. By (1.11) and (1.12), this is achieved if the condition

$$H_1 NTF_1 - H_2 STF_2 = 0 \qquad (1.13)$$

holds. The simplest (and usually most practical) choice for $H_1$ and $H_2$ which satisfies (1.13) is $H_1 = k \cdot STF_2$ and $H_2 = k \cdot NTF_1$, where $k$ is constant chosen to give unity signal gain. Since $STF_2$ is often just a delay, $H_1$ is easily realized.

The overall output is then given by

$$V = H_1 V_1 - H_2 V_2 = k \cdot STF_1 STF_2 U - k \cdot NTF_1 NTF_2 E_2. \qquad (1.14)$$

In a typical case, both stages of the MASH modulator may contain a second-order loop, and their transfer functions may be given by

$$STF_1 = z^{-1} \qquad (1.15)$$

$$STF_2 = 0.5z^{-1} \tag{1.16}$$

and

$$NTF_1 = NTF_2 = (1-z^{-1})^2. \tag{1.17}$$

Choosing $k = 2$, the output is then

$$V = z^{-2}U + 2(1-z^{-1})^4 E_2. \tag{1.18}$$

Thus, the noise-shaping performance is essentially that of a fourth-order single-loop converter, but the stability behavior is that of a second-order one.

If the condition (1.13) is not exactly satisfied, for example due to imperfections in the realization of the analog transfer functions, then $E_1$ will appear at the output multiplied by $k[STF_2 NTF_{1a} - NTF_1 STF_{2a}]$, where subscript "a" denotes the actual value of the analog transfer function. As will be shown in Chapter 4, this may result in a serious deterioration of the noise performance of the converter.

## 1.5 Bandpass ΔΣ Modulators

Up to now, it was assumed that the signal energy was concentrated in a narrow band at low frequencies, centered at dc. In applications such as RF communication systems, the signal is concentrated in a narrow band of width $f_B$ around a center frequency $f_0$, where $f_B$ is much smaller than $f_s$, while $f_0$ is not. In such cases, ΔΣ modulation may still be effective, but now the noise transfer function NTF must have a bandstop, rather than highpass, character, with zeros located at or around $f_0$. Fig. 1.9 compares the conceptual output spectra of a lowpass and a bandpass ΔΣ modulator.

A simple way to obtain the NTF of a bandpass ΔΣ modulator is to find first an appropriate lowpass NTF, and then perform a z-domain mapping on it. For example, the transformation $z \to -z^2$ maps the frequency range around dc (i.e., $z = 1$) to the ranges around $\pm f_s/4$ ($z = \pm j$). Hence, the resulting NTF will have small values near $f_0 = f_s/4$, and will suppress the quantization noise there. This bandstop noise-shaping makes it possible to achieve a high signal-to-noise ratio (SNR) for signals whose energy is restricted to frequencies near $f_s/4$.

Note that the $z \rightarrow -z^2$ mapping doubles the order of the lowpass NTF and transforms the zeros of the NTF from near $z = 1$ to the vicinity of the $z = \pm j$ points, as illustrated in Fig. 1.10.

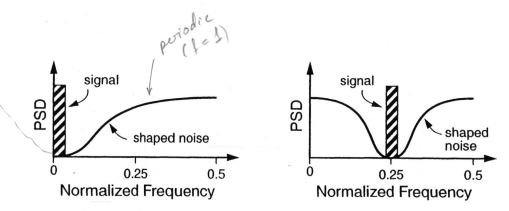

Figure 1.9: Conceptual output spectra for lowpass and $f_s/4$ bandpass modulators.

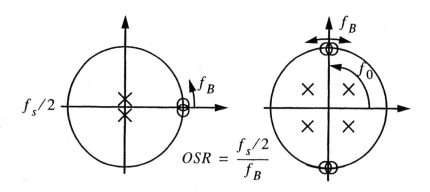

Figure 1.10: Pole-zero locations of a lowpass NTF and a bandpass NTF.

Other techniques for finding the NTF of bandpass $\Delta\Sigma$ modulators will be discussed in detail in Chapter 5 along with circuit design techniques for bandpass $\Delta\Sigma$ modulators.

## 1.6 $\Delta\Sigma$ Modulators with Multi-Bit Quantizers

As explained earlier, nonlinearities in the feedback DAC of a $\Delta\Sigma$ ADC result in comparable nonlinearities for the overall conversion. This occurs because the in-band part of the DAC output signal is forced by the feedback loop to follow the input signal $u$ very accurately. Hence, if the DAC is nonlinear, its input must be distorted to give an accurate output. But the DAC input is the overall output of the loop, which is thus also distorted.

It was this fact which forced early designers of $\Delta\Sigma$ modulators to use single-bit internal ADCs and DACs in the $\Delta\Sigma$ loops. However, single-bit ADCs (which are essentially comparators) have an ill-defined gain factor, as will be shown in Chapter 2. Also as Chapters 3 and 4 will show, loops containing one-bit quantizers must remain stable over a wide range of loop gains. This consideration results in a reduction of the allowable input signal swing, and hence a reduction in the achievable SNR.

For a multi-bit quantizer, the loop is inherently more stable since the quantizer gain is well-defined, and the no-overload range of the quantizer is increased. In fact, linear analysis can be used to design the modulator so that its stability is guaranteed. Furthermore, since the quantization noise decreases by 6 dB for each bit added to the quantizer and since aggressive high-order noise-shaping functions may be used, multi-bit modulators may have very high ENOB even at low *OSR* values. Hence, there is strong motivation to solve the problem of DAC nonlinearity inherent in the use of multi-bit quantization. While brute-force techniques, such as element trimming, have been used earlier, the techniques currently in favor use auxiliary digital circuitry to manipulate the elements of the DAC so as to reduce the in-band portion of the error signal introduced by DAC nonlinearities. These techniques are conceptually very similar to the noise shaping used in $\Delta\Sigma$ modulators, and are often described with the term *mismatch shaping*. As with noise shaping, the effectiveness of mismatch shaping increases with increasing *OSR*. For very low *OSR* values ($OSR < 8$), digital techniques can be used to acquire and then correct the nonlinearity parameters of the DAC.

The topic of multi-bit $\Delta\Sigma$ modulators will be covered in detail in Chapter 4.

## 1.7 Delta-Sigma Digital-to-Analog Converters

The motivation for using $\Delta\Sigma$ modulation to realize high-performance DACs is the same as for ADCs: it is difficult if not impossible to achieve a linearity and accuracy better than about 14 bits for DACs operated at Nyquist rate. Using $\Delta\Sigma$ modulation, this task becomes feasible. A $\Delta\Sigma$ DAC system is illustrated in Fig. 1.11. By operating a fully digital $\Delta\Sigma$ modulator loop at an oversampled clock rate, a data stream with (say) 18-bit word length may be changed into a single-bit digital signal such that the baseband spectrum is preserved. The large amount of truncation noise generated in the loop is shaped in order to make the in-band noise negligible. The single-bit digital output signal can then be converted with high (ideally, perfect) linearity into an analog signal using a simple two-level DAC circuit. The out-of-band truncation noise can be subsequently removed using analog lowpass filters.

As in the case of analog $\Delta\Sigma$ loops, using single-bit truncation may lead to instability, and hence limits the effectiveness of the noise shaping. Using multi-bit (typically, 2-5 bit) truncation improves the noise shaping and makes the task of the analog post filter much easier. The linearity of the DAC for in-band signals can be achieved by using the same mismatch shaping techniques used in the internal DACs of analog multi-bit $\Delta\Sigma$ ADCs.

Also as in the case of ADCs, bandpass $\Delta\Sigma$ DACs can be designed. Noise shaping now suppresses the truncation noise in a narrow frequency band located around a

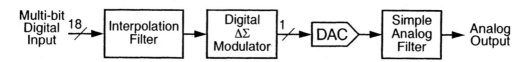

**Figure 1.11:** A $\Delta\Sigma$ DAC system.

nonzero center frequency $f_0$, which need not be much smaller than the clock frequency $f_s$.

$\Delta\Sigma$ DACs will be discussed in detail in Chapter 7.

## 1.8 History; Performance and Architecture Trends

Although the basic idea of using feedback to improve the accuracy of data conversion has been around for about 50 years, the concept of noise shaping was probably first proposed (along with the name delta-sigma modulation) in 1962 by Inose et al. [1]. They described a system containing a continuous-time integrator as the loop filter, and a Schmitt trigger as the quantizer, which achieved (nearly) 40 dB SNR, and had a signal bandwidth of about 5 kHz. Since the trade-off between analog accuracy and higher speed plus additional digital hardware was not particularly attractive at the time, further research on this topic was relatively sparse for a while.

Twelve years later, Ritchie proposed the use of higher-order loop filters [2], and in 1986 Adams described an 18-bit $\Delta\Sigma$ ADC which used a third-order continuous-time loop filter, and a 4-bit quantizer with trimmed resistors performing as the DAC [3]. Useful theory, as well as analysis and design techniques were developed by Candy and his collaborators at Bell Laboratories [4]-[8]. Candy and Huynh also proposed the MASH concept for the digital modulators used in $\Delta\Sigma$ DACs [9]. MASH was first applied in $\Delta\Sigma$ ADCs by Hayashi et al. [10] in 1986.

Using a multi-bit internal quantizer in a $\Delta\Sigma$ loop with digital linearity correction was proposed by Larson et al. [11] in 1988; the use of dynamic matching (randomization) was also introduced for the internal DAC of a $\Delta\Sigma$ ADC by Carley and Kenney in 1988 [12]. Various mismatch-shaping algorithms were suggested subsequently by Leung and Sutarja [13], Story [14], Redman-White and Bourner [15], Jackson [16], Adams and Kwan [17], Baird and Fiez [18], Schreier and Zhang [19], and Galton [20].

Bandpass $\Delta\Sigma$ modulators were motivated for their potential applications in wireless communications, and emerged in the late 1980s [21]-[23].

Current design trends in ΔΣ converters are aimed at extending the signal frequency range without any reduction in SNR. This will open up new applications in digital video, wireless and wired communications, radar, etc. Higher speed can often be achieved by using high-resolution (typically, 5-bit) internal quantizers, and a multistage (2- or 3-stage) MASH architecture. To correct for the nonlinearity of the internal DAC and for quantization noise leakage, digital correction algorithms have been proposed [24] for ΔΣ ADCs. A great deal of effort is also being applied to improving the performance of bandpass ΔΣ ADCs [25]-[28].

Technological trends (finer line widths, accompanied with lower breakdown voltages) stimulated research into ΔΣ modulators needing only low supply voltages [29]. Also, applications opening up in portable devices motivated the development of low-power design techniques for ΔΣ data converters [30],[31]. As noise-shaping theory and practice continue to mature, ΔΣ data converters can be expected to expand their range of application further.

## References

[1] H. Inose, Y. Yasuda and J. Murakami, "A telemetering system by code modulation– Δ-Σ modulation," *IRE Trans. Space Electron. Telemetry*, vol. 8, pp. 204-209, Sept. 1962.

[2] G. R. Ritchie, J. C. Candy and W. H. Ninke, "Interpolative digital to analog converters," *IEEE Transactions on Communications*, vol. 22, pp. 1797-1806, Nov. 1974.

[3] R. W. Adams, "Design and implementation of an audio 18-bit analog-to-digital converter using oversampling techniques," *Journal of the Audio Engineering Society*, vol. 34, pp. 153-166, March 1986.

[4] J. C. Candy, "A use of limit cycle oscillations to obtain robust analog-to-digital converters," *IEEE Transactions on Communications*, vol. 22, no. 3, pp. 298-305, March 1974.

[5] J. C. Candy, B. A. Wooley and O. J. Benjamin, "A voiceband codec with digital filtering," *IEEE Transactions on Communications*, vol. 29, no. 6, pp. 815-830, June 1981.

[6] J. C. Candy and O. J. Benjamin, "The structure of quantization noise from sigma-delta modulation," *IEEE Transactions on Communications*, vol. 29, no. 9, pp. 1316-1323, Sept. 1981.

[7] J. C. Candy, "A use of double integration in sigma-delta modulation," *IEEE Transactions on Communications*, vol. 33, no. 3, pp. 249-258, March 1985.

[8] J. C. Candy, "Decimation for sigma-delta modulation," *IEEE Transactions on Communications*, vol. 34, no. 1, pp. 72-76, Jan. 1986.

[9] J. C. Candy and A. Huynh, "Double integration for digital-to-analog conversion," *IEEE Transactions on Communications*, vol. 34, no. 1, pp. 77-81, Jan. 1986.

[10] T. Hayashi, Y. Inabe, K. Uchimura and A. Iwata, "A multistage delta-sigma modulator without double integration loop," *ISSCC Digest of Technical Papers*, pp. 182-183, Feb. 1986.

[11] L. E. Larson, T. Cataltepe and G. C. Temes, "Multi-bit oversampled ΣΔ A/D converter with digital error correction," *Electronics Letters,* vol. 24, pp. 1051-1052, Aug. 1988.

[12] R. Carley and J. Kenney, "A 16-bit 4'th order noise-shaping D/A converter," *IEEE Proceedings of the Custom Integrated Circuits Conference*, pp. 21.7.1-21.7.4, 1988.

[13] B. H. Leung and S. Sutarja, "Multi-bit Σ–Δ A/D converter incorporating a novel class of dynamic element matching," *IEEE Transactions on Circuits and Systems II*, vol. 39, pp. 35-51, Jan. 1992.

[14] M. J. Story, "Digital to analogue converter adapted to select input sources based on a preselected algorithm once per cycle of a sampling signal," U.S. patent number 5138317, Aug. 11 1992 (filed Feb. 10 1989).

[15] W. Redman-White and D. J. L. Bourner, "Improved dynamic linearity in multi-level ΣΔ converters by spectral dispersion of D/A distortion products," *IEE Conference Publication European Conference on Circuit Theory and Design*, pp. 205-208, Sept. 5-8 1989.

[16] H. S. Jackson, "Circuit and method for cancelling nonlinearity error associated with component value mismatches in a data converter," U.S. patent number 5221926, June 22 1993 (filed July 1 1992).

[17] R. W. Adams and T. W. Kwan, "Data-directed scrambler for multi-bit noise-shaping D/A converters," U.S. patent number 5404142, April 4 1995 (filed Aug. 1993).

[18] R. T. Baird and T. S. Fiez, "Linearity enhancement of multibit ΔΣ A/D and D/A converters using data weighted averaging," *IEEE Transactions on Circuits and Systems II*, vol. 42, no. 12, pp. 753-762, Dec. 1995.

[19] R. Schreier and B. Zhang, "Noise-shaped multibit D/A convertor employing unit elements," *Electronics Letters*, vol. 31, no. 20, pp. 1712-1713, Sept. 28 1995.

[20] I. Galton, "Noise-shaping D/A converters for ΔΣ modulation," *IEEE Transactions on Circuits and Systems II, Proceedings of the 1996 IEEE International Symposium on Circuits and Systems,* vol. 1, pp. 441-444, May 1996.

[21] T. H. Pearce and A. C. Baker, "Analogue to digital conversion requirements for HF radio receivers," *Proceedings of the IEE Colloquium on system aspects and applications of ADCs for radar, sonar and communications*, London, Nov. 1987, Digest No 1987/92.

[22] P. H. Gailus, W. J. Turney and F. R. Yester, Jr., "Method and arrangement for a sigma delta converter for bandpass signals," US Patent number 4,857,928, Aug. 1989 (filed Jan. 1988).

[23] R. Schreier and W. M. Snelgrove, "Bandpass sigma-delta modulation," *Electronics Letters*, vol. 25, no. 23, pp. 1560-1561, Nov. 9 1989.

[24] X. Wang, U. Moon, M. Liu and G. C. Temes, "Digital correlation technique for the estimation and correction of DAC errors in multibit MASH ΔΣ ADCs," *2002 IEEE International Symposium on Circuits and Systems*, vol. 4, pp. 691-694, May 2002.

[25] W. Gao and W. M. Snelgrove, "A 950-MHz IF second-order integrated LC bandpass delta-sigma modulator," *IEEE Journal of Solid-State Circuits,* vol. 33, no. 5, pp. 723-732, May 1998.

[26] G. Raghavan, J.F. Jensen, J. Laskowski, M. Kardos, M. G. Case, M. Sokolich and S. Thomas III, "Architecture, design, and test of continuous-time tunable intermediate-frequency bandpass delta-sigma modulators," *IEEE Journal of Solid-State Circuits*, vol. 36, no. 1, pp. 5-13, Jan. 2001.

[27] P. Cusinato, D. Tonietto, F. Stefani and A. Baschirotto, "A 3.3-V CMOS 10.7-MHz sixth-order bandpass ΣΔ modulator with 74-dB dynamic range," *IEEE Journal of Solid-State Circuits*, vol. 36, no. 4, pp. 629-638, April 2001.

[28] R. Schreier, J. Lloyd, L. Singer, D. Paterson, M. Timko, M. Hensley, G. Patterson, K. Behel, J. Zhou and W. J, Martin, "A 50 mW Bandpass ΣΔ ADC with 333 kHz BW and 90 dB DR," *International Solid-State Circuits Conference Digest of Technical Papers*, pp. 216-217, Feb. 2002.

[29] M. Keskin, Un-Ku Moon and G. C. Temes, "A 1-V 10-MHz clock-rate 13-bit CMOS ΔΣ modulator using unity-gain-reset opamps, "*IEEE Journal of Solid-State Circuits*, vol. 37, no. 7, pp. 817 -824, July 2002.

[30] E. van der Zwan and E. Dijkmans and J. Huijsing, "A 0.2 mW ΣΔ modulator for speech coding with 80 dB dynamic range," *IEEE Journal of Solid-State Circuits*, vol. 31, no. 12, pp. 1873-1880, Dec. 1996.

[31] M. Annovazzi, V. Colonna, G. Gandolfi, F. Stefani and A. Baschirotto, "A low-power 98-dB multibit audio DAC in a standard 3.3-V 0.35μm CMOS technology," *IEEE Journal of Solid-State Circuits*, vol. 37, no. 7, pp. 825-834, July 2002.

# CHAPTER 2 — *The First-Order Delta Sigma Modulator*

After a brief review of quantizers and quantization noise, this chapter examines the first-order modulator, which we designate MOD1, from a variety of perspectives. MOD1 exhibits many of the features found in more complex systems and thus a good grasp of its properties provides the background necessary for understanding many important aspects of the most advanced ΔΣ architectures.

## 2.1 Quantizers and Quantization Noise

To convert a continuous-time analog signal into a digital one, two operations need to be performed: sampling the analog signal (usually with a constant sampling period $T$) and quantizing its amplitude so that it assumes one of a finite number of allowable values. Quantization is usually uniform, so that any two adjacent quantized values differ by a fixed level spacing $\Delta$. The device carrying out the quantization (Fig. 2.1) is called a *quantizer* or ideal A/D converter. It is assumed to be a memoryless nonlinear device completely defined by its static input/output characteristics, i.e., by its $y$-$v$ transfer curve[†]. An example of such a curve is given in

---

[†] A mnemonic for the notation that $y$ is the quantizer input and $v$ is its output is to think of the quantizer as losing information and to picture the letter $v$ as being the letter $y$ minus its tail.

Fig. 2.2, where it is assumed that only positive input values need to be processed (hence the name *unipolar quantizer*) and where the step size, $\Delta$, is 1. A *bipolar* transfer curve, suitable for negative as well as positive inputs, is shown in Fig. 2.3. In this curve, $y = 0$ coincides with a step (rise) of $v$, and hence it is called a *mid-rise* characteristic. In the curve of Fig. 2.4, by contrast, $y = 0$ occurs in the middle of a flat portion (tread) of the curve. Such devices are called *mid-tread* quantizers. The step size is $\Delta = 2$ in both cases. This common value for $\Delta$ allows the quantization levels of both quantizers to be integer values: odd integers for the mid-rise quantizer and even integers for the mid-tread quantizer. Unless otherwise noted, the quantizers considered in this text will be symmetric bipolar quantizers with $\Delta = 2$ and a sampling period of $T = 1$ second.

It is often desirable to approximate the transfer curve with a straight line $v = ky$, where $k$ is the equivalent gain of the quantizer. The deviation of the actual charac-

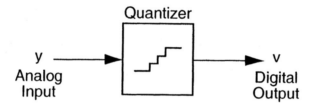

**Figure 2.1: Quantization.**

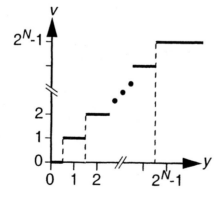

**Figure 2.2: Transfer curve of a unipolar $N$-bit quantizer.**

teristics from the approximate one is called the *quantization error*, or (not entirely correctly) the *quantization noise*. Figs 2.3 and 2.4 illustrate the straight-line approximation and the quantization error $e$. Here, $k = 1$ was used, and thus the difference between input thresholds, also known as the *least-significant bit size* or LSB size, is also 2.

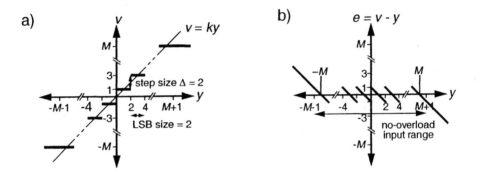

**Figure 2.3:** (a) Transfer curve and (b) error function of a symmetric $M$-step mid-rise quantizer. ($M$ is odd.)

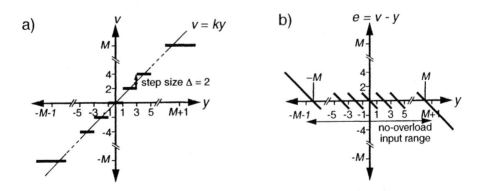

**Figure 2.4:** (a) Transfer curve and (b) error function of a symmetric $M$-step mid-tread quantizer. ($M$ is even.)

## 2 The First-Order Delta Sigma Modulator

It can be seen from the figures that as long as $y$ is between $-(M+1)$ and $(M+1)$, the error $e$ is between -1 and 1. We call the range of $y$ where this condition is satisfied the *no-overload input range*, or simply *input range*. The difference between the lowest and highest levels is called the *full scale (FS)* of the quantizer. Table 2.1 summarizes these and other properties of the quantizers of Figs 2.3 and 2.4.

The ideal quantizer is a deterministic device; $v$, and hence also the error $e$, are fully determined by the input $y$. However, if $y$ stays within the input range of the quantizer, and changes by sufficiently large amounts from sample to sample so that its position within a quantization interval is essentially random, then it is permissible to assume that $e$ is a white noise process with samples uniformly distributed between $-\Delta/2$ and $\Delta/2$. From the assumed uniform distribution, it is easy to derive that $e$ has mean zero and mean-square value $\sigma_e^2 = \Delta^2/12$ [1]. This "noise" is simply added to the scaled input $y$ to give the output

$$v = ky + e \tag{2.1}$$

| Parameter | Value |
|---|---|
| input step size (LSB size) | 2 |
| output step size | 2 |
| number of steps | $M$ |
| number of levels | $M+1$ |
| number of bits | $\lceil \log_2(M+1) \rceil$ |
| no-overload input range | $[-(M+1), M+1]$ |
| full-scale | $2M$ |
| input thresholds | $0, \pm 2, \ldots, \pm(M-1)$, $M$ odd<br>$\pm 1, \pm 3, \ldots, \pm(M-1)$, $M$ even |
| output levels | $\pm 1, \pm 3, \ldots, \pm M$, $M$ odd<br>$0, \pm 2, \pm 4, \ldots, \pm M$, $M$ even |

Table 2.1. Properties of the symmetric quantizers of Figs. 2.3 and 2.4.

Note that (2.1) is always valid; an approximation is only made when $e$ is assumed to have specified properties such as a uniform distribution or a white spectrum! Again, such approximations are only justifiable under the conditions stated above. Inputs for which these conditions are not satisfied, and hence for which the approximation gives wildly wrong results, include a constant $y$, or a periodic $y$ with a frequency harmonically related to $f_s$, especially if $y$ changes from sample to sample by rational fractions of $\Delta$.

To illustrate these points, Fig. 2.5 shows a full-scale sine wave sampled and then quantized by a 16-step symmetrical bipolar quantizer with $\Delta = 2$. The frequency $f$ of the input is moderately high compared to $f_s$ and does not have a simple harmonic relation to $f_s$. As a result, the quantization error sequence plotted in Fig. 2.6 appears fairly random, although a careful examination of the error sequence as the input passes through its peaks does reveal a non-zero correlation between adjacent samples. The mean-square value of the error in this example is found to be 0.30, which is close to the expected value of $\Delta^2/12 = 1/3$.

A fast Fourier-transform (FFT) of $v$ is plotted in Fig. 2.7. The spectrum consists of one large spike representing the input sine wave, plus many smaller spikes distributed evenly along the frequency axis, representing the frequency components of the quantization error. Judging by these results, the white-noise approximation appears to be reasonable in this situation.

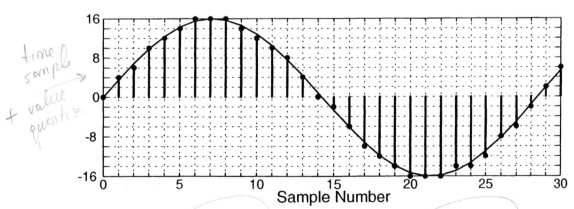

**Figure 2.5: A sine wave quantized by a symmetrical 16-step quantizer.**

## 2 The First-Order Delta Sigma Modulator

Now consider what happens when the full-scale sine wave with frequency $f = f_s/8$ shown in Fig. 2.8 is quantized by the same 16-step quantizer. The quantization error, shown in Fig. 2.9, is now periodic and, since it assumes only 3 values, its distribution is far from uniform. The mean-square value of this error sequence is found to be 0.23, or only about 70% of the expected value. The FFT shown in Fig. 2.10 represents an even more serious departure from our normal assumptions in that the spectrum now consists of only two spikes! The quantization noise energy is fully concentrated in one tone at $f_s/8$ (coincident with the signal itself) and a second tone at $3f_s/8$ (the third harmonic of the signal).

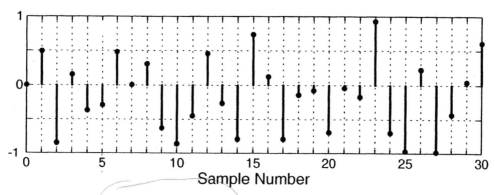

Figure 2.6: Quantization error sequence for the sine wave of Fig. 2.5.

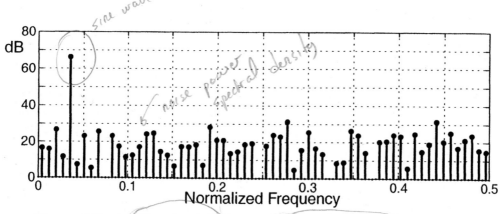

Figure 2.7: A 256-point FFT of the quantized sine wave of Fig. 2.5.

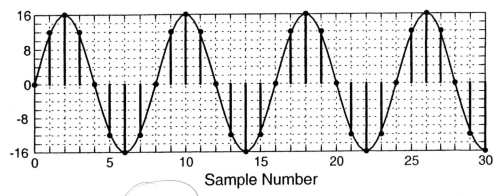

**Figure 2.8:** An $f = f_s/8$ sine wave quantized by a 16-step quantizer.

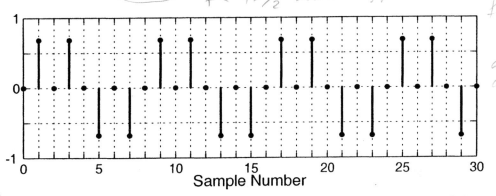

**Figure 2.9:** Quantization error sequence for the sine wave of Fig. 2.8.

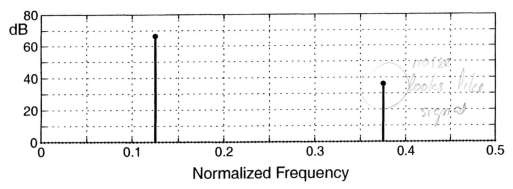

**Figure 2.10:** A 256-point FFT of the quantized sine wave of Fig. 2.8.

## 2.1.1 Binary Quantization

As illustrated in Figs. 2.3-2.4, the gain $k$ of the linear model for multi-bit quantizers is naturally determined by the ratio of the step size to the distance between adjacent thresholds. However, as depicted in Fig. 2.11, the gain of a binary quantizer is not defined so easily because a binary quantizer has only one threshold. The two approximating lines shown are both possible, and they both result in the same maximum error $\Delta/2 = 1$, although with different no-overload ranges.

If the statistical properties of $y$ are known, an obvious optimality criterion for $k$ is to minimize the *mean square value* (also called the *average power*) of the error sequence $e$. This is defined as the expected (or mean) value of $e^2$:[†]

$$\sigma_e^2 = \lim_{N \to \infty} \frac{1}{N} \sum_{n=0}^{N} e(n)^2 \qquad (2.2)$$

A convenient notation is possible by introducing the *inner product*, or *scalar product*. For real sequences $a$ and $b$, the inner product is defined as

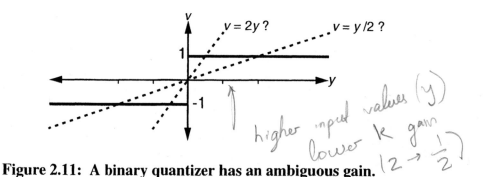

**Figure 2.11:** A binary quantizer has an ambiguous gain.

---

[†] Since we assume that $e$ has zero mean, the variance of $e$ equals its mean-square value, and we are justified in using $\sigma_e^2$ to represent the mean-square value of $e$. If $e$ has a non-zero mean, but is otherwise white, the variance of $e$ is the appropriate quantity to use in the formulae relating the in-band noise power to the quantization noise power.

$$\langle a, b \rangle = \lim_{N \to \infty} \left[ \frac{1}{N} \sum_{n=0}^{N} a(n)b(n) \right] = E[ab] \quad (2.3)$$

Since $e = v - ky$, the average power of $e$ can be written as

$$\sigma_e^2 = \langle e, e \rangle$$
$$= \langle v - ky, v - ky \rangle$$
$$= \langle v, v \rangle - 2k \langle v, y \rangle + k^2 \langle y, y \rangle \,. \quad (2.4)$$

This is minimized for

$$k = \frac{\langle v, y \rangle}{\langle y, y \rangle} = \frac{E[|y|]}{E[y^2]}. \quad (2.5)$$

The last part of (2.5) follows since $v(n)y(n) = \text{sign}[y(n)]y(n) = |y(n)|$.

Clearly, the optimal value of $k$ for the linear model of a binary quantizer is dependent on the statistics of its input $y$. As a sanity check, let $k$ be the gain associated with an input $y$ according to (2.5). If $y' = 10y$, then $E[|y'|] = 10E[|y|]$ and $E[y'^2] = 100(E[y^2])$, so that $k' = k/10$. Thus, the effective gain of a binary quantizer is reduced by a factor of 10 when its input is amplified by a factor of 10. This makes physical sense, since $v$ remains the same while $y$ is increased tenfold.

When a system containing a binary quantizer is replaced by its linear model, the estimate of the quantizer gain $k$ should be found from extensive numerical simulations. Otherwise, misleading results may be obtained from the linear model.

## 2.2 MOD1 as an ADC

As already introduced in Fig. 1.4 (repeated here as Fig. 2.12) the simplest analog $\Delta\Sigma$ modulator is a first-order loop containing an integrator, a 1-bit ADC and a 1-bit DAC. We use the term *delta-sigma* ($\Delta\Sigma$) *modulator* to describe a system (analog or digital) which transforms its input, be it continuous or finely-quantized, into a coarsely-quantized output possessing a noise-shaped spectrum. As described in Chapter 1, $\Delta\Sigma$ modulators can be used to create ADCs or DACs, but since the underlying dynamics and describing equations are essentially independent of whether the modulator is being used to create an ADC system or a DAC system, we use the term "modulator" to encompass both applications. Thus the first-order modulator, MOD1, has a variety of uses and implementations and so the use of MOD1 as an ADC is just the first of several which we shall study.

Fig. 2.13 shows a possible *continuous-time* (CT) realization of MOD1, while Fig. 2.14 shows a *switched-capacitor* (SC) one. In the CT system, the 1-bit ADC is simply a comparator followed by a D flip-flop (DFF), while the DAC function is performed by the $R_f$ resistors which connect the DFF outputs to the opamp inputs. Note that since the reference voltage for the DAC is thus the (noisy) power supply of the DFF, this circuit is intended for low-performance applications, where circuit simplicity is more important than modulation accuracy.

In the SC realization, the 1-bit ADC is implemented with a clocked comparator while the DAC is formed by the $\phi_2$ switches and the $C_1$ capacitors. Since the DAC in this circuit uses a dedicated reference voltage and operates in discrete time, this implementation is inherently more accurate than the simple CT implementation shown in Fig. 2.13.

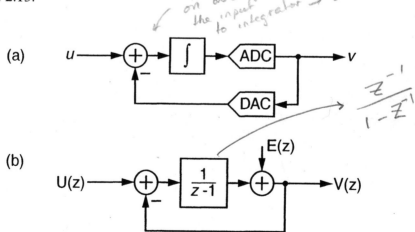

Figure 2.12: (a) A delta-sigma modulator used as an ADC and (b) its linear z-domain model.

Consider the operation of the CT circuit. In order to simplify the discussion, the high and low output voltages of the flip-flop will be assumed to be +1 and −1 V, respectively, so that the output $v$ equals sgn($y$) in volts. The clocked comparator feeds back a positive charge through $R_f$ in those clock periods when the sampled output voltage $y(n)$ of the integrator is positive, thereby driving the output of the integrator down. If $y(n)$ is negative, a negative charge is fed to the virtual ground and the output of the integrator rises.

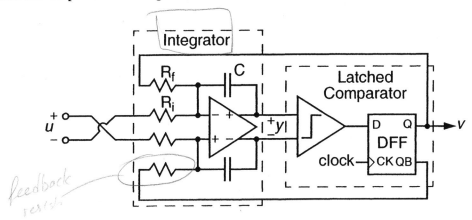

Figure 2.13: An active-RC (continuous-time) implementation of MOD1.

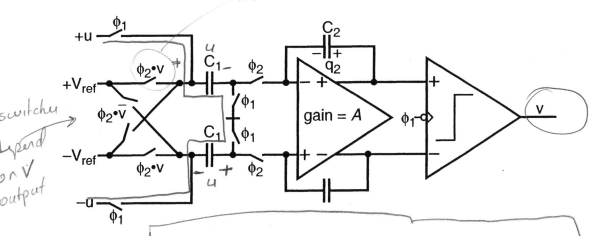

Figure 2.14: A switched-capacitor (discrete-time) implementation of MOD1.

Fig. 2.15 shows representative waveforms for a zero input ($u = 0$) and a slightly positive input ($u = 0.06$). For the zero-input case, the integrator output alternately ramps up and down, and the output $v$ merely toggles between +1 and −1. With a positive input, the integrator output primarily ramps up and then down, but tends to drift high over time. After 15 clock cycles, the integrator output is so high that two adjacent +1s are produced by the comparator and the integrator output takes two successive downward steps. Clearly, the feedback is acting to regulate the output of the integrator towards zero, and the only way that can happen over time is if the average current into the integrating capacitor $C$ equals zero. Thus, the circuit forces the average of $v/R_f$ to equal the average of $u/R_i$. Similar conclusions can be reached regarding the operation of the SC circuit of Fig. 2.14.

The remarkable feature of both of these simple circuits is that for constant inputs arbitrarily high accuracy can be achieved, at least in principle. Furthermore, a common source of nonlinearity in a Nyquist converter, namely component mismatch, is virtually immaterial to a $\Delta\Sigma$ converter. To see this, observe that changing the value of the integrating capacitor in Fig. 2.13 (or changing either $C_1$ or $C_2$ in Fig. 2.14) simply scales the output of the integrator slightly but does not affect its polarity, which is the only thing a 1-bit ADC detects. For this reason, component-value shifts do not affect the operation, and hence the accuracy, of the converter. Similarly, comparator hysteresis, opamp offset and (in the case of the CT circuit)

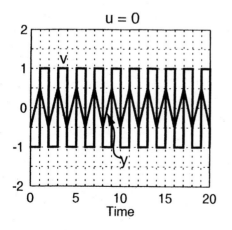 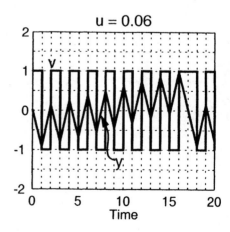

**Figure 2.15: Waveforms for the CT system depicted in Fig. 2.13.**

resistor mismatch, have virtually no impact on converter linearity. At worst, the input full-scale range will shift, and/or the input-referred offset will be non-zero.

The results reached above are readily confirmed by a simple mathematical analysis. Let the circuits in Fig. 2.13 and Fig. 2.14 be modeled by the block diagram of Fig. 2.16[†], where the quantizer Q is assumed to be have the simplified characteristic

$$v(n) = \text{sgn}[y(n)]. \quad (2.6)$$

From the diagram,

$$y(n) = y(n-1) + u(n) - v(n-1) \quad (2.7)$$

and hence, combining equations for $n = 1, 2, \ldots N$

$$y(N) - y(0) = \sum_{n=0}^{N} [u(n) - v(n-1)]. \quad (2.8)$$

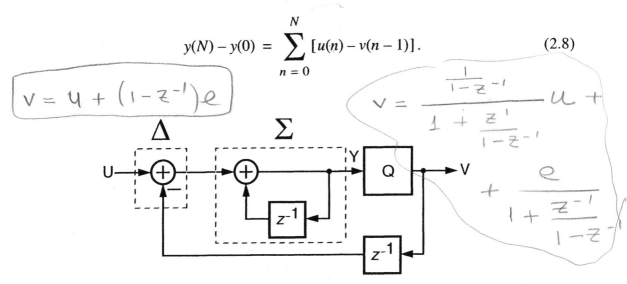

Figure 2.16: A z-domain block diagram of MOD1.

---

† Fig. 2.12b shows a delaying integrator while Fig. 2.16 uses a non-delaying integrator with a delay element in the feedback. Adding a delay to the $u$ input of Fig. 2.16 makes Fig. 2.16 equivalent to Fig. 2.12b.

Assume that the loop is stable, so that $y(n)$ is bounded (i.e., finite). Then

$$\lim_{N \to \infty} \frac{y(N) - y(0)}{N} = 0. \tag{2.9}$$

From (2.8) and (2.9) it follows that the average of the input samples, defined as

$$\bar{u} = \lim_{N \to \infty} \frac{1}{N} \sum_{n=0}^{N-1} u(n), \tag{2.10}$$

equals that of the digital output $v$.

Thus, for a dc input $u$, if we allow the circuit to operate for a sufficiently long time, the average value of the digital output $v$ will become an arbitrarily good approximation of the input. This average value can be recovered by cascading a digital lowpass filter with the $\Delta\Sigma$ loop.

Note that MOD1 uses a single-bit (two-level) quantizer, and hence ideally its DAC is capable of perfectly linear operation. In practice there are several effects which may compromise this linearity, such as changes in the value of the reference or supply voltages with any of the signals, or memory effects in the DAC. These will be briefly covered in Chapter 6.

## 2.3 MOD1 as a DAC

Consider the problem of driving an 8-bit DAC from a 16-bit data stream. One could simply discard the 8 LSBs of the original data, but doing so would of course entail a loss of resolution. An alternative method involves trying to make up for the error caused by truncation in one cycle by introducing the opposite error in the next cycle. By doing so, errors caused by truncation should average out over time.

Fig. 2.17 shows a system which embodies this principle. In this system, wide input data (of width $m_1 + m_2 = m$ bits) is added to the contents of the truncation error register. The output of the accumulator is truncated to $m_1$ bits to drive an $m_1$-bit DAC while the $m_2$-bit truncation error is stored in the error register for use in the next cycle. We will shortly see that this system is another incarnation of the first-order $\Delta\Sigma$ modulator, albeit with a multi-bit "quantizer."

Since the truncation operation can be viewed as subtraction of the LSBs, we may write $v = y - y_{LSB}$, where $y_{LSB}$ is the truncation error consisting of the $m_2$ LSBs of $y$. From the diagram,

$$y(n) = u(n) + y_{LSB}(n-1). \qquad (2.11)$$

Since $y_{LSB}(n-1) = y(n-1) - v(n-1)$, (2.11) may be rewritten as

$$y(n) = u(n) + y(n-1) - v(n-1), \qquad (2.12)$$

which is identical to (2.7).

Thus the system shown in Fig. 2.17 is equivalent to MOD1. As a result of this equivalence, the system shown in Fig. 2.17 responds to a dc input by alternating the DAC input code between adjacent levels in such a way that the average of the analog output corresponds to the $m$-bit input. Furthermore, it is easy to show that the accumulated difference between the input and the output is always less than one DAC LSB.

Digital modulators will be discussed at length in Chapter 7.

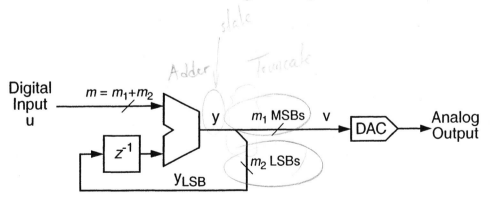

Figure 2.17: A first-order digital modulator used in a $\Delta\Sigma$ DAC.

## 2.4 MOD1 Linear Model

The two preceding sections presented several implementations of MOD1, focussing on its time-domain operation and its response to dc inputs. To provide insight into the behavior of MOD1 with time-varying inputs, this section studies the z-domain model of Fig. 2.18, in which the quantizer has been replaced by its linear model.

From the diagram,

$$Y(z) = z^{-1}Y(z) + U(z) - z^{-1}V(z). \quad (2.13)$$

Thus

$$\begin{aligned} V(z) &= Y(z) + E(z) = z^{-1}Y(z) + U(z) - z^{-1}V(z) + E(z) \\ &= U(z) + E(z) - z^{-1}(V(z) - Y(z)) \\ &= U(z) + E(z) - z^{-1}E(z) \\ &= U(z) + (1 - z^{-1})E(z). \end{aligned} \quad (2.14)$$

*noise shaping (error minus previous error)*

The dc values of $u$ and $v$ (i.e. their average values) can be obtained by setting $z = 1$ in (2.14). If the dc value of $e$ is finite, $V(1) = U(1)$ follows directly from (2.14). This confirms the conclusion reached earlier in Section 2.2 regarding the ability of MOD1 to provide arbitrarily high resolution for a dc input.

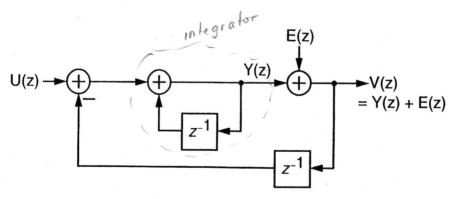

Figure 2.18: A z-domain linear model for MOD1.

# MOD1 Linear Model

Eq. (2.14) can be written in the general form

$$V(z) = STF(z)U(z) + NTF(z)E(z), \qquad (2.15)$$

where in this case the signal transfer function (STF) is unity, and the noise transfer function is $NTF(z) = 1 - z^{-1}$. In order to estimate the in-band power of the quantization noise, it is useful to find the squared magnitude of NTF in the frequency domain, by setting $z = e^{j2\pi f}$. This gives[†]

$$|NTF(e^{j2\pi f})|^2 = [2\sin(\pi f)]^2 \qquad (2.16)$$

For frequencies which satisfy $f \ll 1$, $|NTF|^2 \approx (2\pi f)^2$. Fig. 2.19 (earlier shown as Fig. 1.5) illustrates the frequency response of the NTF. It is clearly a highpass response, which suppresses the quantization noise at and near dc (where the signal energy is) and amplifies it out of band, at and near 0.5 (i.e., $f_s/2$). This noise-shaping action is the key to the effectiveness of $\Delta\Sigma$ modulation.

If the conditions discussed in Section 2.1 (large, randomly-varying $y$) hold, the error $e$ can be treated as a white noise with a mean-square value $\sigma_e^2 = 1/3$. (Recall that we assume $\Delta = 2$.) The 1-sided power spectral density of $e$ is therefore

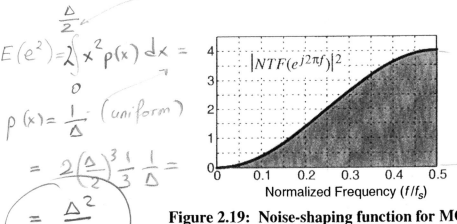

Figure 2.19: Noise-shaping function for MOD1.

---

[†] Here, as before, $f_s = 1$ is used.

## 2 The First-Order Delta Sigma Modulator

$OSR = \dfrac{f_s/2}{f_B}$

$S_e(f) = 2\sigma_e^2 = 2/3$. Then, as already shown in Section 1.2, the in-band noise power in the output $v$ is given approximately by

$f_B = \dfrac{f_s/2}{OSR}$ , $f_s = 1$

$$\sigma_q^2 = \int_0^{1/(2 \cdot OSR)} [2\pi f]^2 S_e(f) df = \dfrac{\pi^2}{9(OSR)^3}. \tag{2.17}$$

Assume that the input signal is a full-scale sine wave with a peak amplitude $M$. Since the STF is unity, the output signal power is $\sigma_u^2 = M^2/2$. Hence, the in-band *signal-to-quantization-noise ratio* (SQNR) is approximately

$$SQNR = \dfrac{\sigma_u^2}{\sigma_q^2} = \dfrac{9M^2(OSR)^3}{2\pi^2}. \tag{2.18}$$

As already stated in Section 1.2, (2.18) shows that the SQNR increases by 9 dB for each doubling of the OSR (i.e., for each octave increase of $f_s$ with $f_B$ fixed). This results in a relatively low SQNR, less than 70 dB, even for a relatively high value of OSR such as OSR = 256. $= 2^8$

### 2.5 Simulation of MOD1

The behavior of MOD1 is easily simulated in the time-domain using the difference equations (2.7) and (2.6). All that is needed to perform such calculations are the samples of the input signal and the initial condition of the integrator. Once the output sequence has been computed, its spectrum can be found using an FFT. However, to achieve the high numerical accuracy needed, several precautionary measures must be used; these are discussed in Appendix A. Simulation is an important tool in $\Delta\Sigma$ modulator design because the linear model is imperfect and can hide important effects which only become apparent when the true nonlinear nature of the modulation process is taken into account.

As a demonstration that MOD1 really does shape quantization noise, Fig. 2.20 plots the spectrum of the output of MOD1 when simulated with a full-scale sine-wave input. This figure clearly shows a noise-shaped characteristic, the hallmark of $\Delta\Sigma$ modulation, and the 20 dB/decade slope of the noise is consistent with first-order shaping. Furthermore, computing the optimal quantizer gain using (2.5) yields $k = 0.90$, which is fairly close to the value of 1 assumed in the analysis. However, the simulated SQNR of 55 dB for $OSR = 128$ is 5 dB less than the value of 60 dB predicted by (2.18) with $M = 1$.

Fig. 2.21, which plots the simulated SQNR vs. the input amplitude for two different test frequencies, gives a clue as to the source of this SQNR discrepancy. As this figure demonstrates, the simulated SQNR of MOD1 is an erratic function of the input amplitude and frequency. Clearly, the dynamics of MOD1 are not as simple as the linear model would lead us to believe.

As a further demonstration of the complexity lurking in MOD1, Fig. 2.22 plots the simulated in-band noise power as a function of the value of a dc input for two val-

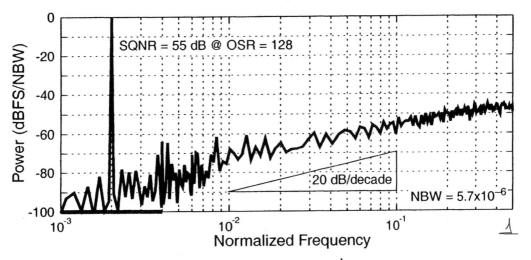

**Figure 2.20: Simulated MOD1 output spectrum† with a full-scale sine-wave input.**

---

† As with all spectral density plots in this text, the vertical axis of Fig. 2.20 is the power relative to a full-scale sine wave contained in a certain bandwidth, the *noise bandwidth* (NBW). Thus the units on the vertical axis are dB relative to a full-scale sine wave, per bandwidth NBW, which is abbreviated dBFS/NBW. Appendix A describes the process of spectral estimation in more detail and explains the origin and significance of NBW.

ues of *OSR*. Also shown are the mean-square levels of the in-band noise, averaged over all input values. Both plots show that MOD1 exhibits increased in-band noise in the vicinity of certain input values, in particular $\pm 1$ and 0. Comparing Fig. 2.22a with Fig. 2.22b, we see that as *OSR* is increased, the widths and absolute heights of the anomalous regions decrease, but relative to the mean noise level the noise spikes are higher. Candy and Benjamin showed that the central noise peaks have a height $-20\log(\sqrt{2}OSR)$ dB and width $(OSR)^{-1}$ [2]. Later, Friedman showed that the dominant pattern consists of two large spikes surrounding a mound of smaller spikes, and that this pattern is duplicated between adjacent pairs of the smaller spikes in an endless recursion [3].

This and other exotic behavior is a direct consequence of the nonlinear difference equations which govern MOD1. Although a thorough study is beyond the scope of this text, it is possible to state some important results and give some indications of the underlying mechanisms.

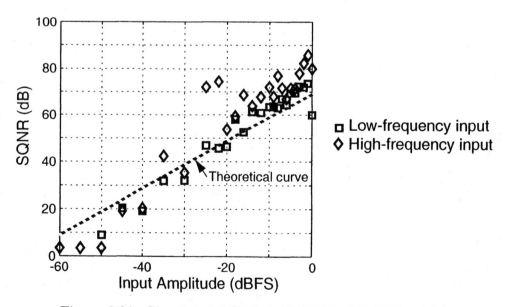

Figure 2.21: **Simulated SQNR for MOD1 with OSR = 256.**

## 2.6 MOD1 under DC Excitation

As mentioned earlier, the linear model used in Section 2.4 is valid only under specific conditions (large and fast random variations) on the input $y$ to the quantizer, which can only be satisfied if the input $u$ to the loop meets similar conditions. In this section, we shall discuss the behavior of MOD1 for an important case in which these conditions are violated, namely the case of a constant input signal.

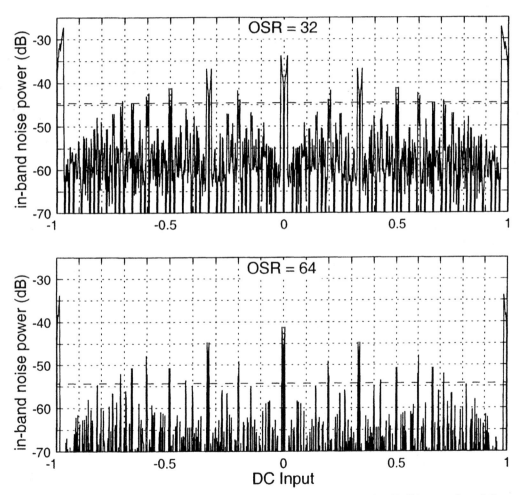

**Figure 2.22:** In-band quantization noise power vs. the DC input level for MOD1. The broken line marks the noise power averaged over all input values.

## 2.6.1 Idle Tone Generation

Consider MOD1 with a dc input $u$. From the block diagram in Fig. 2.18,

$$y(n) = y(n-1) + u - v(n-1) \quad (2.19)$$

and

$$y(n) = u - (v(n-1) - y(n-1))$$

$$v(n) = \operatorname{sgn}(y(n)). \quad (2.20)$$

To avoid ambiguity, for $y(n) = 0$, we assign $v(n) = 1$. It should also be noted that by regarding the DAC as a unity-gain element, as is implied in (2.19), we effectively normalize all analog signals ($y$ and $u$) to the reference voltage $V_{ref}$ of the DAC.

Combining these relations,

$$y(n) = y(n-1) + u - \operatorname{sgn}(y(n-1)) \quad (2.21)$$

results. Although (2.19) and (2.20) can then be used to generate the sequences $y(n)$ and $v(n)$ from a given $u$ and the initial condition $y(0)$, (2.21) shows that the exotic behavior of MOD1 can be captured in a single first-order nonlinear difference equation.

Now assume that $u = 1/2$ and also $y(0) = 1/2$. Then $v(0) = 1$, $y(1) = 1/2 + 1/2 - 1 = 0$ and $v(1) = 1$. The $y$ and $v$ sequences are summarized in Table 2.2 for $n = 0$ to $4$. Clearly, for $n = 4$, the same conditions exist as for $n = 0$. Hence, the output is periodic with a period of 4.

| $n$ | 0 | 1 | 2 | 3 | 4 |
|---|---|---|---|---|---|
| $y(n)$ | $\frac{1}{2}$ | 0 | $-\frac{1}{2}$ | 1 | $\frac{1}{2}$ |
| $v(n)$ | 1 | 1 | $-1$ | 1 | 1 |

Table 2.2. MOD1 operation for $u = 1/2$.

The average value of $v$ for a full period is $(1 + 1 - 1 + 1)/4 = 1/2$, which is identical to the input $u$. Thus, as already shown in Section 2.2, MOD1 is capable of converting a dc input signal with unlimited accuracy, provided it is allowed to operate for unlimited time, and is followed by a perfect lowpass filter. For the input under consideration, the output is periodic with period 4, and so contains tones at $f_s/4$ and its harmonics, which will be removed by the lowpass filter.

The reader should repeat the calculation for different values of $y(0)$ to verify that the output will again be periodic with a period of 4, and that $|y(n)| \leq 2$[†]. Over a period of 4 samples, the output will always contain three +1s and one −1. The order in which they occur will depend on $y(0)$.

Consider now the more general case where the input is a constant rational number: $u = a/b$. Assume $0 < a < b$, and that $a$ and $b$ are odd positive integers with no common factor. The initial value $y(0)$ is given, with $|y(0)| < 1$. Let the first $b$ output samples of $v$ contain $(a+b)/2$ samples of +1, and $(b-a)/2$ samples of −1. Then the average value of $v$ in this set is $a/b$, the same as $u$. The total net input to the integrator for the first $b$ samples is zero. Hence, after the $b^{th}$ sample, the value of $y$ will be again back at $y(b) = y(0)$. The $v$ sequence will therefore be repeated during the next $b$ samples. Therefore, a periodic sequence with a period $b$ will be generated.

If $u = a/b$ with one of $a$ or $b$ even so that $(a+b)$ is odd, it can be easily shown that again a periodic sequence results, with a period $2b$ containing $(a+b)$ +1s and $(b-a)$ −1s. For a negative input $u$, the output is the negative of the above given the positive input $-u$.

Assume now that MOD1 has an output which is periodic with a period $p$, containing $m$ samples of +1 and $p-m$ samples of −1 in each period. The average output for each period will be $(2m-p)/p$, a rational number. Hence, the (constant) input must equal this value, and must also be rational.

---

[†] A tight bound on the input to the quantizer for a dc input $u$ is $-1 + u \leq y \leq 1 + u$, for $|u| \leq 1$ and $|y(0)| \leq 1$.

## 2 The First-Order Delta Sigma Modulator

Thus, a periodic output with a constant input $u$ ($|u| < 1$) implies that $u$ is rational. It follows therefore that if $u$ ($|u| < 1$) is constant but irrational, the output of MOD1 cannot be periodic.

The periodic sequences generated by rational dc inputs are sometimes called *pattern noise*, *idle tones*, or *limit cycles*. They do not represent instability of the loop; their amplitude does not change with time, but is a complicated function of $u$ [2]. Their frequency also depends on the input, as discussed above.

It is important to examine the effect of limit cycles on the in-band noise of MOD1. In the numerical example discussed above, the output spectrum contains a line at dc, corresponding to the average value of the $v$ samples. This represents the digital equivalent of the dc input $u$, and is thus the desired signal. In addition, there will be spectral lines at $f_s/4$, $2f_s/4$ and $3f_s/4$ which represent noise. Since typically the cut-off frequency $f_B$ of the digital lowpass filter following the loop satisfies $f_B = f_s/(2 \cdot OSR) \ll f_s$, these lines are in the stopband of the filter, and hence are harmless.

Unfortunately, the situation is not always so rosy. Assume that the input is the rational dc value $u = 1/100$. Now the argument presented above shows that the output signal will contain tones at $f_s/200$ and its harmonics. Several of these will usually be within the passband of the digital filter, and hence deteriorate the SNR. (A detailed analysis shows that the output sequence will contain an alternation of the values +1 and −1 for 101 periods; the −1 which would have occurred at sample 102 is changed to +1 due to the accumulation of $u$ in the integrator so that two +1s occur in a row. Then the alternation resumes until sample 199, when again a −1 becomes a +1 so that two +1s occur in succession. A quick calculation confirms that the average of the output over 200 samples is (101-99)/200 = 2/200 = $u$, as expected. The output spectrum can be visualized by observing that the output is equivalent to a sampled sine wave of frequency $f_s/2$ modulated by a square wave with a frequency of $f_s/200$ and a 50.5% duty cycle.)

As mentioned earlier, both the frequency and power of the tones are functions of the dc input $u$. Hence, so is the in-band noise which they introduce. Fig. 2.22 shows the variation of in-band noise power for MOD1 with $u$ for $OSR = 32$ and for $OSR = 64$. As the figure shows, large peaks occur near simple rational values, such as $u = 0, \pm 1/2, \pm 1/3$, etc.

In some applications, such as in digital audio, idle tones cannot be tolerated, since the human hearing apparatus can detect tones even 20 dB below the level of any white noise present. The prevention of tone generation is therefore an important aspect of $\Delta\Sigma$ modulator design, one that often motivates the use of high-order modulators or *dither*, an added pseudo-random signal.

Idle tones may also be generated in the presence of slowly varying input voltages, which stay near a critical level (a simple rational value) long enough for a limit cycle to become apparent. This is particularly likely to occur for low-order modulators, such as MOD1.

*2.6.2 Graphical Visualization*

In this section, MOD1 is viewed from the perspective of nonlinear dynamics. Proofs are avoided in order to maintain focus on the results themselves. The interested reader is encouraged to consult the references for complete derivations.

Fig. 2.23 shows a geometric interpretation of the operation of MOD1. The horizontal axis represents time and the vertical axis is the accumulated output

$$s(n) = \sum_{i=1}^{n} v(i) \qquad (2.22)$$

The possible values of $s$ from time 0 to time $n$ form a triangular lattice of points. Also plotted is a line $L$ of slope $u$ passing through the origin. At each time step, $s$ is either incremented or decremented by 1, depending on which choice results in a sum which is closest to the line $L$.

If the input $u$ is rational, $L$ periodically passes through one of the lattice points and the $v$ sequence repeats. If the input $u$ is irrational, $L$ threads its way through the lattice but does not go through any lattice point other than (0,0) and so the $v$ sequence does not follow a repetitive pattern. However, suppose that $L$ comes very close to a lattice point at time $n$ and let $\varepsilon$ denote the distance between $L$ and this lattice point. If $\varepsilon$ is small enough, the modulator will repeat the sequence of decisions leading up to time $n$. After this repetition, the distance between $L$ and the nearest lattice point will be $2\varepsilon$. If this quantity is still small, the modulator will repeat the sequence once again, and the distance will increase to $3\varepsilon$. After $k$ repetitions, the distance will be $k\varepsilon$, and so it will eventually grow large enough that the modulator output is forced to deviate temporarily from a repeating pattern.

A procedure for computing the output of MOD1 given a DC input $u$, without resorting to step-by-step simulation of the difference equations, is available in the literature [3][4], and is summarized here without proof:

**1.** Map $u$ to $x = (1+u)/2$. This transforms the range [-1,1] to [0,1].

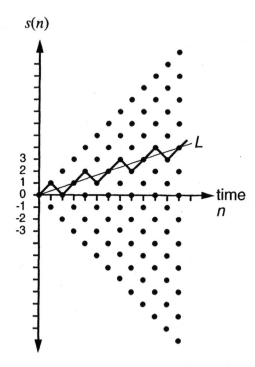

Figure 2.23: A geometric interpretation of the operation of MOD1.

For example, $u = 0.4$ maps to $x = 0.7$.

2. Determine the *continued-fraction expansion* (CFE) of $x$.
   For $x = 0.7$,
   $$x = \frac{7}{10} = \frac{1}{1+\frac{3}{7}} = \frac{1}{1+\frac{1}{2+\frac{1}{3}}} \quad (2.23)$$
   which will be abbreviated as $0.7 = CFE(1, 2, 3)$.

   It can be shown that
   i) for rational $x$, the CFE terminates after a finite number of steps;
   ii) truncating the CFE early results in a *rational of best approximation*[†] to $x$.

3. Next, carry out the recursion
   $$S_n = S_{n-2}(S_{n-1})^{\alpha_n}, \quad (2.24)$$
   with starting conditions
   $$S_0 = 0 \text{ and } S_1 = 1(S_0)^{\alpha_1 - 1}. \quad (2.25)$$
   Here, $S^k$ means the $k$-fold repetition of the sequence $S$ and $\alpha_n$ is the $n^{th}$ argument in $CFE(\alpha_1, \alpha_2, \ldots)$.

   For $x = 0.7$, $\alpha = (1, 2, 3)$ and so
   $S_0 = 0$
   $S_1 = 1(0)^{1-1} = 1$
   $S_2 = 0(1)^2 = 011$
   $S_3 = 1(011)^3 = 1011011011$

---

[†] $a/b$ is said to be a rational of best approximation to $x$ if, for all rationals $p/q$ with $q<b$, $|x - a/b| < |x - p/q|$.

Since the CFE terminates, MOD1 produces the $S_3$ sequence repeated ad infinitum. Note that in this interpretation, the modulator output is taken as 0 or 1, rather than our more usual $-1$ or $+1$. This difference in convention is the reason for the transformation of step 1. Furthermore note that the starting point within the sequence is dependent on the initial conditions of the integrator. For the example sequence, the length is 10 and there are 7 ones, so the average is 7/10 (which equals $x$). Interpreting the 0s as $-1$s yields an average of 4/10 (the value of $u$).

The above procedure gives several insights into the structure of MOD1's output given a DC input. First, since each subsequence $S_n$ corresponds to a rational of best approximation, the 0s and 1s are distributed as uniformly as possible in the output bit stream. Second, if the input is close to a rational $r$, then one of the $\alpha$ arguments in $CFE(\alpha_1, \alpha_2, ...)$ will be large, and thus the pattern which corresponds to $r$ will be repeated in the output data stream many times in a row. If $r$ has a small denominator, then MOD1's output will appear periodic over short time spans. Of course, if the input is equal to $r$, the output of MOD1 will be truly periodic.

Taking a slightly different perspective, the above procedure shows that if $x < 1$ is a rational $a/b$, taken in lowest terms, then MOD1 will produce a period-$b$ output. Since we already know that irrational inputs result in aperiodic behavior, we conclude that MOD1 is like an oscillator whose period can be anything, including infinity! As mentioned in Section 2.6.1, the periodic sequences generated by rational dc inputs are sometimes called *pattern noise*, *idle tones*, or *limit cycles*. They are *not* a result of non-ideal operation of the loop; they are a direct consequence of MOD1's dynamics.

Clearly, a rational input yields an output spectrum consisting of a finite number of discrete spectral lines between 0 and $f_s/2$. It is somewhat surprising that irrational inputs also yield an output spectrum consisting solely of discrete spectral lines, although the number of such lines between 0 and $f_s/2$ is infinite. This remarkable result was established in [5] by a study of MOD1 from the nonlinear dynamical system perspective. This work showed that the tones in the output spectrum have frequencies (normalized to $f_s$)

$$f_n = nx \qquad (2.26)$$

where $n$ is a natural number and, as before, $x = (1+u)/2$. These tones alias to the $[0, 0.5]$ Nyquist range and have an amplitude of

$$a_n = 2|NTF(e^{j2\pi f_n})|/(\pi n). \tag{2.27}$$

For example, assume that a small dc input $u = 1/(2 \cdot OSR)$ is applied to MOD1, and use the above formulas. The largest tone in the output is at a frequency $f_1 = x = (1+u)/2 = 0.5 + 1/(4 \cdot OSR)$, which is very close to 0.5, (i.e., to $f_s/2$). This fits our intuition because we expect the output of MOD1 to alternate between $-1$ and $+1$ most of the time when the input is small. The frequency of the $n = 2$ tone is $f_2 = 2x = 1 + 1/(2 \cdot OSR)$, which aliases to the edge of the passband at $f'_2 = 1/(2 \cdot OSR)$. Since $|NTF(e^{j2\pi f_2})| \approx \pi/(OSR)$, the power associated with this tone is $a_n^2/2 \approx 1/(2 \cdot OSR^2)$. Increasing $u$ would move this second tone out of the band-of-interest, resulting in a sharp drop in the in-band noise, while decreasing $u$ would lower the frequency and hence increase the attenuation provided by the $|NTF(e^{j2\pi f_n})|$ factor in $a_n$. This argument accounts for the shape of the noise peaks in Fig. 2.22 around $u = 0$, which resemble the letter M, and also agrees with the results quoted on page 39 regarding the width and height of these noise peaks.

## 2.7 Stability of MOD1

Linear analysis based on Bode plots would predict unconditional stability for MOD1, since the loop gain decreases by $-6$ dB/octave and the loop phase is $-90°$ at all frequencies. This prediction, however, does not take into account the actual signal processing performed by the quantizer. Hence, time-domain considerations, taking into account nonlinearities, are required.

Consider the stability of the loop under dc excitation. It is obvious that loop becomes unstable, and specifically $y$ becomes unbounded, if $|u| > 1$. For example, if $u = 1.3$, the DAC will try desperately to balance $u$ by feeding back a signal $-1$ every time. Even so, a net input of 0.3 will enter the integrator in every clock period, until $y$ grows so large that the circuit becomes dysfunctional.

Vice versa, if $|u| \leq 1$ and the initial value of $y$ satisfies $|y(0)| \leq 2$, then the loop will remain stable, with $|y|$ bounded by 2. This stability condition is sufficient even for time-varying $u$ and is readily established by induction as follows. Replacing the dc input term $u$ in (2.21) with $u(n)$ gives

$$y(n) = y(n-1) + u(n) - \text{sgn}(y(n-1)). \tag{2.28}$$

If $|y(0)| \leq 2$, then $|y(0) - \text{sgn}\, y(0)| \leq 1$. Thus $|y(1)| = |u(1) + y(0) - \text{sgn}\, y(0)| \leq 2$, since $|u(1)| \leq 1$. Continuing, we can show that $|y(2)|$, $|y(3)|$, etc. all are similarly bounded, provided $|u(n)| \leq 1$ for all $n$.

If $|y(0)| > 2$ and $|u(n)| \leq 1$, then the modulator output will contain a string of +1s (if $y(0) > 2$) or −1s (if $y(0) < -2$) and $|y|$ will monotonically decrease until it is less than 2, at which point the condition of the preceding paragraph will hold and $|y|$ will remain bounded by 2. Thus, it is clear that MOD1 is stable with arbitrary inputs less than or equal to 1 in magnitude, and that it is able to recover from any initial condition.

## 2.8 The Effects of Finite Op-Amp Gain

The foregoing analyses have all assumed ideal operation of MOD1. In this section, we take a brief look at the effect of an important non-ideality: finite op-amp gain. Both linear and nonlinear system perspectives are used, since each provides worthwhile insights.

### 2.8.1 Linear Systems Perspective– Degraded Noise Shaping

If the gain of the op-amp in Fig. 2.14 is $A$, then the difference equation which applies to the charge $q_2(n)$ on integrating capacitor $C_2$ is

$$q_2(n) = q_2(n-1) + C_1\left(u(n) - v(n-1) - \frac{q_2(n)}{C_2(A+1)}\right) \tag{2.29}$$

Assuming $C_1 = C_2$ and $A \gg 1$, the z-transform of the voltage across $C_2$ is

$$Y(z) = p\frac{zU(z) - V(z)}{z - p} \quad (2.30)$$

where the pole $p = 1 - 1/A$ is slightly less than 1. The integrator is therefore *lossy* or *leaky*; its dc gain $Y(1)$ is no longer infinite.

The noise transfer function is then found to be

$$NTF(z) = 1 - pz^{-1}. \quad (2.31)$$

From this expression, it is apparent that the NTF's zero shifts from $z = 1$ to $z = p$, inside the unit circle. Thus the NTF gain at dc changes from its ideal value of zero to $1 - p = 1/A$ and MOD1 loses its ability to achieve infinite precision with dc signals. For busy input signals, the additional noise power resulting from the "filling in" of the noise notch at low frequencies can be estimated by integrating $|NTF(e^{j\omega})|^2 \approx A^{-2} + \omega^2$ across the $\omega \in [0, \pi/(OSR)]$ band of interest and comparing the result against the $A = \infty$ case. If $A > OSR$, the additional noise is less than 0.2 dB, and hence this effect is rarely serious. Although this argument suggests that high op-amp gain is not a critical requirement, the reader should be aware that the above argument assumes linear op-amp gain. As will be discussed in Chapter 6, low op-amp gain can be problematic if the gain is sufficiently nonlinear. Also, the finite-gain effect has serious consequences for the cascade (MASH) architectures discussed in Chapter 5.

### 2.8.2 Nonlinear Systems Perspective—Dead Zones

Assume next that a small positive dc input $u$ is applied to MOD1 under ideal conditions ($A = \infty$), and that the initial condition is $y(0) = 0$. By (2.28), the first few values of the quantizer input are

$$\begin{aligned} y(1) &= y(0) + u - \text{sgn}(y(0)) = u - 1 < 0 \\ y(2) &= u - 1 + u + 1 = 2u > 0 \\ y(3) &= 2u + u - 1 = 3u - 1 < 0 \end{aligned} \quad (2.32)$$

and so on.

## 2 The First-Order Delta Sigma Modulator

Clearly, $v$ alternates: $v(0) = 1$, $v(1) = -1$, $v(2) = 1$, $v(3) = -1$ etc., at least initially. This pattern would hold for all time if the input were exactly zero.

For the $k^{th}$ sample, we have

$$y(k) = \begin{cases} ku - 1, & \text{if } k \text{ is odd} \\ ku, & \text{if } k \text{ is even} \end{cases} \quad (2.33)$$

The even-indexed $y(k)$ will always be positive. The effect of $u > 0$ will occur in the first odd-indexed sample satisfying $ku \geq 1$. At this point, the sign of $y(k)$ becomes positive, and hence $v(k) = 1$ occurs twice in a row after $k \geq 1/u$ periods. The frequency with which this happens is a measure of the value of $u$.

If the integrator is lossy, such that its transfer function becomes $H(z) = \dfrac{p}{z-p}$ as in (2.30), then the recursion relation (2.28) becomes

$$y(n) = py(n-1) + u - \text{sgn}(y(n-1)). \quad (2.34)$$

Assuming again a small positive dc input $u$, and $y(0) = 0$, we now get

$$y(1) = y(0) + u - \text{sgn}(y(0)) = u - 1 < 0$$
$$y(2) = pu - p + u + 1 = (1+p)u + (1-p) > 0$$
$$y(3) = p(1+p)u + p(1-p) + u - 1 = (1 + p + p^2)u - (1 - p + p^2) < 0$$
$$\ldots$$

The general formula is clearly

$$y(k) = \sum_{i=0}^{k-1} p^i u + (-1)^k \sum_{i=0}^{k-1} (-p)^i \quad (2.35)$$

In order for $u$ to have an effect on the output $v$, $\text{sgn}[y(k)]$ must stop alternating for some odd value of $k$. For this to happen, the magnitude of the first term in (2.35) must become larger at some point than that of the second term. For $k \to \infty$, this requires that

$$\frac{u}{1-p} > \frac{1}{1+p} \tag{2.36}$$

or

$$u > \frac{1-p}{1+p} = \frac{1/A}{2-1/A} \approx \frac{1}{2A} \tag{2.37}$$

where $A \gg 1$ is the opamp gain. A similar result can easily be found for $|u|$ if $u < 0$.

The conclusion is that with finite op-amp gain, inputs smaller than $1/(2A)$ in normalized value will have no effect on the output. For example, if $A = 1000$ and $V_{ref} = 1$ V, then dc signals less than 0.5 mV will not register in the output. There is thus a *dead zone* or *dead band* around $u = 0$ if MOD1 uses a leaky integrator. It can be shown that similar dead zones exist around all rational values of $u$, and that with the exception of the dead zones around $\pm 1$, these other dead zones are narrower than the one around $u = 0$. Fig. 2.24 plots the simulated average output, $\bar{v}$, of MOD1 for dc values around zero when $p = 10^{-3}$. This figure shows that the width of the dead zone is as predicted, and that the error quickly decays to a value much less than 0.5 mV for inputs outside the dead zone.

Another consequence of the finite amplifier gain concerns the limit cycles of MOD1. The limit cycles displayed by an ideal MOD1 with a dc input are unstable, or *non-attracting*, because arbitrarily small shifts in the input eventually lead to large changes in the integrator state and hence a different output pattern. However, if the NTF zero is inside the unit circle, the resulting limit cycles are stable or *attracting*, because sufficiently small shifts in the input lead to small changes in the integrator state and no change in the output pattern. This is a detrimental effect, since limit cycles are usually undesirable. We note in passing that if the NTF zero is placed outside the unit circle, all limit cycles are *repelling* and the modulator forms a *chaotic system* [6-8]. This effect may be used to suppress idle tones.

## 2.9 Decimation Filters for MOD1

As shown in Sec. 2.5, the output signal of MOD1 contains very little quantization noise at low frequencies, but the PSD of the output noise grows rapidly with increasing frequencies, and is four times higher at $f_s/2$ than the power of the white quantization noise assumed in the analysis. Hence, the signal band limit $f_B$ must be much smaller than $f_s/2$, and if MOD1 is to be used as an ADC, the out-of-band noise must be removed by a digital lowpass filter (LPF). Afterwards, the LPF's output signal may be decimated, thereby reducing the sampling rate to the Nyquist sampling frequency $2f_B$.

Clearly, the requirements on the LPF are that its gain response be flat and large over the signal band $[0, f_B]$, and very small between $f_B$ and $f_s/2$. Often, it is also desirable to have a flat group delay response in the signal band. The latter condition can be satisfied by using a *linear-phase finite-impulse-response* (FIR) LPF. Since the output signal of MOD1 is a single-bit data stream, it may be practical to use a single-stage high-order linear-phase FIR filter, since there are no actual multiplications required between the samples of the signal and the weights of the taps. In general, however, it is usually more efficient and economical to carry out the fil-

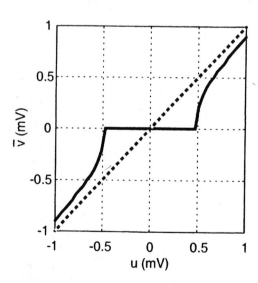

Figure 2.24: Dead-band behavior of MOD1 around $u = 0$ with $A = 1000$.

tering and decimation in stages. The stages most often used are the so-called *sinc filters*. These will be discussed next.

### 2.9.1 The Sinc Filter [9]

A *sinc* filter is an FIR filter with N-1 delays and N equal-valued tap weights, which simply computes a running average of the input data stream $v(n)$. Thus, the output $w(n)$ of a sinc filter is

$$w(n) = \frac{1}{N} \sum_{i=0}^{N-1} v(n-i). \qquad (2.38)$$

Its impulse response is hence given by

$$h_1(n) = \begin{cases} 1/N, & \text{if } (0 \leq n \leq N-1) \\ 0, & \text{otherwise} \end{cases}. \qquad (2.39)$$

The z-domain transfer function is

$$H_1(z) = \frac{1}{N} \frac{1 - z^{-N}}{1 - z^{-1}} \qquad (2.40)$$

and thus its frequency response is

$$H_1(e^{j2\pi f}) = \frac{\text{sinc}(Nf)}{\text{sinc}(f)} \qquad (2.41)$$

where sinc($f$) is defined as $(\sin(\pi f))/(\pi f)$, whence the name of this filter. Figures 2.25 and 2.26 show the impulse response and frequency response, respectively, for a sinc filter with $N = 32$. As Fig. 2.26 shows, the gain is close to 1 at and near $f = 0$, and close to zero near $f_s/N$ and its harmonics. Hence, if the sampling rate at the output of the filter is reduced by $N$, causing the noise around $f_s/N$, $2f_s/N$, $3f_s/N$, etc. to fold into the baseband, the energy of the aliased noise will be reduced by the sinc filter's attenuation. Hence, $N = OSR$ should be chosen for this application.

It is instructive to find out how much residual noise remains at the output of the sinc filter when it is used to suppress the shaped quantization noise $NTF(z)E(z)$ of MOD1, compared to what an ideal LPF with a bandwidth $f_B = f_s/(2N)$ could achieve. The z-transform of the noise at the output of the sinc filter is

$$Q_1(z) = H_1(z)NTF(z)E(z) = \frac{1}{N}(1 - z^{-N})E(z) \qquad (2.42)$$

so that in the time domain

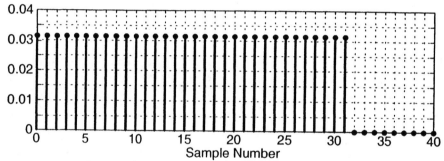

Figure 2.25: Impulse response of a sinc filter with $N = 32$.

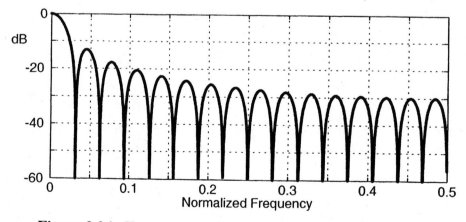

Figure 2.26: Frequency response of a sinc filter with $N = 32$.

$$q_1(n) = \frac{1}{N}[e(n) - e(n-N)]. \qquad (2.43)$$

Assuming $e(n)$ and $e(n-N)$ are uncorrelated, and that each has an rms value $\sigma_e^2$, the quantization noise power at the output of the sinc filter is

$$\sigma_{q_1}^2 = \frac{2\sigma_e^2}{N^2}. \qquad (2.44)$$

The output noise power for an ideal LPF with unit gain at dc is given by Eq (1.7). Using $OSR = N$:

$$\sigma_q^2 = \frac{\pi^2 \sigma_e^2}{3N^3}. \qquad (2.45)$$

Thus, it is clear that the sinc filter is nearly $N$ times less effective than the ideal LPF. Hence, the sinc filter is seldom used as a complete decimation filter. Rather, it normally forms only one stage in a multi-stage filter.

An important advantage of the sinc filter is that for single-bit modulators it can be realized very economically. In the implementation shown in Fig. 2.27, a counter is incremented for each +1 from the modulator. Once in every $N$ clock cycles, the output of the counter is clocked into a register, and the counter is reset. Thus, the output $w(n)$ is a down-sampled count of the number of +1s produced by the modulator over the last $N$ clock cycles. For this reason, a sinc filter with downsampling is also known as an *accumulate-and-dump* stage. If $N = 2^k$, the counter produces a $k$-bit output[†] which may be interpreted as a binary fraction between 0 and 1. In this interpretation, the ADC is considered to be unipolar. For a bipolar ADC, sim-

---

† Strictly speaking, the output of the counter requires k+1 bits in order to accommodate the possibility that the number of +1s accumulated equals $2^k$. To avoid adding an extra bit for this overflow condition, the counter should saturate, instead of wrapping around.

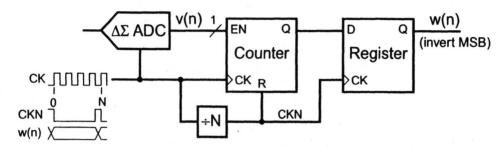

Figure 2.27: Sinc filter realized with a counter.

ply invert the MSB of the counter output, and interpret the result as a signed binary fraction in two's-complement form.

### 2.9.2 The Sinc² Filter

Cascading two sinc filters results in the so-called *sinc²* filter. The impulse response of the sinc² filter is obtained by convolving the rectangular impulse response of the sinc filter with itself. A triangular response results (Fig. 2.28). The z-domain and frequency-domain transfer functions are simply those of the sinc filter, squared:

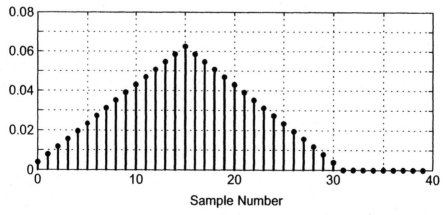

Figure 2.28: Impulse response of an $N = 16$ sinc² filter.

$$H_2(z) = \left[\frac{(1-z^{-N})}{N(1-z^{-1})}\right]^2 \quad (2.46)$$

$$H_2(e^{j2\pi f}) = \left(\frac{\text{sinc}(Nf)}{\text{sinc}(f)}\right)^2 \quad (2.47)$$

The z-transform of the noise at the output of the sinc filter is therefore

$$Q_2(z) = NTF(z)H_2(z)E(z) = \frac{1}{N^2}\frac{(1-z^{-N})}{(1-z^{-1})}(1-z^{-N})E(z)$$

$$= \frac{1}{N}H_1(z)[(1-z^{-N})E(z)]. \quad (2.48)$$

Since $H_1$ implements an $N$-sample average in the time domain and $(1-z^{-N})E(z)$ corresponds to $e(n) - e(n-N)$ in the time domain,

$$q_2(n) = \frac{1}{N^2}\sum_{i=0}^{N-1}[e(n-i) - e(n-N-i)]. \quad (2.49)$$

If $e$ is white noise of power $\sigma_e^2$, the power of the quantization noise at the output of the sinc$^2$ filter is

$$\sigma_{q_2}^2 = \frac{2N\sigma_e^2}{N^4} = \frac{2\sigma_e^2}{N^3}. \quad (2.50)$$

Comparing $\sigma_{q_2}^2$ with the ideal LPF's output $\sigma_q^2$ from (2.45), we see that $\sigma_{q_2}^2$ is actually less than $\sigma_q^2$. However, by taking into account the passband droop, which reduces the signal energy at the LPF output, we find that the SNR is in reality slightly lower with a sinc$^2$ LPF than with an ideal LPF. Since the SNR difference is small, the sinc$^2$ filter is sufficient for the decimation of MOD1's output data. (In general, a sinc$^{L+1}$ LPF is sufficient for an $L^{\text{th}}$-order loop.)

The $sinc^2$ filter can of course be realized by cascading two sinc filters. However, to minimize the hardware requirements, the triangularly-weighted sum needed for the impulse response of a $sinc^2$ filter can be constructed from rectangularly-weighted and ramp-weighted sums, as illustrated in Fig. 2.29a. Fig. 2.29b depicts the realization of a $sinc^2$ filter for single-bit data which implements the concept of Fig. 2.29a. The hardware is made especially simple if $N$ is assumed to be a power of 2. A sinc filter generates the rectangularly-weighted sums, while the ramp-weighted sums are generated by conditionally accumulating the output of a counter which wraps to zero every $N$ clock cycles. The filter output is then an arithmetic combination of these intermediate sums.

It should be pointed out that both sinc and $sinc^2$ filters introduce a droop into the baseband frequency response. If necessary, this droop may be equalized with a post-filter operated at the reduced clock rate.

**2.10 Conclusions**

This chapter discussed quantizers and the first-order $\Delta\Sigma$ modulator MOD1. It was shown that the quantization error of a multi-bit quantizer is usually well-approximated by a white noise $e$ of power $\sigma_e^2 = \Delta^2/12$. For the single-bit quantizers often found in $\Delta\Sigma$ modulators, this approximation is still useful, but must be

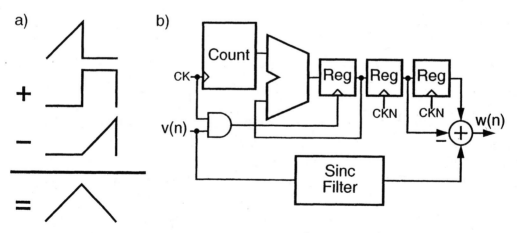

Figure 2.29: A hardware-efficient $sinc^2$ filter implementation.

applied with caution. In particular, the designer should verify that the quantizer gain used in the linear model accurately reflects the actual operation of the quantizer in the $\Delta\Sigma$ loop.

The first-order $\Delta\Sigma$ modulator, MOD1, can be used as an ADC or as a DAC, and although each application can use a variety of topologies and technologies in the implementation, the essential dynamics of all these systems are the same. The most important property of MOD1 in its ideal form is that it can achieve arbitrary accuracy for dc inputs. For low-bandwidth signals, the effective number of bits is approximately $1.5\log_2 OSR$, where $OSR = f_s/(2 \cdot f_B)$ is the oversampling ratio.

When binary quantization is used, MOD1 is immune to a variety of non-idealities, including component errors and misplaced DAC levels. However, an important non-ideality which does have an impact on MOD1 is finite op-amp gain. Finite gain $A$ creates dead zones in the input-output characteristic of MOD1, the width of which are proportional to $1/A$. MOD1 is also prone to limit cycles, an undesirable oscillatory condition. Last but not least, MOD1 requires a very large OSR for accurate data conversion.

# References

[1] D. A. Johns and K. Martin, *Analog Integrated Circuit Design*, John Wiley & Sons, New York, New York, 1997, pp. 450-451.

[2] J. C. Candy and O. J. Benjamin, "The structure of quantization noise from sigma-delta modulation," *IEEE Transactions on Communications*, vol. 29, no. 9, pp. 1316-1323, September 1981.

[3] V. Friedman, "The structure of the limit cycles in sigma delta modulation," *IEEE Transactions on Communications*, vol. 36, no. 8, pp. 972-979, August 1988.

[4] O. Feely and L. O. Chua, "The effect of integrator leak in $\Sigma$-$\Delta$ modulation," *IEEE Transactions on Circuits and Systems*, vol. 38, no. 11, pp. 1293-1305, November 1991

[5] R. M. Gray, "Spectral analysis of quantization noise in a single-loop sigma-delta modulator with dc input," *IEEE Transactions on Communications*, vol. 37, no. 6, pp. 588-599, June 1989.

[6] M. O. J. Hawksford, "Chaos, oversampling, and noise-shaping in digital-to-analog conversion," *Journal of the Audio Engineering Society*, vol. 37, no. 12, December 1989.

[7] O. Feely and L. O. Chua, "Nonlinear dynamics of a class of analog-to-digital converters," *International Journal of Bifurcation and Chaos*, vol. 2, no. 2, June 1992, pp. 325-340.

[8] R. Schreier, "On the use of chaos to reduce idle-channel tones in delta-sigma modulators," *IEEE Transactions on Circuits and Systems I,* vol. 41, no. 8, pp. 539-547, August 1994.

[9] J. C. Candy, "Decimation for sigma-delta modulation," *IEEE Transactions on Communications*, vol. 34, no. 1, pp. 72-76, January 1986.

# CHAPTER 3
# *The Second-Order Delta-Sigma Modulator*

The previous chapter studied MOD1, the first-order $\Delta\Sigma$ modulator. Despite the advantages of simplicity, robustness and stability, the overall performance of MOD1 in terms of resolution and idle-tone generation is inadequate for most applications. This chapter focuses on MOD2, the second-order $\Delta\Sigma$ modulator. We will see that MOD2 overcomes the above disadvantages of MOD1, but does so at the expense of more hardware and a slight reduction in signal range.

## 3.1 The Second-Order Modulator: MOD2

The simplest way to construct a second-order modulator from MOD1 is to replace the quantizer within MOD1 with another copy of MOD1. The resulting structure is shown in Fig. 3.1. Modeling the quantizer in the new system with an additive noise source results in the linear $z$-domain model of Fig. 3.2. From the figure,

$$V(z) = E(z) + \frac{1}{1-z^{-1}}\left(-z^{-1}V(z) + \frac{1}{1-z^{-1}}(-z^{-1}V(z) + U(z))\right)$$

$$= \frac{(1-z^{-1})^2 E(z) - [(1-z^{-1})z^{-1} + z^{-1}]V(z) + U(z)}{(1-z^{-1})^2} \quad (3.1)$$

from which it follows that

$$V(z) = U(z) + (1-z^{-1})^2 E(z). \tag{3.2}$$

Hence, the signal transfer function is $STF(z) = 1$, as for MOD1, but now the noise transfer function is $NTF(z) = (1-z^{-1})^2$. Since this is the square of MOD1's NTF, we expect increased attenuation of quantization noise at low frequencies.

In the frequency domain, the squared magnitude of the NTF is given by

$$\left|NTF(e^{j2\pi f})\right|^2 = (2\sin\pi f)^4 \approx (2\pi f)^4, \text{ for } f \ll 1. \tag{3.3}$$

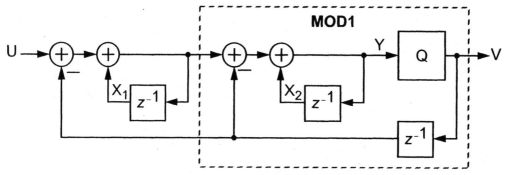

Figure 3.1: MOD2: A second-order $\Delta\Sigma$ modulator obtained by replacing the quantizer in MOD1 with another copy of MOD1.

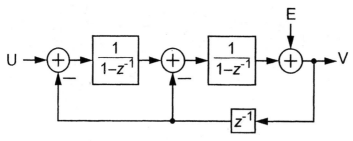

Figure 3.2: Linear model of MOD2.

Fig. 3.3 compares the NTF magnitude of MOD2 and MOD1. At low frequencies, the NTF of MOD1 displays a 20 dB/decade slope, while the NTF of MOD2 has a 40 dB/decade slope. The increased attenuation at frequencies close to dc is desirable because it reduces the amount of quantization noise within the signal band. Unfortunately, since the gain at high frequencies provided by MOD2's NTF is greater than that of MOD1's NTF, the total power of the quantization noise at the output of MOD2 is more than that of MOD1.

Following the same steps as in the derivations for MOD1, we can calculate the power of the in-band quantization noise contained in $v$:

$$\sigma_q^2 \approx \int_0^{1/(2 \cdot OSR)} (2\pi f)^4 \cdot 2 \cdot \sigma_e^2 df = \frac{\pi^4 \sigma_e^2}{5(OSR)^5}. \quad (3.4)$$

Comparing this with the power $M^2/2$ of a sine wave of amplitude $M$, and assuming that $\sigma_e^2 = 1/3$, we find the SQNR of MOD2

$$SQNR = \frac{M^2/2}{\sigma_q^2} = \frac{15M^2(OSR)^5}{2\pi^4}. \quad (3.5)$$

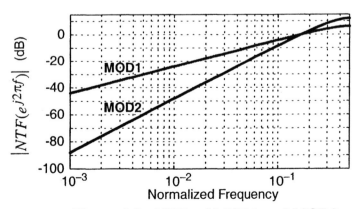

**Figure 3.3: NTFs of MOD1 and MOD2.**

## 3 The Second-Order Delta-Sigma Modulator

Fig. 3.4 compares the theoretical SQNR vs. OSR performance of MOD2 with that of MOD1 for $M = 1$. The SQNR of MOD2 is proportional to the fifth power of OSR, rather than third power, as in the case of MOD1. Hence, doubling OSR increases SQNR by a factor of 32 (15 dB), and thus increases the resolution by 2.5 bits.

As an example of the achievable ENOB, let $OSR = 128$ and $M = 1$. Then, (3.5) gives $SQNR = 94.2$ dB, which corresponds to nearly 16-bit resolution. If the ADC is used to convert audio signals, so that $f_B = 20$ kHz, the clock rate needed will be $f_s = 2 \cdot OSR \cdot f_B = 5.12$ MHz, a highly practical value. To achieve the same resolution with MOD1, (2.18) predicts that $OSR = 1800$ would be needed. This necessitates the use of $f_s = 72$ MHz, which would make the implementation unnecessarily difficult.

If binary quantization is used, MOD2 has the same inherent linearity property as MOD1. Like MOD1, the linearity of MOD2 is not jeopardized by comparator non-idealities, such as offset or hysteresis, since these errors are injected into the loop at the same point as the quantization noise and are thus also attenuated by the NTF. Slight shifts in the loop filter coefficients are similarly benign, since these translate into small shifts in the pole and zero locations of the NTF and STF, rather than into nonlinearities.

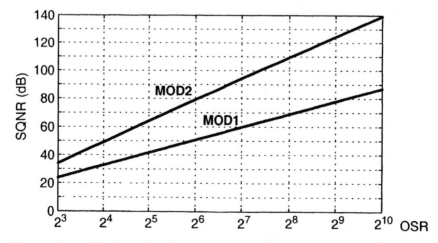

**Figure 3.4:** SQNR vs. OSR for MOD1 and MOD2 with $M=1$.

## 3.2 Simulation of MOD2

As with MOD1, the validity of the assumptions which lead to analytical SQNR predictions for MOD2 are best checked by performing simulations based on the modulator's difference equations. Fig. 3.5 illustrates the input and output waveforms for MOD2 with a half-scale sine-wave input. As is typical in $\Delta\Sigma$ systems, these waveforms provide very little insight into the system's behavior when viewed in the time domain. Only crude observations can be made, such as the increased tendency for the output to be +1 when the input is positive and −1 when the input is negative. However, by viewing the modulator output in the frequency domain, a much clearer picture emerges.

The spectrum shown in Fig. 3.6 clearly exhibits noise shaping, and the 40 dB per decade slope confirms the second-order nature of the shaping. For an oversampling ratio of 128, integrating the PSD yields $SQNR = 86$ dB and extrapolating from this value to the SQNR for a full-scale input results in a predicted peak SQNR of 92 dB. This is in close agreement with the theoretical result of 94 dB predicted by (3.5).

However, Fig. 3.6 displays two features which are not in agreement with our model. First, the $3^{rd}$ and $5^{th}$ harmonics of the signal are distinctly visible, having amplitudes of −88 dBFS[†] and −90 dBFS, respectively. Since white quantization

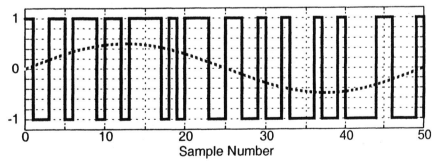

**Figure 3.5: MOD2 input and output waveforms.**

---

†. "dBFS" means "dB with respect to a full-scale sine wave."

noise cannot produce harmonics of the signal, this simulation result is inconsistent with the white noise model. Second, Fig. 3.6 shows a plot of MOD2's NTF, scaled by $2\sigma_e^2 NBW$ so that it should align with the observed PSD[†]. The theoretical curve is similar to the observed PSD, but has a corner frequency that is higher than that of the observed PSD. The higher corner frequency makes the theoretical PSD somewhat lower at low frequencies and somewhat higher at high frequencies than the observed PSD.

To account for the harmonics, a nonlinear quantizer model is needed. Such a model will be presented in the next section. To account for the shift in NTF shape, it is sufficient to evaluate the effective quantizer gain from simulation. Applying (2.5) to the simulation data results in $k = 0.63$, and recomputing the NTF for this value of $k$ results in the pole-zero and PSD plots shown in Fig. 3.7. Now the agreement between the observed PSD and that "predicted" by the linear model is very good.

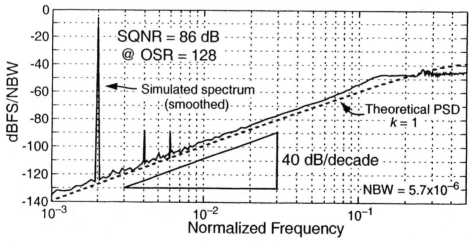

Figure 3.6: Output spectrum of MOD2 with a −6-dBFS sine-wave input.

---

† NBW is an abbreviation for *noise bandwidth*. See Appendix A for a discussion of the importance of NBW in the context of a PSD plot.

When the input amplitude is lowered, simulations show that the optimal value of the quantizer gain changes slightly, and is approximately $k = 0.75$ for inputs that are less than $-12$ dBFS. Thus, simulations indicate that a more accurate NTF for MOD2 with small inputs can be derived if it is assumed that $k = 0.75$, rather than one. As will be shown in Section 4.2, the following formula offers a convenient way to calculate the NTF for any value of the quantizer gain $k$ ($NTF_k$), once the NTF associated with a quantizer gain of 1 ($NTF_1$) is known:

$$NTF_k(z) = \frac{NTF_1(z)}{k + (1-k)NTF_1(z)}. \tag{3.6}$$

Applying (3.6) with $k = 0.75$ yields the following improved estimate of MOD2's NTF.

$$NTF(z) = \frac{(z-1)^2}{z^2 - 0.5z + 0.25} \tag{3.7}$$

Since this NTF has an in-band gain 2.5 dB above that of $NTF_1(z) = (1 - z^{-1})^2$, the in-band noise power seen in simulation should be about 2.5 dB higher than that predicted from $NTF_1$.

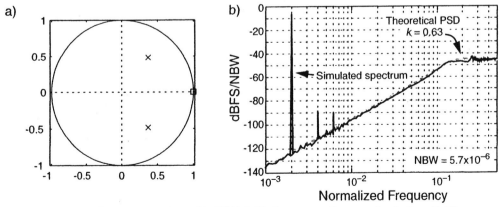

**Figure 3.7:** (a) NTF of MOD2 if the quantizer gain is $k = 0.63$; (b) Comparison of corresponding PSD with simulation.

## 3 The Second-Order Delta-Sigma Modulator

When the input amplitude exceeds half of full-scale, the optimal quantizer gain decreases, degrading the noise-shaping further and increasing the difference between the expected and simulated SQNR. To illustrate this, Fig. 3.8 plots the SQNR of MOD2 as a function of the input amplitude, for two different test frequencies, and compares the simulation results against the predictions of the unity-gain white-noise quantizer model. The simulation data follows (3.5) quite closely over the middle portion of the amplitude range. For small inputs, the observed SQNR is somewhat lower than predicted, with the result that 0 dB SQNR requires a larger input than expected. For large input amplitudes, the SQNR of MOD2 saturates, peaking for inputs in the vicinity of −5 dBFS, and then dropping abruptly as the input amplitude approaches full-scale. The largest degradation in SQNR occurs for low-frequency full-scale signals because these signals apply large values for extended periods of time to the input of MOD2. Comparing Fig. 3.8 with the corresponding one for MOD1 (Fig. 2.21), we see that the SQNR of MOD2 is more well-behaved than MOD1, except for the aforementioned saturation under large-signal conditions.

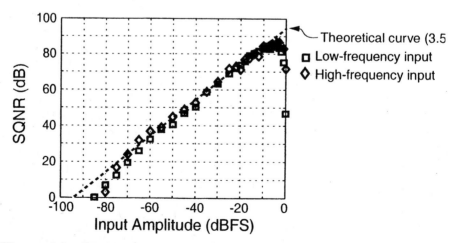

**Figure 3.8:** Simulated SQNR of MOD2 with $OSR = 128$.

## 3.3 Nonlinear Effects in MOD2

The simulations of MOD2 presented above confirm the essential insights provided by the linear model, but also exhibit anomalies which can only be explained with nonlinear models. Unfortunately, the dynamics of MOD2 with a binary quantizer are much more complex than those of MOD1 and defy exact analysis[†]. In light of the intractable dynamics, it is reasonable to use approximate and/or empirical techniques to explain the observed behavior of MOD2.

### 3.3.1 Signal-dependent quantizer gain

As Section 2.1.1 described, the gain of a binary quantizer is indeterminate. Moreover, the simulations of the previous section suggest that, in the case of MOD2, the quantizer's gain is in fact signal-dependent. A model for MOD2 which includes this signal-dependent gain (a nonlinear effect) is depicted in Fig. 3.9. In this figure, the quantizer is replaced by a weak nonlinearity, the *quantizer transfer curve* (QTC), plus additive noise. The QTC is determined by computing the average quantizer output as a function of the average quantizer input, while the dc input signal is swept across its range. Fig. 3.10 shows the QTC determined by simulations. This figure shows that the QTC is *compressive*, i.e. it exhibits reduced gain when the magnitude of its input is large. Fig. 3.10 also plots a cubic approximation to the quantizer nonlinearity, $\bar{v} = k_1 y + k_3 y^3$, and lists the $(k_1, k_3)$ coefficients of this approximation. We can use these coefficients to estimate the distortion induced by the QTC as follows.

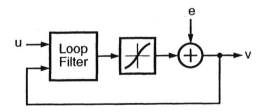

**Figure 3.9: A modulator model incorporating quantizer nonlinearity.**

---

[†]. For MOD2 with 4 quantization levels and dc inputs less than 0.5 LSB in magnitude, exact analysis is possible and has shown that the quantization noise is in fact white [2].

First, determine the effective NTF of the modulator by using $k = k_1$ as the effective the quantizer gain:

$$NTF(z, k_1) = \frac{(z-1)^2}{z^2 - 0.775z + 0.3875}. \qquad (3.8)$$

Next, since the distortion term $k_3 \bar{y}^3$ is added at the same place in the loop as the quantization error, we conclude that the spectrum of the distortion is shaped by $NTF(z, k_1)$. Thus, distortion is greatest for frequencies where the NTF gives the least protection, i.e. when the distortion term lies at the edge of the passband. For the spectrum of Fig. 3.7, the input signal is located at $f = f_s/500$ and thus the third harmonic is at a normalized frequency of $3f = 0.006$, where the attenuation provided by $NTF(z, k_1)$ is about 53 dB.

For a small low-frequency sine-wave input of amplitude $A$, the local average of the output follows the input and thus, according to the linear model, the local average of the quantizer input is also a low-frequency sine-wave, but of amplitude $A/k_1$. The amplitude of the third harmonic induced by the QTC is $k_3(A/k_1)^3/4$ [†], from which it follows that the third harmonic distortion is

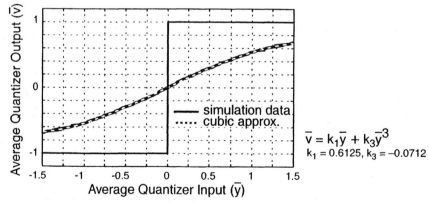

Figure 3.10: Empirically-determined quantizer transfer curve for MOD2.

---

†. The third harmonic component of $(A\sin\omega t)^3$ is $A^3/4$. [3]

$$HD_3 = \left| \frac{k_3 A^2}{4k_1^3} NTF(z, k_1) \right|, \tag{3.9}$$

where $z = e^{j2\pi(3f)}$. Evaluating (3.9) under the conditions of Fig. 3.7 ($A = 0.5$, $f = 1/500$) yields $HD_3 = -87$ dB. Since this calculated value is about 5 dB short of the observed value of $-82$ dB, the distortion estimate provided by the above calculation can only be considered a rough approximation. The discrepancy is greatest when the input is large and the distortion is severe.

Returning now to our empirical study of MOD2's behavior as its input is varied, Fig. 3.11 plots the in-band quantization noise power of MOD2 (with binary quanti-

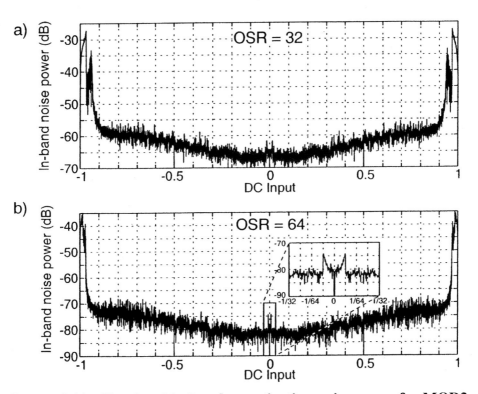

Figure 3.11: Simulated in-band quantization noise power for MOD2 as a function of the DC input level.

zation) vs. the dc input level for oversampling ratios of 32 and 64. Comparing these plots with those for MOD1 (Fig. 2.22), we immediately note that the numerous spikes associated with periodic behavior in MOD1 are now absent, and thus conclude that MOD2 is more resistant to tones than MOD1. However, as the expanded view around $u = 0$ in Fig. 3.11b illustrates, MOD2 still exhibits vestiges of this undesirable behavior.

A second feature apparent in the plots of Fig. 3.11 is the U-shape of the in-band noise power, which culminates in very large noise peaks as $|u| \rightarrow 1$. Even over the range $|u| < 0.7$ (i.e. for inputs less than −3 dBFS), the in-band noise varies by about 6 dB. The bulk of this behavior is accounted for by the signal-dependent quantizer gain (and hence signal-dependent NTF) of MOD2. As the input increases in magnitude, the quantizer gain decreases, reducing the loop gain and hence the effectiveness of the noise-shaping. As a result, the noise at the output of MOD2 increases with signal level. In the terminology of stochastic processes, this property makes the quantization noise of MOD2 with a large deterministic input *non-stationary*. Specifically, when the absolute value of the signal is large, the quantization noise power is also large, and thus the statistics of the quantization noise are time-varying.

### 3.3.2 Stability of MOD2

The preceding section used a combination of simulation and quasi-linear modeling to explain several of the nonlinear aspects of MOD2. In this section, results from the literature regarding the stability of MOD2 with a dc input are presented.

Hein and Zakhor [4,5] have shown that if MOD2 is constructed as in Fig. 3.1, so that the difference equations are

$$v(n) = \text{sgn}[x_2(n)] \tag{3.10}$$

$$\begin{bmatrix} x_1(n+1) \\ x_2(n+1) \end{bmatrix} = \begin{bmatrix} 1 & 0 \\ 1 & 1 \end{bmatrix} \begin{bmatrix} x_1(n) \\ x_2(n) \end{bmatrix} + \begin{bmatrix} -1 \\ -2 \end{bmatrix} v(n) + \begin{bmatrix} 1 \\ 1 \end{bmatrix} u, \tag{3.11}$$

then for dc inputs $u$ satisfying $|u| < 1$ the following bounds apply:

$$|x_1| \leq |u| + 2, \qquad (3.12)$$

$$|x_2| \leq \frac{(5-|u|)^2}{8(1-|u|)}. \qquad (3.13)$$

Fig. 3.12 plots these bounds along with values found by simulation. As this figure shows, the analytic bounds are fairly tight for $|u| \leq 0.7$, but the analytic bound on $x_2$ "blows up" more quickly than do the simulation results as $|u| \rightarrow 1$. Since $x_2$ is (a delayed version of) the quantizer input, Fig. 3.12 once again demonstrates that the input to the quantizer in MOD2 becomes large as the input approaches full-scale.

According to (3.12) and (3.13), the internal states of MOD2 are guaranteed to be bounded for dc inputs less than 1 in magnitude, although the bound on $x_2$ does become arbitrarily large as $|u| \rightarrow 1$. Since MOD2 tracks the low-frequency content of its input, one might assume that arbitrary time-varying inputs satisfying $|u(n)| < 1$ for all $n$ would lead to bounds similar to those for dc inputs. However, as Fig. 3.13 illustrates, an input waveform with $|u(n)| < 0.3$ can lead to large internal states if the waveform is chosen appropriately. The input waveform shown in Fig. 3.13 predominantly oscillates between +0.3 and –0.3, and it does so with just the right phase and period so as to drive the modulator state into ever-increasing oscillations. Furthermore, the values at the transitions are chosen so as to maximize the excursion of the internal oscillations. Fortunately, such a waveform is

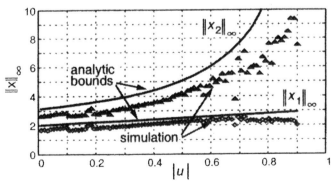

Figure 3.12: Comparison of analytic state bounds with simulation for MOD2.

unlikely to be encountered in practice, and even if it were, it is even more unlikely that it would be precisely synchronized with the modulator's internal dynamics the way the waveform in Fig. 3.13 is. It is known that MOD2 is stable for arbitrary inputs less than 0.1 in magnitude, but the upper limit on the input amplitude for which stable operation is guaranteed is not known.

The preceding discussion shows that MOD2 is "less stable" than MOD1. Although the stability of MOD2 with dc inputs less than 1 in magnitude has been rigorously established, the states may become arbitrarily large as $|u| \rightarrow 1$. It is therefore wise to limit the input of MOD2 to $|u| < 0.8$ or 0.9, if possible, so that the state of the second integrator is not unduly large. Unfortunately, even though such an input limit will keep the modulator state reasonable for dc and slowly-varying inputs, it is possible for the modulator state to become much larger than intended. It is therefore important to include means for detecting overly large states and for placing the modulator in a "good" state.

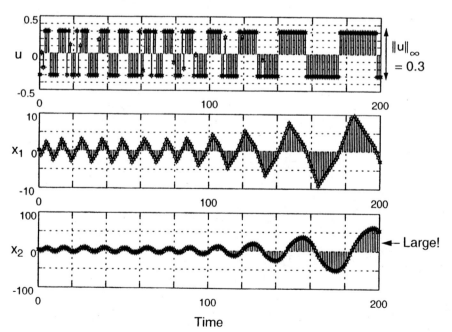

**Figure 3.13:** A hostile input with $|u| = 0.3$ can drive MOD2 "unstable."

### 3.3.3 Dead-band behavior

Section 2.8.2 showed that when the gain of the op amp in the integrator is finite, the zero in MOD1's NTF shifts from $z = 1$ to $z = 1 - \mu$, where $\mu$ is inversely proportional to the op-amp gain, $A$. For simplicity, we will assume as before that the proportionality constant is 1, i.e. $\mu = 1/A$[†]. This shift in the NTF zero creates dead bands in the response of MOD1 to dc inputs; the most troublesome of these dead bands is the one centered on $u = 0$, the width of which was calculated to be $1/A$.

In MOD2, we will see that the impact of finite op-amp gain is drastically reduced because the open-loop gain of the loop filter is proportional to $A^2$. Assuming as above that finite op-amp gain results in integrator poles at $z = 1 - \mu$, with $\mu = 1/A$, the two zeros in the NTF of MOD2 under finite gain are both located at $z = 1 - 1/A$.

To determine the width of the dead-band around $u = 0$ which results from finite amplifier gain, we will calculate the largest dc input which can maintain the period-2 limit cycle tabulated below in Table 3.1. Assume that the $v$ sequence is as

| $n$ | 0 | 1 |
|---|---|---|
| $x_1(n)$ | 0.5 | −0.5 |
| $x_2(n)$ | 0.75 | −0.75 |
| $v(n)$ | 1 | −1 |
| $x_1(n+1)$ | −0.5 | 0.5 |
| $x_2(n+1)$ | −0.75 | 0.75 |

**Table 3.1. A length-2 limit-cycle for MOD2 with $u = 0$.**

---

†. In practice, the proportionality constant depends on the values of the components used in the integrator.

shown in the third row, and that the op-amp gain is high enough that the response of MOD2 to $u = 0$ is close to that given in Table 3.1. Applying a small input $u \neq 0$ to MOD2 changes the dc component of the output of the first integrator to $Au$. Similarly, the dc component of the output of the second integrator becomes $A^2 u$. In order for this dc component to leave the sign of $x_2$ unchanged, we need $|A^2 u| < 0.75$. Thus, the expected width of the dead band centered on $u = 0$ is $(2 \cdot 0.75)/A^2 = 1.5/A^2$. For reasonable values of $A$, the dead band will therefore be very small.

To verify the preceding calculation, Fig. 3.14 plots the average value of the output of MOD2 for small dc inputs; the poles of the loop filter are located at $z = 1 - 1/A$ with $A = 100$. The observed width of the dead band is identical to our $1.5/A^2$ estimate. Furthermore, Fig. 3.14 shows that, as with MOD1, inputs outside the dead band are accurately represented by the average output of MOD2.

The gain-squaring nature of the two-integrator cascade found in MOD2 is responsible for MOD2's increased tolerance of finite op-amp gain and reduced susceptibility to tones relative to MOD1.[†] Also, as noted earlier, MOD2 possesses a vastly

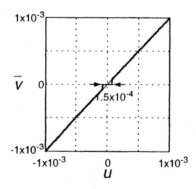

Figure 3.14: Dead-band behavior of MOD2 around $u = 0$ with $A = 100$.

---

[†]. An alternative perspective regarding tonal behavior in MOD1 and MOD2 considers the dimensionality of the state-space for each system. Since the state-space of MOD2 is 2-dimensional, whereas the state-space of MOD1 is only 1-dimensional, it is far more unlikely for a particular input to result in equal initial and final states in MOD2 than in MOD1. Thus repetitive behavior in MOD2 is less probable than in MOD1.

superior SQNR/OSR trade-off. Aside from the added analog complexity, the main drawback of MOD2 is reduced stability, which is manifested in a practical input range that is about 80% to 90% of the potential full-scale range of the converter. The following section presents alternative realizations of MOD2, as well as a modified version of MOD2 that offers improved SNR and better stability.

## 3.4 Alternative Second-Order Modulator Structures

A number of alternative structures exist which can perform a second-order modulation, and which also give a unity-gain STF (albeit with a delay of one or two clock periods), as well as the same $(1-z^{-1})^2$ NTF as the structure shown in Fig. 3.1. In devising such structures, care must be taken to avoid delay-free loops (which are not realizable), and to preserve reasonable robustness against the unavoidable nonideal practical effects such as element inaccuracies, finite op-amp gain, etc.

Second-order modulator structures which implement lowpass STFs exist, as do structures which implement NTFs other than plain second-order differentiation. In dealing with such NTFs, care must also be taken to ensure that the resulting modulator is stable. For example, converting the delay-free integrators of Fig. 3.1 into delaying-integrators and removing the feedback delay alters the NTF and yields a marginally-stable system.

### 3.4.1 The Boser-Wooley Modulator

A second-order modulator which contains two delaying integrators [6] is shown in Fig. 3.15. Using delaying integrators is desirable because it allows the op amps in each integrator to settle independently of each other, thereby relaxing their speed requirements [7].

Presuming unity gain for the quantizer Q, linear analysis shows that the STF is

$$STF(z) = \frac{a_1 a_2 z^{-2}}{D(z)} \qquad (3.14)$$

and the NTF is

## 3 The Second-Order Delta-Sigma Modulator

$$NTF(z) = \frac{(1-z^{-1})^2}{D(z)}, \tag{3.15}$$

where

$$D(z) = (1-z^{-1})^2 + a_2 bz^{-1}(1-z^{-1}) + a_1 a_2 z^{-2}. \tag{3.16}$$

To achieve $STF(z) = z^{-2}$ and $NTF(z) = (1-z^{-1})^2$, the conditions $a_1 a_2 = 1$ and $a_2 b = 2$ must be satisfied. Since we have 3 parameters and only 2 constraints, there are many possible solutions. For example, $a_1 = a_2 = 1$, $b = 2$, or $a_1 = 0.5$, $a_2 = 2$ and $b = 1$ can be used. In the actual design process, dynamic range scaling (to be discussed in Section 8.3) removes any ambiguity in finding the parameters needed to implement a given NTF and STF.

### 3.4.2 The Silva-Steensgaard Structure

Another useful second-order structure is shown in Fig. 3.16 [8][9]. The distinguishing features of this circuit are the direct feedforward path from the input to the quantizer and the single feedback path from the digital output. Linear analysis confirms that the output is given in the z-domain by

$$V(z) = U(z) + (1-z^{-1})^2 E(z) \tag{3.17}$$

as before. The input signal to the loop filter is, however, different: it contains only the shaped quantization noise:

$$U(z) - V(z) = -(1-z^{-1})^2 E(z) \tag{3.18}$$

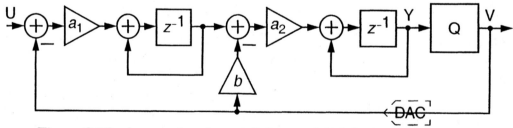

Figure 3.15: A second-order modulator with delaying integrators.

Since the loop filter thus does not need to process the signal, the requirements on its linearity may be greatly reduced, which is a significant practical advantage. Also, as can be seen from (3.18), the output of the second integrator will give $-z^{-2}E(z)$ directly. This is advantageous if the modulator is the input stage of a MASH structure.

### 3.4.3 The Error-Feedback Structure

As an example of a circuit which looks simple and hence attractive, but which is impractical for analog $\Delta\Sigma$ loops, we shall next consider the so-called *error-feedback structure* shown in Fig. 3.17. Here, the quantization error $e$ is obtained in analog form by subtracting the internal ADC's input from the DAC output; $e$ is then fed back to the input through a filter $H_f$. The output signal in the z-domain is

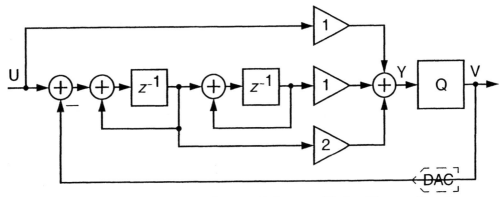

Figure 3.16: A second-order modulator with feedforward paths.

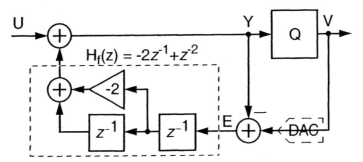

Figure 3.17: Error-feedback implementation of MOD2.

$$V(z) = E(z) + U(z) + H_f(z)E(z) \qquad (3.19)$$

Hence, $STF(z) = 1$ and $NTF(z) = 1 + H_f(z)$. To obtain $NTF(z) = (1-z^{-1})^2$ as in (3.2), $H_f(z) = (1-z^{-1})^2 - 1 = z^{-2} - 2z^{-1}$ is needed. This is readily realized as shown in Fig. 3.17.

In spite of its attractive simplicity, the structure of Fig. 3.17 is not practical for analog implementation, since it is very sensitive to variations of its parameters. Assume, e.g., that the element (say, a capacitor) realizing the multiplication by 2 has a +0.5% mismatch error, so its actual value is 2.01. Then the actual NTF will be $H'(z) = (1-z^{-1})^2 - 0.01z^{-1}$. At very low frequencies, instead of 0, the magnitude of $H'$ will then equal 0.01, or −40 dB. Hence, even with careful analog design, the achievable ENOB will typically be less than 10 bits for ADCs with a single-bit quantizer, even for high OSR. By contrast, comparable parameter changes still allow over 18-bit resolution for all second-order structures discussed earlier.

Note that the error feedback structure is very useful, and is often applied, in the digital loops required in $\Delta\Sigma$ DACs. There, e.g., the parameter values 1 and 2 are easily and exactly realized. More general values can also be accurately implemented. This topic will be discussed in detail in Chapter 7.

### 3.4.4 Generalized Second-Order Structures

The structure of Fig. 3.1 gives $STF(z) = 1$ and $NTF(z) = (1-z^{-1})^2$. More general STFs can be obtained by feeding $u$ not only to the input of the first integrator, but also to the inputs of the second integrator and the quantizer (Fig. 3.18). The STF then becomes

$$STF(z) = b_1 + b_2(1-z^{-1}) + b_3(1-z^{-1})^2 \qquad (3.20)$$

The STF now has two zeros, and a double pole at $z = 0$. In this way, a "free" second-order FIR signal pre-filter can be incorporated into the ADC.

Similarly, by feeding the output signal $v$ back to all three blocks in the forward path (Fig. 3.19), two nonzero poles can be generated in both the STF and the NTF. Thus, more general STFs and NTFs may be obtained. The new functions are

$$STF(z) = \frac{B(z)}{A(z)} \qquad (3.21)$$

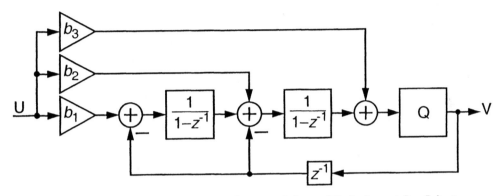

**Figure 3.18:** A second-order modulator with multiple input feed-in terms.

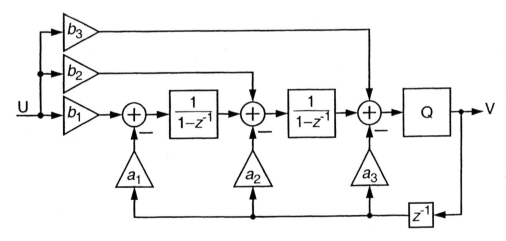

**Figure 3.19:** A second-order modulator with arbitrary feed-in and feed-back coefficients.

and

$$NTF(z) = \frac{(1-z^{-1})^2}{A(z)}, \quad (3.22)$$

where

$$B(z) = b_1 + b_2(1-z^{-1}) + b_3(1-z^{-1})^2 \quad (3.23)$$

and

$$A(z) = 1 + (a_1 + a_2 + a_3 - 2)z^{-1} + (1 - a_2 - 2a_3)z^{-2} + a_3 z^{-3}. \quad (3.24)$$

Since the $a_3$ feedback term increases the NTF order to 3, but does not increase the number of in-band NTF zeros, this term is rarely used. (It may be useful if the modulator is continuous-time, rather than discrete-time as assumed here.)

By incorporating both multiple feedforward and feedback features into the second-order structure, more flexibility is obtained for enhancing stability and improving dynamic range.

### 3.4.5 Optimal Second-Order Modulator

Given the variety of modulator architectures shown thus far, the reader might well wonder "Which architecture is the best?" One part of the answer to this question deals with the optimal NTF, while the second part of the answer deals with the optimal topology. The STF is a secondary consideration because the STF merely filters the signal; it plays no role in determining the peak SQNR. Also, since the choice of the topology is tied more closely to practical considerations than to fundamental mathematical limits, this section will only consider the problem of optimizing the NTF.

Thus, the first step in finding an optimal second-order modulator is to find the second-order NTF which yields the highest SQNR, or equivalently, the NTF which minimizes the in-band noise. For high values of OSR, the magnitude of $NTF(z) = (1-z^{-1})^2/A(z)$ in the passband is approximately $K\omega^2$, where

$K = 1/|A(1)|$. By shifting the NTF zeros from $z = 1$ to $z = e^{\pm j\alpha}$, the magnitude of the NTF in the passband becomes $K(\omega - \alpha)(\omega + \alpha) = K(\omega^2 - \alpha^2)$. The integral of the square of this quantity over the passband is a measure of the in-band noise, and can be minimized by choosing $\alpha$ such that

$$I(\alpha) = \int_0^{\omega_B} (\omega^2 - \alpha^2)^2 d\omega \tag{3.25}$$

is minimized. The solution to this optimization problem can be obtained by differentiating $I(\alpha)$ with respect to $\alpha$, and equating the result to 0. This gives

$$\alpha_{opt} = \omega_B/\sqrt{3}. \tag{3.26}$$

Since the ratio $I(0)/I(\alpha_{opt}) = 9/4$, the expected SQNR improvement is $10\log_{10}(9/4) = 3.5$ dB. Note that these results assume that the quantization noise is white, and that $|A(z)| \approx |A(1)|$ in the 0-$\omega_B$ frequency range.

An exhaustive search of the NTF design space for the NTF with the highest peak SQNR yields the NTF whose pole-zero plot and SQNR curve are depicted in Fig. 3.20. The denominator of this optimal NTF is

$$A_{opt}(z) = 1 - 0.5z^{-1} + 0.16z^{-2}. \tag{3.27}$$

Compared to the SQNR curve of MOD2 shown in Fig. 3.8, the SQNR curve associated with this NTF is more linear and supports signals closer to full-scale without saturating. As a result, the peak SQNR for this NTF is about 94 dB, which is about 6 dB higher than that of MOD2.

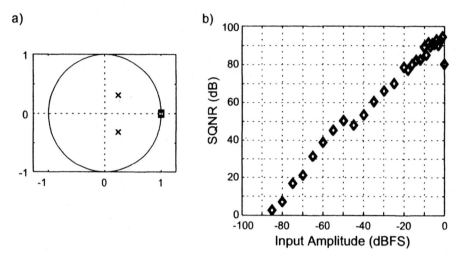

**Figure 3.20:** Optimal second-order NTF for OSR = 128;
(a) pole-zero plot, (b) SQNR curve.

## 3.5 Decimation Filtering for Second-Order ΔΣ Modulators

As with MOD1, efficient decimation filtering can be based on cascaded sinc filters for higher-order modulators also. In general, the determining factors in finding the order $K$ of the $\text{sinc}^K$ filter for an $L^{\text{th}}$-order ΔΣ modulator are

(a) the filter should cut off at a faster rate around $f_B$ than the NTF of the ΔΣ modulator rises there, and

(b) its gain response should be flatter around $f_s/OSR$ and its harmonics than the NTF is around dc.

Condition (a) insures that very little out-of-band noise is left unsuppressed around $f = f_B$ after decimation, while Condition (b) guarantees that the folding of the noise from frequency bands around $f_s/OSR$, $2f_s/OSR$, etc. after decimation adds little to the in-band noise. Both conditions require that $K > L$; usually, $K = L + 1$ is adequate.

If high resolution is required, a two-stage approach is efficient (Fig. 3.21). Here, a $\text{sinc}^K(\pi f/f_D)$ filter, clocked at $f_s$, is used first to reduce the sampling rate from $f_s$ to the intermediate clock rate $f_D$ [10, Ch.1]. The sinc filter is followed by a sharp lowpass filter clocked at the reduced rate $f_D$. The second filter, which may have either a *finite impulse response* (FIR) or an *infinite impulse response* (IIR), suppresses the remaining noise in the range between $f_B$ and $f_D/2$, allowing the further reduction of the clock frequency to its decimated value close to $2f_B$. The second filter may be implemented with a cascade of *halfband FIR filters* [11,12] and may also incorporate compensation for the droop introduced by the $\text{sinc}^K$ filter. Fig. 3.22 illustrates the spectra and filter responses for the two-stage approach with $f_D = 4f_B$. In the case of MOD2, $L = 2$, and thus a $\text{sinc}^3$ filter is adequate for the first stage.

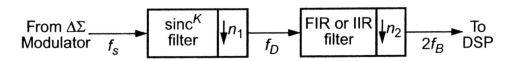

Figure 3.21: A two-stage decimation filter.

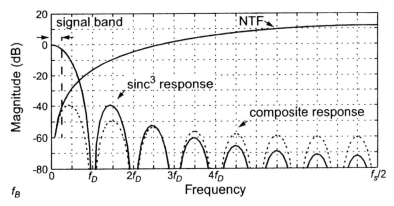

Figure 3.22: Decimating with a $\text{sinc}^3$ filter; $f_s$ is the modulation rate, $f_D$ is the intermediate decimation frequency and $[0, f_B]$ is the signal band.

## 3 The Second-Order Delta-Sigma Modulator

In selecting the intermediate frequency $f_D$, a compromise must be reached. Lowering $f_D$ allows the reduction of the speed and complexity of the second filter. On the other hand, as $f_D$ approaches $2f_B$ so that the intermediate OSR ($OSR_1 = f_D/(2f_B)$) drops below 4, condition (b) may not hold even for $K = L + 1$; furthermore, the droop of the $sinc^K$ filter at $f_B$ becomes large. Hence, an intermediate OSR of 4 seems to be about optimal. In [13], useful graphical design information is given for the added noise and droop introduced by this approach.

A $sinc^3$ filter may be constructed from a combination of three of the accumulator sections already introduced in Section 2.9. However, a more economical realization of $sinc^K$ filters for $K > 2$ is provided by the *Hogenauer structure* [14] shown in Fig. 3.23a. This structure is obtained by factoring the transfer function of the $sinc^3$ filter as follows

$$H(z) = \left(\frac{1}{1-z^{-1}}\right)\left(\frac{1}{1-z^{-1}}\right)\left(\frac{1}{1-z^{-1}}\right)\left(\frac{1-z^{-N}}{N}\right)\left(\frac{1-z^{-N}}{N}\right)\left(\frac{1-z^{-N}}{N}\right) \quad (3.28)$$

and realizing each factor by an accumulator or differencing stage. By performing the decimation just before the differencer stages, the $N$-period delays needed to

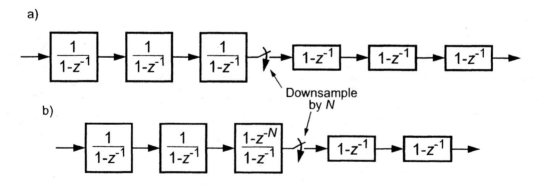

Figure 3.23: Hogenauer structure of a $sinc^3$ filter.

realize the $z^{-N}$ factors can be implemented by single-delay blocks. The constant factor $N$ is usually an integer power of 2, and hence division by $N$ does not require any arithmetic operation.

In the Hogenauer structure, the accumulators will tend to overflow, but if an arithmetic such as *two's complement* or *modulo arithmetic* with "wrap-around" characteristics is used, then the correct result will be obtained at the output, in spite of the intermediate overflows. For an $M$-bit input, using $M + K\log_2 N$ bits throughout is sufficient to realize the filter exactly. Fewer bits can be used in internal states since the associated truncation errors are noise-shaped in the final output. Also, the structure can be simplified somewhat by combining the last accumulator and the first differencer into an accumulate-and-dump stage as shown in Fig. 3.23b [10, Ch. 1].

## 3.6 Conclusions

This chapter examined the second-order modulator, MOD2 and several of its variants. Like MOD1, MOD2 is theoretically able to achieve arbitrarily high resolution for dc inputs, and is immune to a variety of imperfections. In contrast to MOD1, which displays a 9 dB increase in SQNR for every doubling of OSR, the SQNR of MOD2 increases at 15 dB per octave. As a result, MOD2 is able to achieve a given level of performance with a lower sampling rate than MOD1 would require. Due to the gain-squaring provided by the two-integrator cascade, MOD2 is also more robust in the face of finite op-amp gain than MOD1. Furthermore, the quantization noise at the output of MOD2 is less likely to contain tones than the noise at the output of MOD1. In all these areas, MOD2 is markedly superior to MOD1. The primary drawbacks associated with MOD2 are increased hardware complexity (both analog and digital) and a slight decrease in the allowable signal range.

## References

[1] J. C. Candy, "A use of double integration in sigma-delta modulation," *IEEE Transactions on Communications*, vol. 33, no. 3, pp. 249-258, March 1985.
[2] N. He, F. Kuhlmann and A. Buzo, "Double-loop sigma-delta modulation with dc input," *IEEE Transactions on Communications*, vol. 38, no. 4, pp. 487-495, April 1990.
[3] D. Johns and K.W. Martin, *Analog Integrated Circuits*, Wiley, New York, 1997.

[4] S. Hein and A. Zakhor, "On the stability of interpolative sigma delta modulators," *Proceedings of the 1991 IEEE International Symposium on Circuits and Systems,* vol. 3, pp. 1621-1624, June 1991.

[5] S. Hein and A. Zakhor, "On the stability of sigma delta modulators," *IEEE Transactions on Signal Processing,* vol. 41, no. 7, pp. 2322-2348, July 1993.

[6] B. E. Boser and B. A. Wooley, "The design of sigma-delta modulation analog-to-digital converters," *IEEE Journal of Solid-State Circuits,* vol. 23, pp. 1298-1308, December 1988.

[7] R. Gregorian and G. C. Temes, *Analog MOS Integrated Circuits,* John Wiley & Sons, New York, New York, 1986, pp. 491-496.

[8] J. Silva, U. Moon, J. Steensgaard and G. C. Temes, "Wideband low-distortion delta-sigma ADC topology," *Electronics Letters,* vol. 37, no. 12, pp. 737-738, June 7 2001.

[9] J. Steensgaard, "High performance data converters," Ph.D. thesis, Technical University of Denmark, Department of Information Technology, March 8, 1999. Available on the worldwide web at www.steensgaard.org.

[10] S. R. Norsworthy, R. Schreier and G. C. Temes, Eds., *Delta-Sigma Data Converters,* Piscataway NJ: IEEE Press, 1997.

[11] T. Saramäki, T. Karema, T. Ritoniemi and H. Tenhunen, "Multiplier-free decimator algorithms for superresolution oversampled converters," *Proceedings of the 1990 IEEE International Symposium on Circuits and Systems,* vol. 4, pp. 3275-3278, May 1990.

[12] D. A. Johns and K. Martin, *Analog Integrated Circuit Design,* John Wiley & Sons, New York, New York, 1997, pp. 551-554.

[13] J. C. Candy, "Decimation for sigma-delta modulation," *IEEE Transactions on Communications,* vol. 34, no. 1, pp. 72-76, January 1986.

[14] E. B. Hogenauer, "An economical class of digital filters for decimation and interpolation," *IEEE Transactions on Acoustics, Speech and Signal Processing,* vol. 29, pp. 155-162, April 1981.

[15] K. C. H. Chao, S. Nadeem, W. L. Lee, and C. G. Sodini, "A higher order topology for interpolative modulators for oversampling A/D conversion," *IEEE Transactions on Circuits and Systems,* vol. 37, pp. 309-318, March 1990.

[16] R. Schreier, "An empirical study of high-order single-bit delta-sigma modulators," *IEEE Transactions on Circuits and Systems II,* vol. 40, no. 8, pp. 461-466, August 1993.

# CHAPTER 4
# *Higher-Order Delta-Sigma Modulation*

In this chapter, higher-order modulators are developed as generalizations of the first and second-order modulators studied in earlier chapters. High-order modulators are found to offer improved performance at the expense of more hardware and reduced signal range. This chapter also investigates various architectural options for implementing high-order modulators.

## 4.1 High-Order Single-Quantizer Modulators [1]-[3]

Fig. 4.1 shows the general structure of a single-quantizer $\Delta\Sigma$ modulator. In this diagram, the modulator is divided into two parts: a linear part (the loop filter) containing memory elements, and a memoryless nonlinear part (the quantizer). The loop filter is a two-input system whose single output $Y$ can be expressed as a linear combination of its inputs $U$ and $V$:

$$Y(z) = L_0(z)U(z) + L_1(z)V(z). \tag{4.1}$$

The operation of the quantizer is, as usual, described as the addition of an error signal:

$$V(z) = Y(z) + E(z). \tag{4.2}$$

Using these two equations, the output $V$ can be written as a linear combination of two signals, namely the modulator input $U$ and the quantization error $E$:

$$V(z) = STF(z)U(z) + NTF(z)E(z), \tag{4.3}$$

where

$$NTF(z) = \frac{1}{1 - L_1(z)} \quad \text{and} \quad STF(z) = \frac{L_0(z)}{1 - L_1(z)}. \tag{4.4}$$

Conversely, given the desired NTF and STF, one can compute the loop filter transfer functions which are required to implement them, namely

$$L_0(z) = \frac{STF(z)}{NTF(z)} \quad \text{and} \quad L_1(z) = 1 - \frac{1}{NTF(z)}. \tag{4.5}$$

Since these relations apply regardless of the structure of the loop filter, the input-output characteristics of the loop filter (and hence of the modulator) are determined solely by $STF$, $NTF$ and the properties of the quantizer. Structural details of the loop filter are thus unimportant in determining the (ideal) input-output behavior of the modulator.

The reader is reminded that the quantizer Q shown in Fig. 4.1 is in fact an analog-to-digital converter. Hence, $y(n)$ is an analog signal (usually voltage), while $v(n)$ is a digital data stream. Thus, (4.2) implies that the reference voltage $V_{REF}$ of the ADC is assumed to be unity, or equivalently that all analog quantities are in fact normalized to $V_{REF}$. This also means that an ideal DAC using the same $V_{REF}$ is a

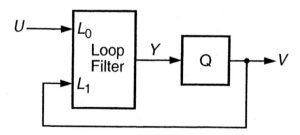

Figure 4.1: General structure of a single-quantizer $\Delta\Sigma$ modulator.

unity-gain block. Hence, the DAC was omitted from the feedback path of the structure of Fig. 4.1.

Equation (4.4) suggests that $L_1$ must be large in the signal frequency range 0 to $f_B$ to reduce the NTF there, and hence $L_0$ must also be large in the same range to allow the STF to remain close to 1. This suggests that both $L_1$ and $L_0$ should have their poles in this range. In fact, as can be foreseen from the structure of Fig. 4.1 and its input-output relations (4.1)-(4.5), due to the shared circuitry used to realize both $L_0$ and $L_1$, these two functions indeed usually have the same poles (which are also the zeros of NTF). $L_0$ and $L_1$ do, however, have in general different zeros.

In the simplest case, the signal is only delayed by $k$ clock periods in the modulator so that the STF satisfies $|STF| = 1$, and the NTF requires the quantization noise to be differentiated $N$ times. Then,

$$STF(z) = z^{-k} \text{ and } NTF(z) = (1 - z^{-1})^N \quad (4.6)$$

which, by (4.5), leads to

$$L_0(z) = z^{-k}(1 - z^{-1})^{-N} \text{ and } L_1(z) = 1 - (1 - z^{-1})^{-N} \quad (4.7)$$

The $N$ poles of both $L_0$ and $L_1$ lie on the unit circle, at $z = 1$. Considering their zeros, $N - k$ of the $N$ zeros of $L_0$ lie (for $N > k$) at $z = 0$, and $k$ zeros at infinity, while those of $L_1$ are located at the roots the equation $(1 - z^{-1})^N = 1$. This gives the following formula for the $N - 1$ zeros of $L_1$

$$z_i = \frac{1}{1 - e^{\frac{j2\pi}{N}i}} = \frac{1}{2}[1 + j\cot(\pi i/N)], \, i = 1, 2, \ldots, N - 1. \quad (4.8)$$

(The $N^{\text{th}}$ zero corresponding to $i = 0$ is at infinity.)

The zeros and poles of $L_0$ and $L_1$ are illustrated for $N = 2, 3, 4$ and 5 in Fig. 4.2. Note that for higher degrees $N$, the finite zeros of $L_1$ may be outside the unit circle, in which case $L_1$ is a *non-minimum-phase* transfer function.

An important special case occurs when the loop filter has a single input, and only the difference $u(n) - v(n)$ enters the loop filter (Fig. 4.3). Then $L_0 = L$ and $L_1 = -L$, and (4.4) becomes

$$NTF(z) = \frac{1}{1 + L(z)} \text{ and } STF(z) = \frac{L(z)}{1 + L(z)}. \qquad (4.9)$$

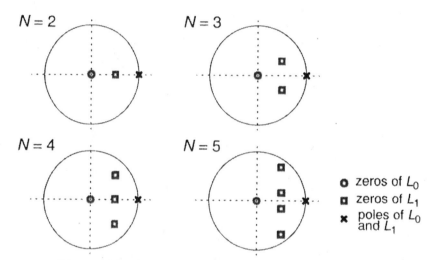

Figure 4.2: Poles and zeros of the $L_0$ and $L_1$ loop filters for $NTF(z) = (1 - z^{-1})^N$.

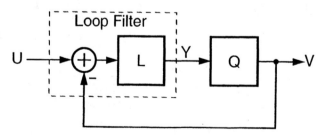

Figure 4.3: The single feedback topology.

where $L$ is the transfer function of the common portion of the loop filter. Now $L$, along with the quantizer, determines all important properties (stability, signal and noise transfer) of the modulator.

In another special situation (Fig. 4.4), a forward signal path is added to the structure of Fig. 4.3 so that $L_0 = L + 1$ while $L_1$ remains unchanged. Hence, the NTF remains as given in (4.9), but the STF becomes

$$STF(z) = \frac{L(z) + 1}{1 + L(z)} = 1. \qquad (4.10)$$

The output thus contains the input without any filtering or delay. It is only scaled by $V_{REF}$, here assumed to be 1. Hence, the input to the loop filter is, in the $z$ domain,

$$U - V = U - (STF \cdot U + NTF \cdot E) = -E/(1 + L). \qquad (4.11)$$

Thus, the input to the loop filter no longer contains the signal, only the filtered quantization noise. This makes the design of the loop filter a much easier task, since its linearity need not be very high [4], [5].

Note that there must be at least one clock-period of delay in the loop containing Q and $L_1$ in Fig. 4.1. Otherwise, a given value $y(n)$ would make $v(n)$ into $y(n) + e(n)$, and this sample would pass through $L_1$, and change $y(n)$ instantly.

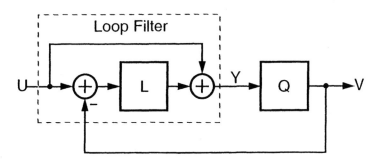

**Figure 4.4: The single feedback topology with a feedforward path.**

Thus the value of $y(n)$ would continuously vary during the same period! In the structure of Fig. 4.1, this delay must exist in the transfer function $L_1$.

From the required loop delay, we can derive an important realizability condition which restricts the choice of the NTF. This condition can be derived from (4.2), which in the time domain translates into $v(n) = y(n) + e(n)$. Assume that there is no input, so $u(n) = 0$, and that initially all signals are zero. Let next the error change to $e(0) = 1$. The new value of the output will then be $v(0) = y(0) + 1 = 1$.[†]

Since, as shown above, there must be at least one delay in the loop filter $L_1$ between its input $v(n)$ and its output $y(n)$ (otherwise, a physically unrealizable delay-free loop would exist in the modulator), $y(n)$ cannot depend on $v(n)$, and for the conditions stated above $v(0) = e(0) = 1$ must hold. This condition is easily translated into a condition on the NTF. Denoting for brevity the NTF by $H(z)$, by the above derivation the impulse response $h(n)$ corresponding to the NTF $H(z)$ must satisfy the condition $h(0) = 1$. From the equation $H(z) = h(0) + z^{-1}h(1) + z^{-2}h(2) + \ldots$, it can also easily be seen that this is equivalent to the condition $H(\infty) = 1$.

Next, let be $H(z)$ a rational function of the form

$$H(z) = \frac{b_m z^m + b_{m-1} z^{m-1} + \ldots + b_1 z + b_0}{a_n z^n + a_{n-1} z^{n-1} + \ldots + a_1 z + a_0} \qquad (4.12)$$

Then it follows from $H(\infty) = 1$ that $m = n$ and $b_m = b_n = a_n$ must hold.

In conclusion, if a physically unrealizable delay-free loop is to be avoided in the modulator, the equivalent conditions

---

[†] Note that in the physical structure $e(n)$ does not change independently of $y(n)$; however, we are discussing here the linearized model where $e(n)$ is an independent input.

$$h(0) = H(\infty) = \frac{b_n}{a_n} = 1 \tag{4.13}$$

on the NTF must hold. This seriously restricts the choice of $H(z)$ available to the designer.

A simpler proof of (4.13), but one without physical insight, is based on (4.4). As discussed above, the loop filter $L_1(z)$ must delay its input signal $v$ by at least one clock period. Thus the first sample of the impulse response of $L_1$ is zero, i.e. $L_1(\infty) = 0$. Then, by (4.4),

$$H(\infty) = \frac{1}{1 - L_1(\infty)} = 1. \tag{4.14}$$

The rest of (4.13) follows directly.

## 4.2 Stability Considerations in High-Order Modulators

The linearized model of the single-quantizer modulator (as described by Fig. 4.1 and Eqs. 4.1-4.5) predicts that the stability of the modulator is determined solely by the loop gain $L_1(z)$, which in turn is determined by the NTF $H(z)$. However, this argument ignores the nonlinear limitations of the quantizer. For example, consider a situation in which the input $u(n)$ is so large that the input to the first integrator is positive at every time step. In this case, the output of the integrator will monotonically increase without bound, and the loop will be unstable. This thought experiment clearly demonstrates that the magnitude of the input is of relevance in determining the stability of the loop. The STF is similarly relevant in that it effectively provides a (frequency-dependent) scaling of the input magnitude.

The range of input magnitudes over which the modulator functions properly is called the *stable input range*. Proper modulator operation is assured if the loop filter remains linear (i.e. internal signals do not grow so large as to saturate the op amps) and if the internal quantizer is not severely overloaded. By virtue of the preceding argument, the stable input range must be less than or equal to the full-scale of the first feedback DAC. In a high-order modulator, especially in one employing single-bit quantization, the stable input range is usually a few dB below the full-scale range of the feedback DAC. This loss in range usually results from the non-linear effects of quantizer overload rather than from insufficient linear range in the loop filter. Since the input to the quantizer is given by

$$Y(z) = STF(z)U(z) + (NTF(z) - 1)E(z) \qquad (4.15)$$

it is clear that when the modulator input $u(n)$ (as amplified by the STF) approaches the edge of the no-overload range of the quantizer, the addition of filtered quantization error may push $y(n)$ into the overload range. This overload will increase $e(n)$, which will aggravate the original overload, and this vicious circle eventually results in severe quantizer overload and hence the saturation of the active blocks. Correction of this runaway behavior often requires some sort of intervention, such as modulator reset, since the return of the input to the stable input range of the modulator may not be sufficient to restore stable operation.

Since the STF acts essentially like a pre-filter, the stable input range of a $\Delta\Sigma$ modulator is primarily determined by the NTF and the number of bits in the quantizer. Single-bit quantization is an important case which is given special attention in the following section. Multi-bit quantization is discussed Section 4.2.2.

### 4.2.1 Single-Bit Modulators

If the NTF is all one needs to know to describe the stability properties of a binary (1-bit) modulator, an important question is "What NTF properties are necessary and sufficient for stable operation?" Unfortunately, a simple and exact answer to this question is not known. The proven results are generally either too restrictive (that is, too conservative), or apply only for specific modulators with constant inputs. The most widely-used approximate criterion is the (modified) *Lee criterion* [1][7]:

A binary $\Delta\Sigma$ modulator with an $NTF = H(z)$ is likely to be stable if $\max_{\omega}|H(e^{j\omega})| < 1.5$.

Note that this criterion is neither necessary (as we have seen in some stable MOD2, $\max_\omega |H(e^{j\omega})| = 4$ was allowed), nor sufficient (the criterion says nothing about a limit on the input signal). Nevertheless, due to its simplicity, it is of some use.

The quantity $\max_\omega |H(e^{j\omega})|$ is the maximum gain of the NTF over all frequencies, also known as the *infinity-norm* of $H$, for which the mathematical notation is $\|H\|_\infty$. In the original statement of the condition, the limit on $\|H\|_\infty$ was given as 2, but as experience with higher-order modulators was gained, the rule of thumb was revised to use a limit of 1.5. For moderate-order modulators (order 3 or 4), slightly higher values may be tolerable, while for very high-order modulators (7 or more) a more conservative $\|H\|_\infty < 1.4$ may be more appropriate.

Although Lee's rule is a helpful rule of thumb for predicting *a priori* instabilities in single-bit modulators, it has no solid theoretical foundations, and needs to be confirmed by extensive simulations.

Note that the maximum of $|H(e^{j\omega})|$ usually occurs at $\omega = \pi$, since this point is farthest from the zeros (which are clustered around $z = 1$) and closest to the poles. An exception may occur if high-Q poles exist in $H(z)$, in which case the peak value may occur near the dominant (highest-Q) pole.

A somewhat more reasoned and satisfying explanation of the runaway behavior associated with instability is based on the linear modulator model shown in Fig. 4.5. Here, the quantization has been replaced by a multiplication with a con-

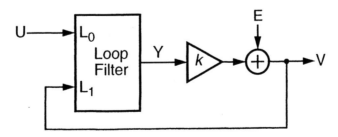

Figure 4.5: **Linear model of a modulator wherein the quantizer is modeled by a gain $k$ and additive error signal $E$.**

stant $k$ and by the addition of the random noise $e(n)$. The gain $k$ can be found by simulation, as described in Section 2.1. The result obtained there was

$$k = \frac{\langle v, y \rangle}{\langle y, y \rangle} = \frac{E[|y|]}{E[y^2]}. \tag{4.16}$$

where the quantization $v(n) = \text{sign}[y(n)]$ was assumed. In the improved linear model which incorporates $k$, the transfer functions $L_0$ and $L_1$ are effectively replaced by $kL_0$ and $kL_1$, respectively. Hence, the new noise transfer function is now

$$NTF_k(z) = \frac{1}{1 - kL_1(z)} = \frac{NTF_1(z)}{k + (1-k)NTF_1(z)} \tag{4.17}$$

as already stated in Section 3.2. Here, $NTF_1(z)$ is the NTF for $k = 1$. The natural modes of the linear model are the roots of the denominator of $NTF_k(z)$. By drawing the locus of these roots for $0 < k < 1$, the stability of the system may be predicted [8]. (In some cases, $k$ may be larger than 1, but in the overload zone $k$ is usually less than one.) Note that for $k = 0$, these roots coincide with the zeros of $NTF_1(z)$; for $k = 1$, with its poles. For stability, all roots must be within the unit circle. As an example, Fig. 4.6 shows the root locus of a fifth-order modulator. For

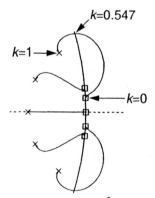

Figure 4.6: Root locus of a $5^{th}$-order modulator as the effective quantizer gain, $k$, varies.

$k > 0.547$ all roots are inside the unit circle, but if $k$ goes below 0.547, one pair of roots moves outside the unit circle.

Note that the root locus indicates a reduced stability for decreasing values of $k$. To understand this phenomenon better, consider the Bode gain plot associated with the modulator loop. The loop gain is $kL_1(z)$, which (as explained in Section 4.1) typically has all its $N$ poles at or near dc, i.e. $z = 1$, on the unit circle. Hence, the log-log gain-vs.-frequency plot at low frequencies is high, but it drops rapidly, with a slope $-20N$ dB/octave. A typical Bode plot for $k = 1$ is as shown in Fig. 4.7. It indicates conditional stability, since there are frequencies at which $\angle L_1 = 180°$

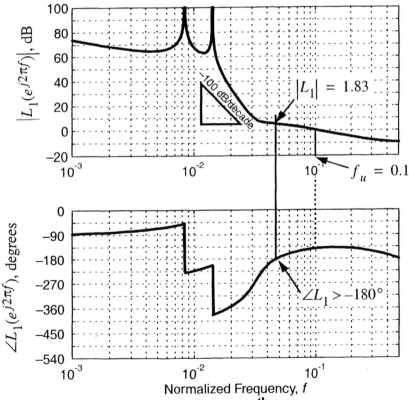

**Figure 4.7: Bode plot for the 5$^{th}$-order loop filter.**

and $|L_1| > 1$. For decreasing $k$ values the $|kL_1|$ curve drops until the marginal instability condition, $|kL_1| = 1$ at $\angle L_1 = 180°$, holds. The value of $k$ for which this occurs, namely $k_{min} = 1/1.83 = 0.547$, is the lowest allowable quantizer gain as indicated in Fig. 4.6.

Since the output amplitude $|v(n)|$ of the single-bit quantizer is fixed, a smaller $k$ by (4.16) is equivalent to larger input signals $|y(n)|$ and hence larger $|u(n)|$, confirming the observation that instability can be prevented by limiting the input signal.

An alternative way to explain the phenomenon of conditional stability is via the Nyquist plot, shown in Fig. 4.8. From linear systems theory [9], the number of clockwise encirclements of the *critical point* at $(-1,0)$ made by the $-L_1(e^{j\omega})$ contour as $\omega$ varies from $-\pi$ to $\pi$ is equal to the number of closed-loop poles which are outside the unit circle, minus the number of open-loop poles which are outside the unit circle. Since $L_1$ has poles which are on the unit circle, the $e^{j\omega}$ contour is modified to have infinitesimal bumps that skirt these singularities on the outside of the unit circle. The number of open-loop poles which are then outside this modified contour is zero, and thus the condition for closed-loop stability is simply that

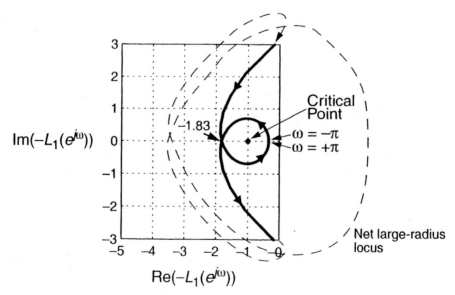

**Figure 4.8**: Nyquist plot for the loop filter of a $5^{th}$-order modulator.

the number of clockwise encirclements of the critical point is zero. In Fig. 4.8, the number of encirclements is zero, so the closed loop system is expected to be stable. However, if the quantizer gain were to drop by a factor of 1.83 or more, the $-L_1(e^{j\omega})$ contour would encircle the critical point twice and the (linear) closed-loop system would be unstable.

The reader should note that the root-locus, Bode and Nyquist techniques for assessing the stability of a $\Delta\Sigma$ modulator are all based upon the same linear modulator model. Any of these techniques can be used to compute the range of quantizer gains over which the linear model is stable, and the results will be identical. Unfortunately, since a $\Delta\Sigma$ modulator is a nonlinear system, the quantizer gain is signal-dependent and this dependency cannot captured by any linear model.

A more sophisticated (and complicated!) technique for modeling one-bit modulators was suggested by Ardalan and Paulos [10]. They proposed splitting the modulator into two linear loops, one processing the signal and the other one the noise, with different equivalent quantizer gains. The quantizer gains are nonlinear functions of the modulator's input. The predictions obtained using this model for the maximum stable input and SNR vs. input characteristics have been found to be very close to the values obtained by simulation. The quasi-linear model advocated by Ardalan and Paulos was extended to non-Gaussian probability density functions by Risbo [11].

In practice, even a very elaborate model cannot predict loop stability with 100% reliability, and the designer of high-order single-bit modulators must resort to extensive simulations. As shown in Fig. 3.13, some rapidly varying input signals can drive a modulator loop into instability, even if their amplitudes are relatively low. Hence, simulations should be performed using realistic worst-case input signals. As R. Adams suggests [6, Sec. 5.4], a representative worst-case signal is a square-wave with the largest permissible amplitude and with a fundamental frequency $\omega_f$ equal to that of the dominant (highest-Q) pole of the NTF. Usually, $\omega_f$ is far outside of the signal band of the modulator, and hence this observation suggests that the stability can be improved by analog prefiltering, which prevents such harmful out-of-band signals from exciting any resonant oscillations in the modulator loop.

## 4.2.2 Multi-Bit Modulators [12]

For multi-bit modulators, the following theoretical result can be useful:

> Consider a modulator with an **$M$-step** (i.e. **($M$+1)-level**) quantizer which has one of the characteristics shown in Fig. 2.3 and 2.4. Let the initial input $y(0)$ to the quantizer be within its linear (no-overload) range. Then, the modulator is guaranteed not to experience overload for any input $u(n)$ such that $\max_n |u(n)| \leq M + 2 - \|h\|_1$, where $\|h\|_1 = \sum_{n=0}^{\infty} |h(n)|$. Here, $h(n)$ is the inverse $z$-transform of the noise transfer function $H(z)$.

For example, if $M = 16$ and $H = (1 - z^{-1})^3$, then $\|h\|_1 = 8$ and thus any input with $\max_n |u(n)| < 10$ is guaranteed to not overload the modulator, i.e. this modulator system is stable for inputs up to 62.5% of the full-scale value 16.

It is easy to use the above rule to establish a more specific condition, namely that modulators employing $N^{th}$-order differentiation for the NTF and $M = 2^N + 1$ steps in the quantizer are stable for arbitrary inputs at least up to half of the quantizer input range. For an $N^{th}$-order differentiating NTF,

$$H(z) = (1 - z^{-1})^N = 1 - Nz^{-1} + -..., \qquad (4.18)$$

indicating that $|h(n)| = (-1)^n h(n)$, and hence, $\|h\|_1 = H(-1) = 2^N$. Thus, the condition $\max_n |u(n)| < 2^{N+1} + 2 - 2^N = 2^N + 2 = M/2 + 2$ guarantees stability. Since the input range of the quantizer spans $-(M+1)$ to $(M+1)$, any input which is less than half of the quantizer input range is sufficiently small to prevent quantizer overload.

To prove the above sufficient (but not necessary!) condition, we first calculate the input $y(n)$ of the quantizer Q (Fig. 4.1). In the $z$-domain

$$Y(z) = V(z) - E(z) = STF(z)U(z) + NTF(z)E(z) - E(z) \qquad (4.19)$$

and hence, assuming that the STF is only a delay of $k$ clock periods,

$$y(n) = u(n-k) + (h*e)(n) - e(n) = u(n-k) + \sum_{i=1}^{\infty} h(i)e(n-i) \quad (4.20)$$

Here, the asterisk ($*$) denotes convolution, and we used the condition $h(0) = 1$, proven in Section 4.1. To avoid overdriving the quantizer, $|y(n)| \leq M + 1$ must hold, as Figs. 2.3 and 2.4 show. Hence, using the worst-case (i.e., largest) values for the terms on the RHS of (4.20), the stability condition becomes

$$\max_n |u(n)| + \max_n |e(n)| \cdot [\|h\|_1 - 1] = \|u\|_\infty + \|h\|_1 - 1 \leq M + 1. \quad (4.21)$$

Here we used $\max_n |e(n)| = 1$ (Figs. 2.3 & 2.4), and $h(0) = 1$. Thus,

$$\max_n |u(n)| \leq M + 2 - \|h\|_1 \quad (4.22)$$

guarantees stability, as stated.

As an example, consider the circuit shown in Fig. 4.9. Clearly,

$$L_0(z) = \frac{z^{N-1}}{(z-1)^N} = \frac{z^{-1}}{(1-z^{-1})^N} \quad (4.23)$$

and

$$L_1(z) = -[I + I^2 + \ldots + I^N]z^{-1} = -I\left(\frac{I^N - 1}{I - 1}\right)z^{-1} \quad (4.24)$$

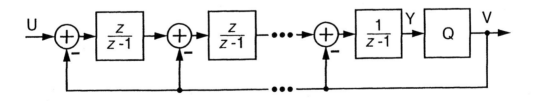

**Figure 4.9:** An implementation of $NTF(z) = (1 - z^{-1})^N$.

where $I(z)$ is the transfer function of the delay-free accumulator:

$$I(z) = \frac{z}{z-1}. \tag{4.25}$$

From the last two equations, it follows that $L_1 = 1 - I^N$ and the NTF is

$$H(z) = \frac{1}{1 - L_1(z)} = I^{-N} = (1 - z^{-1})^N. \tag{4.26}$$

As shown earlier, for this NTF $\|h\|_1 = H(-1) = 2^N$.

The STF is

$$G(z) = H(z)L_0(z) = z^{-1} \tag{4.27}$$

Hence, $\max_n |u(n)|$ appears in $y(n)$ unchanged.

From the theorem given above, therefore, the condition

$$\max_n |u(n)| \leq M + 2 - 2^N \tag{4.28}$$

is sufficient to guarantee the stability of the modulator shown in Fig. 4.9. Recalling that the no-overload range of the quantizer is between $-(M+1)$ and $(M+1)$, clearly the accumulated quantization noise occupies a part $(2^N - 1)$ of the available range, allowing $u(n)$ to utilize the rest.

Equation (4.28) can be used to find the minimum quantizer resolution $M$ for higher-order modulators with the structure of Fig. 4.9. If, e.g., $M = 2^N$ ($N$-bit quantizer) is chosen, the remaining range for signals is only 2— i.e., only one LSB! Using $N + 1$ bits, the signal can be as large as $0.5(M + 1) + 1.5$, more than 50% of the quantizer range; for $N + 2$ bits, over 75% of the range can be utilized for the signal.

Extensive simulations performed for dc, sinusoid and noise input signals [13] showed that for $N = 5$ and $M > 2^5$, the condition (4.28) is a very tight one. Thus, signal levels only slightly higher than the $\max_n |u(n)|$ value given by (4.28) could cause instability. This indicates the practical value of this condition.

In concluding this section, it should be reiterated that while there are impressive research results available on the stability of high-order delta-sigma modulators, extensive behavioral simulations are still advisable before implementing them. The lower the resolution of the quantizer used, the more suspicious the designer should be about unforeseen instability!

## 4.3 Optimization of the NTF Zeros and Poles

Although most examples used earlier in the text dealt with NTFs of the form $(1 - z^{-1})^N$, which had all their zeros at $z = 1$ and all their poles at $z = 0$, significant improvement in the SQNR can be obtained by using more general functions. These are NTFs with separated zeros on the unit circle, spread over the signal range, and with complex poles located within the unit circle surrounding the signal band. Spreading the zeros reduces the total noise power in the signal band, while moving the poles nearer to the zeros reduces the out-of-band NTF gain, resulting in improved stability.

Techniques for constructing such NTFs will be discussed next.

### 4.3.1 NTF Zero Optimization

It was shown in Section 3.4.5 that shifting the two zeros of a second-order NTF from dc ($\omega = 0$) to an optimized value equal to $\omega_B / \sqrt{3}$ resulted in a significant (3.5 dB) SQNR improvement. It is to be expected that even greater benefits can be obtained by optimizing the location of the zeros of higher-order NTFs. The principle of optimization, however, remains the same: the normalized noise power, given by the integral of the squared magnitude of the NTF over the signal band, is minimized with respect to the values of all its zeros. The optimal zeros are found by equating the partial derivatives of the integral to zero.

The resulting values for the zeros (normalized to the signal band limit $\omega_B$) are given, for NTFs with degrees from 1 to 8, in Table 4.1. Note that the optimization process giving these zeros assumed that the quantization noise is white, and that the poles of the NTF have no significant effect on the inband noise. If these conditions do not hold, or if the noise at different frequencies should be weighted differently as is the case, e.g., for A-weighting of audio signals, then the optimization may still be performed, by incorporating these factors into the optimization process in the form of a weight factor under the integral.

As an illustration of the improvements achievable using shifted zeros, Fig. 4.10 compares the frequency responses of two fifth-order NTFs, one with all its zeros at dc, and the other with optimized zeros. The optimization assumed OSR = 32, i.e., a normalized signal band limit equal to 1/64. Fig. 4.11 shows the optimized output spectrum when the input signal of the single-bit modulator is a half-scale sine-

| $N$ | zero locations, normalized to $\omega_B$ | SQNR improvement |
|---|---|---|
| 1 | 0 | 0 dB |
| 2 | $\pm 1/(\sqrt{3})$ | 3.5 dB |
| 3 | $0, \pm\sqrt{3/5}$ | 8 dB |
| 4 | $\pm\sqrt{3/7 \pm \sqrt{(3/7)^2 - 3/35}}$ | 13 dB |
| 5 | $0, \pm\sqrt{5/9 \pm \sqrt{(5/9)^2 - 5/21}}$ | 18 dB |
| 6 | $\pm 0.23862, \pm 0.66121, \pm 0.93247$ | 23 dB |
| 7 | $0, \pm 0.40585, \pm 0.74153, \pm 0.94911$ | 28 dB |
| 8 | $\pm 0.18343, \pm 0.52553, \pm 0.79667, \pm 0.96029$ | 34 dB |

Table 4.1. Zero placement for minimum in-band noise.

wave. The SQNR is 84 dB. The Figure also illustrates the responses predicted by using linear models with quantizer gain $k = 1$ and with the optimized value $k = 1.72$ (the value obtained *a posteriori* from simulation data according to the method described in Section 2.1). Once again, note that once the value of the quantizer gain used for analysis accurately reflects the actual quantizer gain, the predicted PSD matches the observed PSD very well.

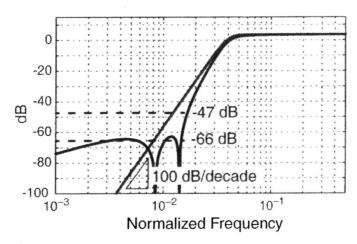

Figure 4.10: Fifth-order NTFs, with and without optimized zeros.

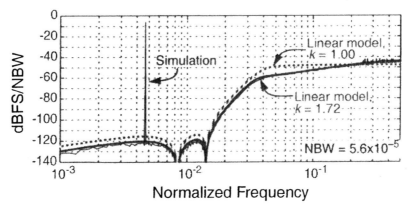

Figure 4.11: Output spectrum of a $5^{th}$-order modulator with a half-scale input.

Fig. 4.12 demonstrates the computed SQNR vs. input amplitude curves for the modulator. Here, the continuous lines show the SQNR predicted by the linear models, while the diamonds illustrate the simulated values for a high-frequency input sine-wave, and the squares the simulation results for a low-frequency one (cf. Fig. 3.8 of Chapter 3). This modulator shows behavior which is less erratic than the earlier low-order modulators, indicating that modeling the quantizer with a white noise source of constant power is more valid in a complex high-order system than it is in a simpler lower-order system. For moderate signal levels (–40 dBFS[†] to –5 dBFS), the predicted behavior matches the simulated behavior quite well. At high signal levels, instability causes the saturation and eventual deterioration of the SQNR. Lastly, at low signal levels (below an input-frequency-dependent threshold) the character of the spectrum changes abruptly (see Fig. 4.13) and the simulated SQNR is about 7 dB higher than predicted from the linear model. As Fig. 4.13 shows, the observed PSD in the low-signal regime is markedly different from the moderate-signal regime– there is a collection of tones in the vicinity of $f_s/4$ and a notch at $f_s/2$. Explanation of these new features requires modifications to our simple linear model which are beyond the scope of this text.

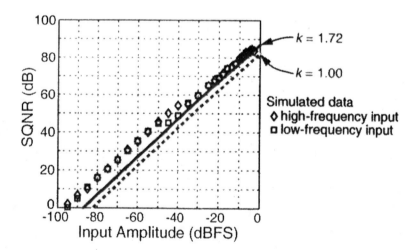

**Figure 4.12: SQNR vs. input amplitude for the $5^{th}$-order modulator.**

---

†. Recall that "dBFS" means "dB with respect to a full-scale sine wave."

For even-order NTFs, the optimized zeros do not provide perfect dc suppression, since there are no zeros at $z = 1$. Often such dc zeros are desirable in order to ensure exact reproduction of the dc input level, and also to reduce the probability of low-frequency tones. In such cases, it is possible to place a double zero at $z = 1$, and optimize the remaining zeros as before so as to minimize the mean-square inband gain of the NTF.

### 4.3.2 NTF Pole Optimization

When the poles of the NTF are determined, a number of constraints must be observed:

(a) As discussed in Section 4.1, the overall noise transfer function, $H(z)$, must satisfy the realizability condition $H(\infty) = 1$.

(b) The out-of-band NTF gain, the loop gain, and hence the stability of the whole modulator are largely determined by the choice of the poles. (The zeros, as explained above, are constrained for efficient noise shaping to the signal-band region of the unit circle.) Hence, the location of the poles should be at least partially based on stability considerations.

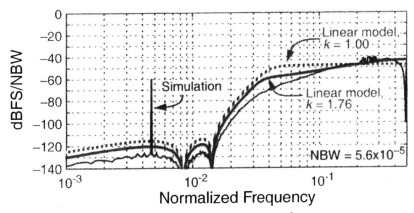

Figure 4.13: Output spectrum of a $5^{th}$-order modulator with a $-60$ dBFS input.

(c) In optimizing the zeros, it was assumed that the influence of the poles on the magnitude of the NTF in the signal band is minimal. Also, for many modulator structures, the STF and NTF have the same poles. For both of these reasons, the denominator of $H(z)$ should have a magnitude which is flat over the signal band.

These constraints make the selection of poles a trade-off between conflicting design considerations, with no unique general solution. Fortunately, there are software tools available to perform this task for the designer. These will be discussed in Sections 8.1 and 8.2. There is also useful information based on extensive computations using these tools [14]. Figs. 4.14-4.16 provide some additional design information. These curves show the achievable peak signal-to-quantization-noise ratio (SQNR) for modulators of orders $N = 1$ to 8 employing optimal zero placement, with 1- to 3-bit internal quantization. The curves include the effects of the reduction of the input $u$ necessary to satisfy the stability conditions. Hence, they accurately predict the actual performance of the real (nonlinear as opposed to linearized) modulator.

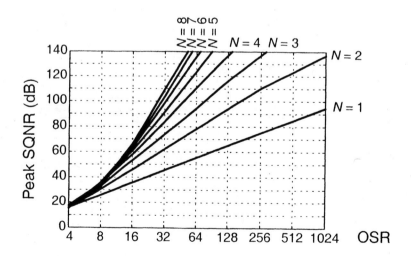

**Figure 4.14:** Empirical SQNR limit for 1-bit modulators of order $N$.

If the appropriate computer tools are not readily available, a "cookbook" design methodology, similar to that described in Section 4.4.1 of [6], may be used to find the complete NTF. It consists of the following steps:

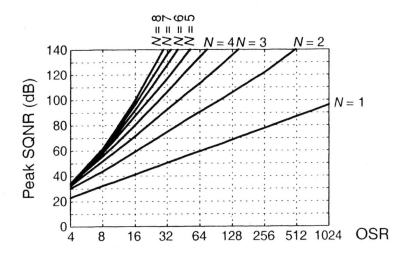

Figure 4.15: Empirical SQNR limit for modulators with 2-bit quantizers.

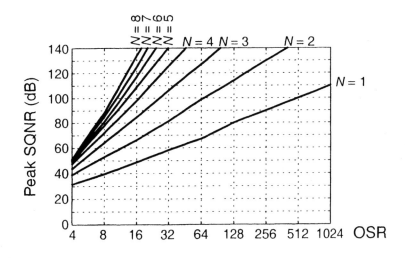

Figure 4.16: Empirical SQNR limit for modulators with 3-bit quantization.

1. Choose the order $N$ of the modulator. This can be done based on the specified SQNR and OSR, using the curves of Figs. 4.14-4.16.
2. Choose the NTF approximation type. Usual choices are Butterworth, inverse Chebyshev, or maximally-flat all-pole highpass transfer functions.
3. Place the 3-dB cutoff frequency $\omega_{3dB}$ of the NTF slightly above the edge of the signal band.
4. Based on the choices made in Steps 1. and 2., find the zeros $z_i$ and poles $p_i$ of the NTF. To satisfy Condition (a) given above, the NTF must be in the form

$$H(z) = \prod_{i=1}^{N} \frac{z - z_i}{z - p_i}. \qquad (4.29)$$

5. Predict the stability of the modulator. For multi-bit quantization, the main Theorem given in Section 4.2.2 may be used to estimate the permissible input signal amplitude for guaranteed stability. For single-bit quantization, Lee's Rule can be utilized. Since, as discussed above, the maximum value of $|H(z)|$ on the unit circle usually occurs at $z = -1$ ($\omega = \omega_s/2$), Lee's Rule requires

$$|H(-1)| = \prod_{i=1}^{N} \frac{1 + z_i}{1 + p_i} < 1.5. \qquad (4.30)$$

(Even if the peak gain occurs elsewhere due to some high-Q poles, its value is usually close to $|H(-1)|$.)

6. Confirm the stability estimation with extensive simulations under all practical conditions.
7. If the predicted stability is unsatisfactory, the poles need to be shifted away from the $z = -1$ point, while maintaining the flat gain in the signal band. This can be achieved by reducing the cutoff frequency $\omega_{3dB}$. As shown in [14], this reduces the peak NTF gain and hence enhances the stability of the modulator.
8. If the stability is robust, but the SQNR does not reach the limit values predictable from Figs. 4.14-4.16, it may be beneficial to make the design more aggressive by increasing $\omega_{3dB}$, and then repeating the stability tests. Steps 6. to 8. can be iterated until the optimum condition (maximum SQNR with adequate stability) is approached.

## 4.4 Loop Filter Architectures

In this Section, some basic configurations will be described for the implementation of the loop filter of Fig. 4.1. Some of these can be considered as the straightforward generalization of the second-order modulator MOD2 discussed in Chapter 3; others are different. Their advantages and disadvantages will also be discussed. [4][5][6, Sec. 5.6][15].

### 4.4.1 Loop Filters with Distributed Feedback and Input Coupling– The CIFB and CRFB Structures

The circuit containing *cascaded integrators with distributed feedback as well as distributed input coupling* (CIFB), shown in Fig. 4.17, is an extension of that shown earlier in Fig. 3.15. It contains a cascade of $N$ delaying integrators, with the feedback signal as well as the input signal being fed to each integrator input terminal with different weight factors $a_i$ and $b_i$. By inspection, the transfer function of the signal filter $L_0$ is now given by

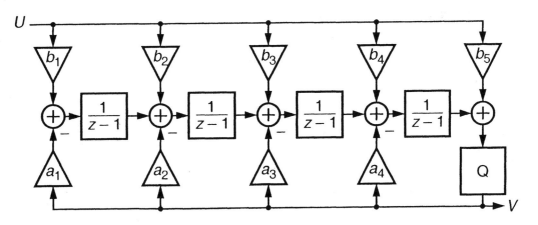

Figure 4.17: Cascade of 4 integrators with distributed feedback and distributed coupling (CIFB structure).

$$L_0(z) = \sum_{i=1}^{N+1} \frac{b_i}{(z-1)^{N+1-i}} = \frac{b_1 + b_2(z-1) + \ldots + b_{N+1}(z-1)^N}{(z-1)^N} \quad (4.31)$$

while the feedback filter $L_1$ has the transfer function

$$L_1(z) = \sum_{i=1}^{N} \frac{-a_i}{(z-1)^{N+1-i}} = -\frac{a_1 + a_2(z-1) + \ldots + a_N(z-1)^{N-1}}{(z-1)^N} \quad (4.32)$$

where $a_1, b_1 > 0$. By (4.4), the NTF for this circuit is of the form

$$H(z) = \frac{1}{1 - L_1(z)} = \frac{(z-1)^N}{D(z)} \quad (4.33)$$

where

$$D(z) = a_1 + a_2(z-1) + \ldots + a_N(z-1)^{N-1} + (z-1)^N. \quad (4.34)$$

Thus, all zeros of the NTF of this structure must lie at $z = 1$ (dc). Note that the realizability condition $H(\infty) = 1$ is satisfied by $H(z)$, as it must be for a physical structure.

The weight factors $a_i$ can be used to introduce finite nonzero poles into the NTF, and also to determine the zeros of $L_1(z)$. The $a_i$ can be found by comparing $D(z)$ to the denominator of the desired NTF, and equating the coefficients of like powers of $z$. (Or, preferably, of $(z-1)$!)

The STF is, also by (4.4),

$$G(z) = \frac{L_0(z)}{1 - L_1(z)} = \frac{b_1 + b_2(z-1) + \ldots + b_{N+1}(z-1)^N}{D(z)} \quad (4.35)$$

where $D(z)$ is given in (4.34). This indicates that the $b_i$ determine the zeros of the STF, and the $a_i$ its poles. As explained before, the $b_i$ can be found by coefficient matching with the numerator of the specified STF. The poles of the STF and the NTF are shared.

It is usually necessary to choose nonzero values for all $a_i$ to realize the prescribed poles suitable for stable operation. There is some latitude, however, in choosing the zeros of the STF, and hence the $b_i$. To simplify the circuit, all $b_i$ except $b_1$ can be chosen as zero. Then all zeros of the STF lie at infinity in the $z$ plane, and the signal transfer function is determined by $b_1/D(z)$. In this case, $|D(e^{j\omega})|$ should be flat in the signal band, to make $|STF|$ constant there.

Another interesting choice is $b_i = a_i$ for all $i \leq N$ and $b_{N+1} = 1$ [15]. Then, by (4.35), the STF is exactly 1, and hence the output of the modulator is

$$V(z) = U(z) + H(z)E(z). \tag{4.36}$$

Hence, the input signal to the $i^{\text{th}}$ integrator is, from Fig. 4.18 with $b_i = a_i$,

$$W_i(z) = X_{i-1}(z) - a_i V(z) + b_i U(z) = X_{i-1}(z) - a_i H(z)E(z) \tag{4.37}$$

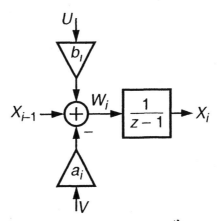

Figure 4.18: Input signals to the $i^{\text{th}}$ integrator.

Thus, the input signal $u(n)$ is not present in any integrator input signal. The loop filter only processes the quantization noise $e(n)$. This usually results in reduced op-amp output swings, especially for multi-bit quantizers.

As discussed in Section 3.4.2, not having to process $u(n)$ in the loop filter has several advantages. One is that the unavoidable nonlinearities of the integrators will not introduce harmonic distortion into the output signal. The other is that, in the absence of $u(n)$, the integrator outputs are considerably reduced, especially if a multi-bit quantizer is used. Thus, this circuit needs less extensive dynamic range scaling[†] and hence has more convenient element values, than other CIFB structures, such as the one with $b_i = 0$ for $i > 1$.

As (4.33) indicated, the structure of Fig. 4.17 can realize NTF zeros only at dc ($z = 1$). We have already seen that much better SQNR performance may achieved if these zeros are located at nonzero frequencies on the unit circle. This requires a modification of the CIFB architecture, as indicated in Fig. 4.19. The circuit shown

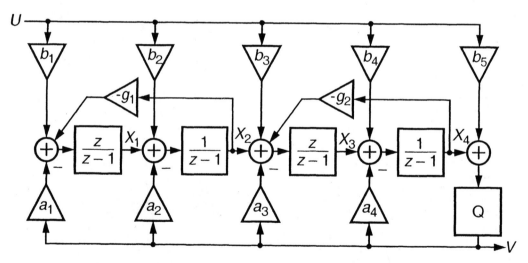

Figure 4.19: Cascade of resonators with distributed feedback and input (CRFB structure).

---

†. Scaling will be discussed in Section 8.3.

there is capable of realizing four NTF zeros as two conjugate complex pairs on the unit circle. The first and second integrators, together with the feedback path $-g_1$, form a resonator with two complex poles which are the zeros of $z^2 - (2-g_1)z + 1$. These poles will be on the unit circle at frequencies $\pm\omega_1$ which satisfy $\cos\omega_1 = 1 - g_1/2$. Similarly, the third and fourth integrators, with feedback branch $-g_2$, form a resonator giving rise to poles at $\pm\omega_2$ such that $\cos\omega_2 = 1 - g_2/2$. For the usual case when the normalized pole frequency $\omega_i \ll 1$, $\omega_i \approx \sqrt{g_i}$ is a good approximation. This configuration will be called *a cascade of resonators with distributed feedback* (CRFB) structure.

One of the integrators in each resonator (here, the first and third integrators) needs to be a delay-free one to insure that the poles stay on the unit circle.

The transfer functions of the resonators can easily be found. For the first resonator, the transfer function from $V$ to $X_2$ is

$$R_1(z) = \left.\frac{X_2(z)}{V(z)}\right|_{U(z)=0} = -\frac{a_1 z + a_2(z-1)}{z^2 - (2-g_1)z + 1}. \qquad (4.38)$$

Note that if the order $N$ is odd, a plain integrator is also needed in the loop filter. In this case, it is customary to make the integrator the input stage in order to minimize the input-referred contribution of noise sources from subsequent stages.

For high-frequency (wideband) ADCs realized using switched-capacitor (SC) integrators, it is advantageous to have a delay associated with every integrator, since this reduces the speed requirements of the amplifiers used. In such situations, both integrator blocks in the resonator have transfer functions $1/(z-1)$. The transfer function of the first resonator then becomes

$$R_1(z) = \left.\frac{X_2(z)}{V(z)}\right|_{U(z)=0} = -\frac{a_1 + a_2(z-1)}{z^2 - 2z + (1+g_1)}. \qquad (4.39)$$

The poles are now outside the unit circle, at $z = 1 + j\sqrt{g_1}$. For $\omega_i \ll 1$, $\omega_i \approx \sqrt{g_i}$ gives again a good approximation.

The resonators by themselves are unstable, as can be inferred from their pole locations. However, they are embedded in a stable feedback system, which prevents local oscillations.

In designing the CRFB circuit, the values of the $g_i$ can immediately be determined from the $\omega_i$ as shown above. The rest of the parameters (the $a_i$ and $b_i$) can readily be found by calculating $L_0(z)$ and $L_1(z)$, first from the specified STF and NTF, then in terms of the $a_i$, $b_i$ and $g_i$ from the circuit diagram, and matching the coefficients of like powers of $z$. A less laborious way is to use the software tools described later in Chapter 8.

To illustrate the analysis of the circuit of Fig. 4.19, we shall express $L_1(z)$ in terms of the resonator transfer functions $R_1(z)$, given by (4.38), and $R_2(z)$, given by

$$R_2(z) = -\frac{a_3 z + a_4(z-1)}{z^2 - (2-g_2)z + 1}. \tag{4.40}$$

If $U(z) = 0$, then, from Fig. 4.19,

$$\frac{X_4(z)}{V(z)} = R_1(z) \cdot \left(\frac{z}{z^2 - (2-g_2)z + 1}\right) + R_2(z) \tag{4.41}$$

and thus

$$L_1(z) = -\frac{(a_1 z + a_2(z-1)) \cdot z + [z^2 - (2-g_1)z + 1] \cdot (a_3 z + a_4(z-1))}{[z^2 - (2-g_1)z + 1] \cdot [z^2 - (2-g_2)z + 1]} \tag{4.42}$$

The expression for $L_0(z)$ is the negative of the above, with the $b_i$ replacing the $a_i$ in $R_1(z)$ and $R_2(z)$, and with $b_5$ added as a constant term.

### 4.4.2 Loop Filters with Distributed Feedforward and Input Coupling– The CIFF and CRFF Structures

The transfer functions $L_0(z)$ and $L_1(z)$ discussed in Section 4.1 can also be realized by using feedforward, rather than feedback, signal paths to create the zeros of the NTF. A circuit constructed from cascaded integrators and feedforward branches is shown in Fig. 4.20. By inspection, the transfer function of the feedback filter is

$$L_1(z) = -a_1 I(z) - a_2 I(z)^2 - \ldots - a_N I(z)^N \qquad (4.43)$$

where $I(z) = 1/(z-1)$ is the delaying integrator's transfer function. Similarly, the signal filter function is

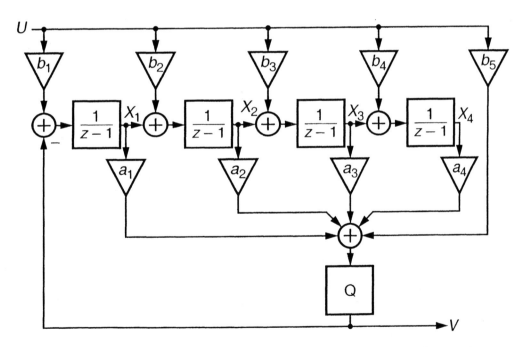

Figure 4.20: Chain of integrators with weighted feedforward summation (CIFF structure).

$$L_0(z) = b_1 \cdot (a_1 I + a_2 I^2 + \ldots + a_N I^N) + b_2 \cdot (a_2 I + \ldots + a_N I^{N-1})$$
$$+ b_3 \cdot (a_3 I + \ldots + a_N I^{N-2}) + \ldots + b_{N+1} \quad (4.44)$$

Consider next the case when $b_2 = b_3 = \ldots = b_N = 0$ and $b_1 = b_{N+1} = 1$. Then, from (4.43) and (4.44), $L_0(z) = 1 - L_1(z)$ holds, and hence from (4.4) it follows that $STF(z) = 1$. Hence, the input to the loop filter satisfies

$$U(z) - V(z) = U(z) - [U(z) + H(z)E(z)] = -H(z)E(z) \quad (4.45)$$

Thus, the loop filter does not process the input signal $u(n)$, and hence this configuration has the low-distortion property discussed earlier for the distributed feedback structures.

As (4.43) shows, all $N$ poles of $L_1(z)$ lie at dc ($z = 1$) for the structure of Fig. 4.20. Hence, so do all zeros of $H(z)$. To obtain optimized zeros for $H(z)$, resonators must be created by internal feedback within the loop filter. The resulting structure is illustrated in Fig. 4.21, where for simplicity only the first and last input weight factors were included, to obtain the low-distortion structure. The analysis of this system is left for the reader, as an exercise.

## 4.5 Multi-Stage Modulators

For low OSR values, as Figs. 4.14-4.16 illustrate, it is no longer possible to obtain high SQNR values in a single-quantizer modulator simply by raising the order of the loop filter, since stability considerations limit the permissible input signal amplitude for higher-order loops, which counteracts the improved noise suppression. The SQNR can still be increased by using more bits in the internal quantizer, but this requires a flash ADC and also means to insure the in-band linearity of the internal DAC, as will be discussed in Section 6.4. As a result, the complexity of the quantizer grows exponentially with the number of bits used in it. Hence, this number can seldom be higher than 4 or 5 bits.

A different strategy, which relies on the cancellation rather than the filtering of the quantization noise, is to use a *multi-stage* (also called *cascade*) structure for the modulator. This section is devoted to a discussion of this useful modulator topology.

### 4.5.1 The Leslie-Singh (L-0 Cascade) Structure [16]

A simple two-stage delta-sigma ADC is illustrated in Fig. 4.22. It contains an $L^{th}$-order delta-sigma modulator as its first stage, and a static (i.e., zero-order) ADC as its second stage. The outputs of the two stages, $v_1$ and $v_2$, are digitally filtered and combined to obtain the overall output $v$.

As shown, the quantization error $e_1(n)$ of the input stage is extracted in analog form by subtracting the input signal $y_1$ of the internal quantizer from its output $v_1$. The error $e_1$ is then converted into digital form by a multi-bit (say, 10-bit) ADC which forms the second stage of the modulator. This introduces another quantization error $e_2(n)$, which however can be much smaller than $e_1(n)$, since the second-stage ADC (not being in a feedback loop) is allowed to have arbitrary latency, and hence it can be realized as a low-complexity multi-bit pipeline structure.

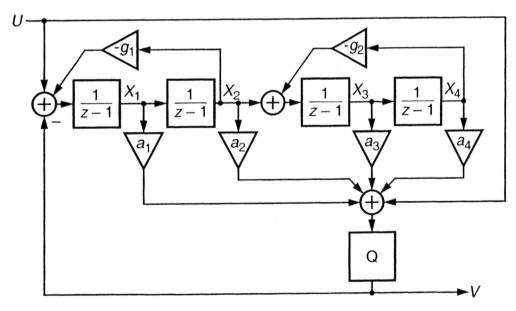

**Figure 4.21: Chain of integrators with feedforward summation and local resonator feedbacks (modified CIFF structure).**

Next, the outputs $v_1$ and $v_2$ of the two stages are filtered by the digital stages $H_1$ and $H_2$, respectively, and added. Usually, $H_1(z) = z^{-k}$ simply implements a delay which equals the latency of the second-stage ADC, while $H_2$ is chosen as the digital equivalent of the NTF of the first stage. Then, subtracting the output of $H_2$ from that of $H_1$ produces the output

$$\begin{aligned}V(z) &= H_1(z) \cdot V_1(z) - H_2(z) \cdot V_2(z) \\ &= z^{-k} \cdot [STF_1(z) \cdot U(z) + NTF_1(z) \cdot E_1(z)] - NTF_1(z) \cdot z^{-k} \cdot [E_1(z) + E_2(z)] \\ &= z^{-k} \cdot [STF_1(z) \cdot U(z) - NTF_1(z) \cdot E_2(z)] \end{aligned} \quad (4.46)$$

Comparing the output $V(z)$ with the first-stage output $V_1(z)$, it is clear that (apart from the delay of $k$ clock periods) the difference is that $E_1(z)$ is replaced by $-E_2(z)$ in $V(z)$. As explained above, $E_2(z)$ may be much smaller than $E_1(z)$ since it is much cheaper to construct a multi-bit pipeline ADC than a multi-bit loop quantizer for the first stage. Hence, this technique can enhance the SQNR by as much as 25~30 dB.

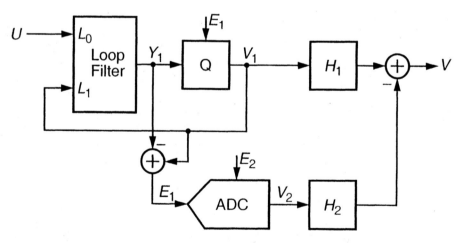

Figure 4.22: The *L-0* cascade (Leslie-Singh) structure.

To obtain $e_1(n)$ by simple subtraction, the operation of the quantizer must be delay-free, which may not be practical. In this case, the signal $y_1$ must be delayed before the subtraction is carried out.

To avoid the subtraction altogether, the input signal of the second stage can be chosen as $y_1(n)$, the input signal of the first-stage ADC, instead of $e_1(n)$. It is given by

$$Y_1(z) = V_1(z) - E_1(z) = STF_1(z) \cdot U(z) + [NTF_1(z) - 1] \cdot E_1(z) \quad (4.47)$$

Keeping $H_1(z) = z^{-k}$, but choosing the other filter function as $H_2(z) = NTF_1(z)/[NTF_1(z) - 1]$,[†] the overall output now becomes

$$V(z) = z^{-k} \cdot [STF_1(z) \cdot U(z) + NTF_1(z) \cdot E_1(z)] \quad (4.48)$$
$$- \frac{NTF_1(z)}{NTF_1(z) - 1} \cdot z^{-k} \cdot \{STF_1(z) \cdot U(z) + [NTF_1(z) - 1] \cdot E_1(z) + E_2(z)\}$$

so, assuming ideal cancellation of like terms,

$$V(z) = \frac{z^{-k} \cdot STF_1(z)}{1 - NTF_1(z)} \cdot U(z) + \frac{z^{-k} \cdot NTF_1(z)}{1 - NTF_1(z)} \cdot E_2(z) \quad (4.49)$$

results. In the signal band, $|NTF_1| \ll 1$, and hence the SQNR obtained with the new $V(z)$ is very close to the one obtainable with the $V(z)$ given in (4.46). A disadvantage of using $y_1(n)$ as the input to the second stage is that it contains $u(n)$ as well $e_1(n)$, and hence the second stage must be able to handle a larger input signal. Also, the second stage must have low distortion, to avoid generating harmonics of $u(n)$.

Consider next one of the low-distortion structures discussed in Section 4.4 used as the first stage. Assume, e.g., that in the CIFB modulator of Fig. 4.17 the conditions

---

[†]. $H_2(z)$, as written, is non-causal. Multiplying $H_2(z)$ and $H_1(z)$ by $z^{-1}$ yields a realizable pair of filters and preserves the desired noise cancellation.

$b_i = a_i$ for all $i \leq N$ and $b_{N+1} = 1$ hold. Then $STF(z) = 1$, and the output signal of the last integrator is

$$X_N(z) = Y(z) - b_N \cdot U(z) = STF(z) \cdot U(z) - [1 - NTF(z)] \cdot E(z) - b_N \cdot U(z)$$
$$= -[1 - NTF(z)] \cdot E(z) \qquad (4.50)$$

This signal can be used as the input signal of the second-stage ADC. It does not contain $u(n)$, and hence the second stage has a smaller input signal, and need not be very linear.

It can easily be seen that similar conclusions apply to the other low-distortion structures: it is possible to extract $y_1(n) - u(n) \approx e(n)$, and use it as the input to the second stage. As an example, Fig. 4.23 shows a second-order low-distortion CIFF modulator. Simple analysis shows that its noise transfer function is $(1 - z^{-1})^2$, its signal transfer function is 1, and the output signal of its second integrator is $X_2(z) = -z^{-2}E_1$. Hence, $X_2(z)$ can be used directly as the input to the second stage of the structure.

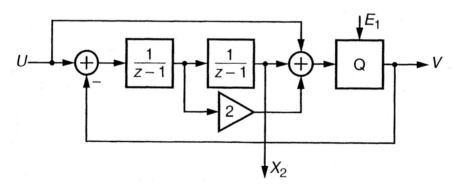

Figure 4.23: Low-distortion CIFF modulator used as first MASH stage.

## 4.5.2 Cascade (MASH) Modulators

An obvious extension of the Leslie-Singh modulator, which historically preceded it, is the cascade modulator, also called *multi-stage* or *MASH* (for Multi-stAge noise-SHaping) modulator [17-19]. Here, the second stage is realized by another delta-sigma modulator. The basic concept is illustrated in Fig. 4.24. The output signal of the first stage is given by

$$V_1(z) = STF_1(z)U(z) + NTF_1(z)E_1(z) \tag{4.51}$$

where $STF_1$ and $NTF_1$ are the signal and noise transfer functions, respectively, of the first stage.

As shown in Fig. 4.24, the quantization error $e_1$ of the input stage is found in analog form by subtracting the input to its internal quantizer from its output. It is then fed to another $\Delta\Sigma$ loop forming the second stage of the modulator, and converted into digital form. Hence, the output signal of the second stage in the z-domain is given by

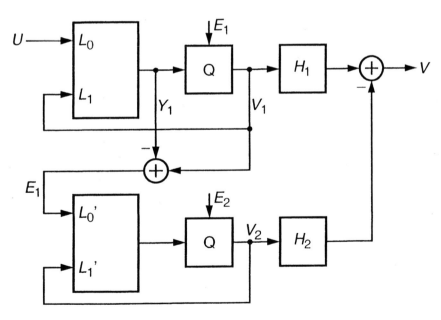

**Figure 4.24: A two-stage MASH structure.**

$$V_2(z) = STF_2(z)E_1(z) + NTF_2(z)E_2(z) \tag{4.52}$$

where $STF_2$ and $NTF_2$ are the signal and noise transfer functions, respectively, of the second stage. The digital filter stages $H_1$ and $H_2$ at the outputs of the two modulator loops are designed such that in the overall output $V(z)$ of the system the first-stage error $E_1(z)$ is cancelled. By (4.51) and (4.52), this is achieved if the condition

$$H_1 \cdot NTF_1 - H_2 \cdot STF_2 = 0 \tag{4.53}$$

holds. The simplest (and usually most practical) choice for $H_1$ and $H_2$ which satisfies (4.53) is $H_1 = STF_2$ and $H_2 = NTF_1$. Since $STF_2$ is often just a delay, $H_1$ is then easily realized.

The overall output is given by

$$V = H_1 V_1 - H_2 V_2 = STF_1 \cdot STF_2 \cdot U - NTF_1 \cdot NTF_2 \cdot E_2 \tag{4.54}$$

In a typical case, both stages of the MASH may contain a second-order loop, and their transfer functions may be given by

$$STF_1(z) = STF_2(z) = z^{-2} \text{ and } NTF_1(z) = NTF_2(z) = (1 - z^{-1})^2 \tag{4.55}$$

Then, the overall output will be

$$V(z) = z^{-4} \cdot U(z) - (1 - z^{-1})^4 \cdot E_2(z) \tag{4.56}$$

Thus, the noise-shaping performance is that of a fourth-order single-loop converter, while the stability is that of a second-order one, since both internal feedback loops are of order 2.[†]

---

[†] In practice, the $E_1$ input to the second modulator stage needs to be scaled to fit within the stable input range. For a second-order, single-bit first stage, the usual scaling factor is 1/4. If multi-bit quantization is used in the first stage, the scaling factor can be greater than 1. The inverse of this scaling factor needs to be included in $H_2$ in order to cancel $E_1$.

If the condition (4.53) is not exactly satisfied due to imperfections in the realization of the analog transfer functions, then $E_1$ will appear at the output multiplied by $STF_2 \cdot NTF_{1a} - NTF_1 \cdot STF_{2a}$, where subscript "$a$" denotes the actual value of the analog transfer function. As will be shown in Section 4.5.3, this may result in a serious deterioration of the noise performance of the converter.

As discussed in the preceding section, it is advantageous for MASH systems as well to use a low-distortion loop filter structure in all stages. This makes it possible to obtain the first-stage error $e_1(n)$ without any subtraction, for entering it into the second stage. In addition, of course, the low-distortion property improves the performance of both stages.

An advantage of the MASH configuration is that the remaining error in the output $V$ is the shaped quantization error $e_2(n)$ of the second stage, operating with an input $e_1(n)$ which is itself noise-like. Hence, the second-stage quantization error $e_2(n)$ is very similar to a true white noise. This remains valid even if the first stage noise contains tones. As an illustration, Fig. 4.25 shows the simulated output spec-

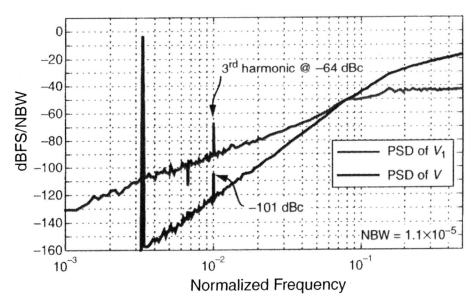

**Figure 4.25: Output spectra for a 2-2 MASH modulator.**

tra of the input stage ($V_1$) and the overall modulator ($V$) for a 2-2 MASH with single-bit quantization. $V_1$ contains 3$^{rd}$ harmonic near $f = 0.01$ which is greatly reduced in $V$. Thus, a MASH modulator is less likely to need dithering than a single-stage one.

Another useful property of the MASH structure is that it often allows the use of a multi-bit quantizer in the second stage, without any dynamic or other correction of the DAC nonlinearity [20]. This is true since the nonlinearity error of the second-stage DAC (as part of $V_2$) is multiplied by $-H_2(z)$ before being added into the output signal $V$. As shown above, $H_2(z)$ contains the NTF of the first stage. Since this $NTF_1(z)$ is a highpass filter function, the nonlinearity error of the second-stage DAC is suppressed in the baseband.

Also, since the input to the second stage contains the quantization error $e_1$ of the first stage rather than the input signal, no harmonic distortion of the signal is generated in the second stage, and (especially for high OSR and small nonlinearity error) the small added noise due to the DAC nonlinearity is usually tolerable.

The principle of quantization error cancellation implemented by the two-stage MASH structure of Fig. 4.24 can be extended. Just as the second stage of the MASH is used to cancel the quantization error $e_1(n)$ of the first stage, a third stage can be added to cancel $e_2(n)$, the quantization error of the second stage (Fig. 4.26). The cancellation conditions can be found exactly as for the two-stage MASH. They are

$$H_1 \cdot NTF_1 - H_2 \cdot STF_2 = 0$$
$$H_2 \cdot NTF_2 - H_3 \cdot STF_3 = 0 \qquad (4.57)$$

Under these conditions, $E_1$ and $E_2$ are cancelled in the overall output signal:

$$V = [STF_1 \cdot U + NTF_1 \cdot E_1] \cdot H_1 - [STF_2 \cdot E_1 + NTF_2 \cdot E_2] \cdot H_2$$
$$+ [STF_3 \cdot E_2 + NTF_3 \cdot E_3] \cdot H_3$$
$$= STF_1 \cdot H_1 \cdot U + NTF_3 \cdot H_3 \cdot E_3 \qquad (4.58)$$

Using (4.57) to express $H_3$, $V$ can be written as

$$V = STF_1 \cdot H_1 \cdot U + \left( \frac{H_1 \cdot NTF_1 \cdot NTF_2 \cdot NTF_3}{STF_2 \cdot STF_3} \right) \cdot E_3 \qquad (4.59)$$

As discussed earlier, $H_1$ and the signal transfer functions usually contain only simple delays, or have flat gain in the signal band, and hence they will not shape either the signal or the noise significantly. However, the NTFs provide suppression over the baseband. Hence, under ideal conditions, the quantization errors of the first two stages are cancelled, while that of the third stage is filtered by the product of the NTFs of all three stages. Thus, if all three stages contain, e.g., second-order loop filters, the overall NTF of the structure will be equivalent to that of a sixth-order modulator, but without the troublesome stability problems inherent in such a high-order loop.

Since the three-stage MASH is normally used only when it is necessary to provide very high SQNR performance, the leakage of the poorly filtered quantization noise of the first stage due to imperfect matching between the analog transfer functions

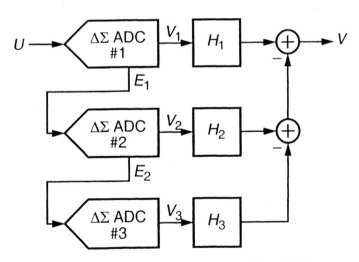

Figure 4.26: A 3-stage MASH ADC.

($NTF_{1,2,3}$, $STF_{1,2,3}$) and the digital ones ($H_{1,2,3}$) is a very critical issue here, and typically limits the practically achievable resolution.

The topic of noise leakage will be discussed in the next section.

### 4.5.3 Noise Leakage in Cascade Modulators

In higher-order single-stage modulators, the imperfect matching of the passive loop filter elements (usually capacitors) and the finite gain of the active ones (usually op amps) will change the coefficients of the NTF and STF, but will usually not affect the SQNR performance significantly. This is because the quantization error is suppressed by filtering, and as long as the gain $L_1$ of the loop filter remains sufficiently large in the signal band, $|NTF| \approx (1/|L_1|) \ll 1$ will continue to hold there. It can easily be shown, e.g., that for op-amp gains as low as $OSR/\pi$ the SQNR decreases by only a few decibels from its ideal value for high-order single-stage ADCs.

In a two-stage MASH structure, by contrast, a large SQNR is achieved by accurate cancellation of the first-stage quantization error $e_1$, which is shaped only by a low-order $NTF_1$. As (4.57) shows, this requires accurate matching between the nominally identical mixed-signal (analog and digital) transfer functions $H_1 \cdot NTF_1$ and $H_2 \cdot STF_2$. For the designer, it is important to know how precise the analog circuit needs to be to obtain good performance for the cascade, i.e. how accurately the components need to be matched, what is the minimum acceptable gain for the op amps, etc., in order to keep the leakage of $e_1$ acceptably low. For a three-stage MASH, the leakage of $e_2$ also needs to be analyzed. Even for relatively simple structures, the equations describing the leakage can become very complex.

As usual for delta-sigma modulators, accurate behavioral simulation is the most reliable technique for predicting the effects of all nonidealities on the SQNR of a MASH modulator. However, under some (usually valid) conditions, valuable and simple results can be obtained using linear analysis and approximations, as will be shown next.

Referring back to (4.58), the transfer functions from $E_1$ and $E_2$ to the overall output $V$ are

$$H_{l1} = H_1 \cdot NTF_1 - H_2 \cdot STF_2$$
$$H_{l2} = H_2 \cdot NTF_2 - H_3 \cdot STF_3 \qquad (4.60)$$

respectively. Ideally, both of these "leakage" transfer functions are identically zero, but since the NTF and STF functions are realized using imperfect analog components, they will be inaccurate. Hence, $H_{l1}$ and $H_{l2}$ will be nonzero, allowing $E_1$ and $E_2$ to leak into $V$. We can usually justify the following simplifying assumptions:

1. The leakage of $E_2$ is less important than that of $E_1$. This is true since the terms in $H_{l2}$ represent higher-order noise shaping than those in $H_{l1}$. As an example, in a 2-2-1 MASH, the noise-shaping due to $H_{l1}$ is at most of order 2, while that of $H_{l2}$ is of order 4. Also, often $E_2$ is smaller than $E_1$ if a multi-bit second-stage quantizer is used.

2. In $H_{l1}$, the effect of an imperfect $NTF_1$ dominates that of the imperfect $STF_2$, even though the gain error is the same for both. This holds since the second stage is followed by the noise-shaping block $H_2 = NTF_1$. Hence, error signals due to imperfect $STF_2$ are inherently noise shaped. This is not true for errors due to the imperfect $NTF_1$, since the $H_1$ block has unity gain in the signal band.

3. In view of assumption 2, we may set $STF_2 = H_1 = 1$ in (4.60), as a first approximation. Then

$$|H_{l1}| \approx |NTF_1 - H_2| = |NTF_{1a} - NTF_{1i}| \qquad (4.61)$$

results, where the subscripts "$a$" and "$i$" denote actual and ideal functions, respectively.

4. From (4.4), $NTF_1 = 1/(1 - L_1)$, where $L_1$ is the gain of the loop filter from the quantizer output to its input (Fig. 4.1). Assuming only small errors, $|L_1| \gg 1$ holds for both the ideal and actual functions. Then, (4.61) can be further approximated by

$$|H_{l1}| \approx |1/L_{1i} - 1/L_{1a}| \qquad (4.62)$$

Eq. (4.62) is much simpler to evaluate than the complete original relations (4.58) or (4.60). It can be used for both two- and three-stage MASH modulators.

## 4 Higher-Order Delta-Sigma Modulation

As an illustration, consider the simple case of a 1-1 or 1-1-1 MASH modulator. The loop filter for the first stage is just a delaying integrator, with an ideal transfer function

$$I_i(z) = \frac{a}{z-1} \tag{4.63}$$

If the integrator is, as usual, realized using switched-capacitor (SC) circuitry, such as shown in Fig. 4.27, then an error $D$ in the nominal capacitance ratio ($C_1/C_2$) will change the factor $a$, and the finite dc gain $A$ of the op amp will change both $a$ and the value of the pole (ideally at $p = 1$). The resulting actual transfer function is

$$I_a(z) = \frac{a'}{z-p'}, \tag{4.64}$$

where, as a simple analysis shows, for $D \ll 1$ and $a/A \ll 1$,

$$a' \approx a[1 - D - (1+a)/A] \tag{4.65}$$

and

$$p' \approx 1 - a/A. \tag{4.66}$$

Since here $L_1(z) = -I(z)$, by (4.62),

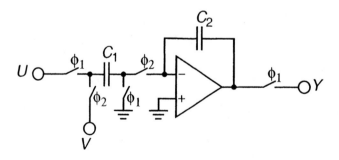

**Figure 4.27:** A delaying switched-capacitor integrator as used in a first-order $\Delta\Sigma$ modulator.

$$|H_{l1}| = |(z-1)/a - (z-p')/(a')|$$
$$\approx |1/a'| \cdot |(a/A) + (z-1) \cdot [D + (1+a)/A]|$$
$$\approx |(1/A) + (z-1) \cdot [D/a + (1+1/a)/A]| \quad (4.67)$$

As (4.67) shows, there is an unfiltered leakage component approximately equal to $E_1/A$, and a first-order-filtered component given approximately by $(z-1) \cdot [D/a + (1+1/a)/A]E_1$. For high specified SQNR, a very-high-gain op amp with excellent settling is required to reduce the unfiltered leakage to a sufficiently low level. If the OSR is low, then the second component will also be significant, requiring $D \ll 1$. Hence, the matching accuracy of the capacitors must also be very high.

Errors in the path coupling the first and second stages will also add to $H_{l1} \cdot E_1$. However, their effects on the output $V$ will be at least first-order filtered, since error signal will pass through the $H_2$ filter.

For a second-order first stage, the leakage of $E_1$ can be reduced. The calculations, however, become much more complicated. Consider, e.g., the 2-0 MASH (Leslie-Singh) modulator (Fig. 4.22), with the ideal output given in (4.46). Assume that the first stage is realized by the low-distortion modulator structure shown in Fig. 4.23, built from two cascaded integrators. The ideal transfer functions of the integrators are given by (4.63) and their actual transfer functions by (4.64)-(4.66).

As a result of these changes, and the inaccuracies in the other coefficients $b_i$ in the loop, there will be leakage of the first-stage quantization error $E_1$ to the output $V$ of the overall ADC system. It is useful to represent the parasitic leakage transfer function by its Taylor-series expansion around $z = 1$ [21]:

$$H_{l1}(z) = A_0 + A_1 \cdot (1-z^{-1}) + A_2 \cdot (1-z^{-1})^2 + \ldots \quad (4.68)$$

where, assuming $A \gg 1$ and $D \ll 1$, the coefficient values are

$$A_0 = 1/A^2$$
$$A_1 = [1/a_1 + 1/a_2]/A$$
$$A_2 = 1/(a_1 \cdot a_2) - 1 + 2[1 - 1/(a_1 \cdot a_2) - 1/a_2^2]/A + 2D/(a_1 \cdot a_2) \quad (4.69)$$

The first term in the series expansion of $H_{l1}$ represents unfiltered leakage. Since it is inversely proportional to $A^2$, it is usually very small. The second term gives the linearly filtered error leakage, the third the quadratically filtered leakage, etc. For $OSR \gg 1$, op-amp gains and matching errors, the linear and quadratic terms containing $A_1$ and $A_2$ tend to dominate $H_{l1}$, since $A_0$ is normally very small, and since higher-order filtering suppresses the terms beyond the quadratic one.

The derivation given above ignored leakage due to the errors in the coupling branch and in the second stage. In this case, these errors contribute only to the quadratic and higher-order terms ($A_2$, $A_3$, etc.), since $H_2$ is here a second-order highpass filter.

As an illustration, for $A = 1000$ and $D = 0.5\%$, we find that $A_0 = 10^{-6}$ and the values of coefficients $A_1$-$A_4$ are between 0.001 and 0.2. The multipliers $(1 - z^{-1})^L$ introduce highpass filtering into the terms in $H_{l1}$, which reduces their effects on the inband noise. The reduction increases rapidly with increasing $L$ and $OSR$. For, e.g., an $OSR = 64$, the linear term ($L = 1$) is reduced by about a factor 1/30, the quadratic term by about 1/1000, and the cubic one by about 1/30,000. Hence, only the first three terms are significant.

## 4.6 Conclusions

In this chapter, higher-order modulators were discussed. Special attention was devoted to the stability of high-order delta-sigma loops, both with single-bit and with multi-bit quantizers. For single-bit loops, the equivalent gain of the quantizer varies strongly with the value of its input, so that linearized stability analysis becomes a difficult task. The stability of the loop may be established using root locus techniques which examine the pole locations for all possible quantizer gains, or by extensive simulation. For multi-bit loops, the quantizer's gain varies only slightly with its input signal, and hence tight theoretical bounds can be found for the signal range which insure stable loop operation.

The optimization of the noise transfer function zeros and poles, already discussed for second-order loops in Chapter 3, was generalized here for higher-order modulators.

The most commonly used loop architectures were also described, analyzed and compared for single-stage modulators. Finally, multi-stage (cascade, MASH) modulators were discussed, and their advantages and drawbacks relative to the single-stage ones analyzed.

## References

[1] W.L. Lee, "A novel higher order interpolative modulator topology for high resolution oversampling A/D converters," Master's Thesis, Massachusetts Institute of Technology, Cambridge, MA. June 1987.

[2] D.R. Welland, B.P. Del Signore, E.J. Swanson, T. Tanaka, K. Hamashita, S. Hara, and K. Takasuka, "Stereo 16-bit delta-sigma A/D converter for digital audio," *Journal of the Audio Engineering Society*, vol. 37, pp. 476-486, June 1989.

[3] R.W. Adams, P.F. Ferguson, A. Ganesan, S. Vincelette, A. Volpe and R. Libert, "Theory and practical implementation of a fifth-order sigma-delta A/D converter," *Journal of the Audio Engineering Society*, vol. 39, pp. 515-528, July 1991.

[4] J. Steensgaard, "High-Performance Data Converters," Ph.D. dissertation, Technical University of Denmark, Dept. of Information Technology, pp. 170-171, March 8, 1999.

[5] J. Silva, U. Moon, J. Steensgaard and G.C. Temes, "Wideband low-distortion delta-sigma ADC topology," *IEE Electronic Letters*, vol. 37, no. 12, pp. 737-738, June 7, 2001.

[6] S.R. Norsworthy, R. Schreier and G.C. Temes, *Delta-Sigma Data Converters*2, IEEE Press, 1997.

[7] K.C.H. Chao, S. Nadeem, W.L. Lee, and C.G. Sodini, "A higher order topology for interpolative modulators for oversampling A/D conversion," *IEEE Transactions on Circuits and Systems*, vol. 37, pp. 309-318, March 1990.

[8] T. Ritoniemi, T. Karema and H. Tenhunen, "The design of stable high order 1-bit sigma-delta modulators," *Proceedings of the 1990 IEEE International Symposium on Circuits and Systems*, vol. 4, pp. 3267-3270, May 1990.

[9] H. Kwakernaak and R. Sivan, *Modern Signals and Systems*, pp. 682, Prentice Hall, Englewood Cliffs, NJ, 1991.

[10] S.H. Ardalan and J.J. Paulos, "Analysis of nonlinear behavior in delta-sigma modulators," *IEEE Transactions on Circuits and Systems*, vol. 34, pp. 593-603, June 1987.

[11] L. Risbo, "Σ-Δ modulators– stability analysis and optimization," Ph.D. thesis, Technical University of Denmark, DK-2800 Lyngby, Denmark, 1994.

[12] J.G. Kenney and L.R. Carley, "Design of multibit noise-shaping data converters," *Analog Int. Circuits Signal Processing Journal*, vol. 3, pp. 259-272, 1993.

[13] Y. Yang, G.C. Temes and R. Schreier, "A tight sufficient condition for the stability of high-order multibit delta-sigma modulators," Oregon State University research report, 1991.

[14] R. Schreier, "An empirical study of high-order single-bit delta-sigma modulators," *IEEE Transactions on Circuits and Systems II*, vol. 40, no. 8, pp. 461-466, August 1993.

[15] P. Benabes, A. Gauthier and D. Billet, "New wideband sigma-delta convertor," *IEE Electronic Letters*, vol. 29, no. 17, pp. 1575-1577, August 19, 1993.

[16] T.C. Leslie and B. Singh, "An improved sigma-delta modulator architecture," *Proceedings of the 1990 IEEE International Symposium on Circuits and Systems*, vol. 1, pp. 372-375, May 1990.

[17] T. Hayashi, Y. Inabe, K. Uchimura and A. Iwata, "A multistage delta-sigma modulator without double integration loop," *ISSCC Digest of Technical Papers*, pp. 182-183, February 1986.

[18] Y. Matsuya, K. Uchimura, A. Iwata, T. Kobayashi, M. Ishikawa and T. Yoshitome, "A 16-bit oversampling A-to-D conversion technology using triple-integration noise shaping," *IEEE Journal of Solid-State Circuits*, vol. 22, pp. 921-929, December 1987.

[19] J.C. Candy and A. Huynh, "Double integration for digital-to-analog conversion," *IEEE Transactions on Communications*, vol. 34, no. 1, pp. 77-81, January 1986.

[20] B.P. Brandt and B.A. Wooley, "A 50-MHz multibit sigma-delta modulator for 12-b 2-MHz A/D conversion," *IEEE Journal of Solid-State Circuits*, vol. 26, no. 12, pp. 1746-1756, December 1991.

[21] P. Kiss, J. Silva, A. Wiesbauer, T. Sun, U.K. Moon, J.T. Stonick and G.C. Temes, "Adaptive Digital Correction of Analog Errors in MASH ADCs - Part II," *IEEE Transactions on Circuits and Systems II*, vol. 47, no. 7, pp. 629-638, July 2000.

# CHAPTER 5
# *Bandpass and Quadrature Delta-Sigma Modulation*

Previous chapters describe the operation of lowpass $\Delta\Sigma$ modulators, for which the highest frequency-of-interest is a relatively small fraction of the sample rate. Advantages of the $\Delta\Sigma$ approach include simplified anti-alias filtering, inherent linearity, robust operation and low power consumption, whereas the primary disadvantage is the need to sample at a rate many times higher the highest frequency-of-interest. This chapter shows how $\Delta\Sigma$ modulation can also be applied to narrowband signals whose highest frequency-of-interest is an appreciable fraction of the sampling rate. The resulting *bandpass* and *quadrature* modulators preserve many of the advantages of lowpass $\Delta\Sigma$ modulation, and are particularly useful in modern receiver systems where high-frequency narrowband signals are converted to digital form without prior translation to baseband.

## 5.1 The Need for Bandpass and Quadrature Modulation

Fig. 5.1 shows a signal processing chain which is found in a variety of wired and wireless receivers. As depicted in the figure, the incoming signal is repeatedly filtered, amplified and mixed down to a lower frequency before finally being digitized and sent to a digital signal processor (DSP). The number and location of the amplifiers, mixers, bandpass filters and lowpass filters, as well as the gain distribution and gain control strategy, are determined by the system designer based on the requirements of the system and on the capabilities of available building blocks. For example, a receiver with modest requirements for instantaneous dynamic range will typically rely more on AGC (automatic gain control) than a system which needs to attain high instantaneous dynamic range, and so will be able to accommodate an ADC with only modest dynamic range. Similarly, increased filtering (possibly combined with an increased number of mixing operations) can relax both the bandwidth and dynamic range requirements of the ADC.

However, since the current trend is toward multi-standard receivers which can handle a variety of modulation formats, receiver specifications must increasingly accommodate the specifications of several standards. Also, since broadcast bandwidth is a limited resource, bandwidth-efficient modulation schemes are increasing in popularity. The result is that bandwidth, linearity and noise requirements are all becoming more difficult and that receiver flexibility is a must. In a similar vein, cost, size and power pressures have put increased emphasis on system simplicity and efficiency. How does the modern receiver designer deal with these conflicting requirements?

Thus far, the answer has been to remove as many analog processing stages as possible, and to put as much of the signal-processing burden on the ADC and DSP as possible. Performing analog-to-digital conversion early in the signal chain requires conversion of either narrowband bandpass or narrowband quadrature signals, while the high dynamic range requirements mandate the use of $\Delta\Sigma$ techniques.

Fig. 5.2 depicts a bandpass $\Delta\Sigma$ ADC system as well as the spectra of its key signals. As the figure shows, the input to the ADC is either an IF (intermediate frequency) or RF (radio frequency) signal. This signal may have already been heavily

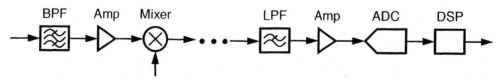

Figure 5.1: **A signal processing chain found in many receiver systems.**

filtered (so that the primary component is the desired signal), but it is more likely to contain undesired signals which are many tens of decibels stronger than the desired signal. The output of the converter contains the desired signal surrounded by shaped quantization noise, plus the interfering signals. The primary feature distinguishing a bandpass modulator from a lowpass modulator is that the quantization noise stopband of a bandpass modulator is not centered on dc. Instead, the stopband is placed at some non-zero frequency $f_0$ that is well removed from dc, such as $f_0 = f_s/4$. The digital output of the modulator is mixed to dc by a digital quadrature mixer, and then lowpass-filtered and decimated by a quadrature lowpass digital decimation filter, to remove out-of-band signals and noise and thereby to produce baseband digital data.

At the system level, the two primary advantages of bandpass conversion are the elimination of one or more analog downconversion operations, and the preservation of the spectral separation between the signal and various low-frequency noise and distortion components. At the circuit level, bandpass conversion is more efficient than converting the entire frequency band between 0 and $f_0 + f_B/2$, since no power is wasted on the accurate conversion of unwanted signals between 0 and $f_0 - f_B/2$. At present, the only converter architecture that is able to focus its operation on a particular frequency band in this way is the bandpass $\Delta\Sigma$ converter.

Fig. 5.3 contrasts the pole/zero placement of the NTFs of hypothetical lowpass and bandpass $\Delta\Sigma$ modulators. The lowpass modulator has NTF zeros at or near $z = 1$

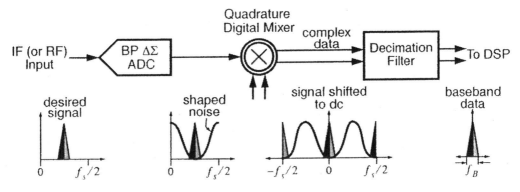

Figure 5.2: A bandpass $\Delta\Sigma$ ADC system.

(which corresponds to dc), whereas the bandpass modulator has NTF zeros elsewhere on the unit circle, such as $z = \pm j$ (which corresponds to $f_0 = f_s/4$). Note that since the NTF zeros occur in complex conjugate pairs, realizing $n$ NTF zeros in the passband of a bandpass modulator requires a $(2n)^{th}$-order converter.

The oversampling ratio of a bandpass system is defined in the same manner as in a lowpass system, namely $OSR = f_s/(2f_B)$, so that $OSR = 1$ corresponds to Nyquist-rate sampling. For a bandpass signal, $f_B$ represents the *two-sided* bandwidth.

In a lowpass system, the modulator output can be decimated by a factor of $OSR$ without loss of information, since the minimum sample rate at the output of the decimation filter is $2f_B$. In a bandpass system, however, the minimum sample rate at the output of the decimation filter is only $f_B$ because the output of the decimation filter is complex data. (Section 5.5 provides a brief overview of complex, or quadrature, signals.) Thus, the data from a bandpass ADC can be decimated by a factor as high as $2 \times OSR$ without loss of information.

As a numerical example, consider a system in which the signal band is centered at 10 MHz and has a bandwidth of 200 kHz (the channel spacing in the popular GSM cellular telephone standard). Sampling at 40 MHz implies that a bandpass modulator with $f_0 = f_s/4$ is needed. The OSR corresponding to this sampling rate is $OSR = f_s/(2f_B) = (40 \text{ MHz})/(2 \times 200 \text{ kHz}) = 100$, even though the sampling rate is only 4 times the center frequency of the desired signal. This example

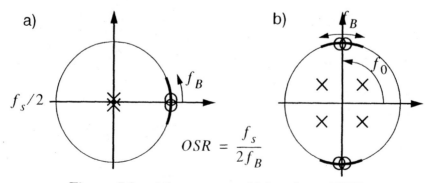

**Figure 5.3:** a) Lowpass vs. b) bandpass NTFs.

makes it clear that while a bandpass modulator oversamples the *bandwidth* by a large factor, the ratio $f_s/f_0$ need not be especially large. Since a $3^{rd}$-order binary lowpass modulator can achieve approximately 100-dB SQNR with $OSR = 100$, similar performance should be expected from a $6^{th}$-order binary bandpass modulator with the same OSR.

The lowest frequency which aliases into the passband of an $OSR = 100$ lowpass modulator as a result of a 40-MHz sample rate is 39.8 MHz. Since this alias frequency is nearly 200 times the upper frequency of the passband, anti-alias filtering for such a modulator is virtually trivial. However, in the case of our example bandpass modulator, the lowest frequency which aliases into the passband is 29.9 MHz. Since this frequency is only about a factor of 3 higher than the desired signal, the anti-alias filter will most likely need to be a bandpass filter in order to provide sufficient alias protection. The Q of this filter is dictated by the required alias suppression and the $f_0/f_s$ ratio; the filter is easier to realize if $f_0$ is a small fraction of $f_s$. As will be discussed in Section 5.3, and demonstrated in Section 5.4, the anti-alias filter can be eliminated entirely if the modulator makes use of a continuous-time loop filter.

Just as a bandpass modulator can exploit the narrowband character of its input, a *quadrature $\Delta\Sigma$ modulator* can exploit the additional information available in a quadrature signal[†]. Fig. 5.4 illustrates the main signal-processing operations that occur within a quadrature $\Delta\Sigma$ ADC system. A quadrature signal, such as that produced by a quadrature mixer, is applied to a quadrature $\Delta\Sigma$ modulator which outputs a digital quadrature signal containing the desired signal and the shaped quantization noise. The distinguishing feature of a quadrature modulator is that its quantization-noise stopband need only exist for positive (or negative) frequencies. In a sense, a quadrature converter is more efficient than a bandpass converter because no power is wasted digitizing the negative-frequency content of the input.

---

† A quadrature signal consists of two real signals, commonly denoted either by *I* (for in-phase) and *Q* (for quadrature phase), or by *re* (for real) and *im* (for imaginary). The key difference between a real signal and quadrature signal is that the spectrum of a quadrature signal need not be symmetric about zero frequency, i.e., for a quadrature signal, positive frequencies are distinct from negative frequencies. Section 5.5 discusses quadrature signals and quadrature filters in more detail.

The modulator output is mixed to baseband by a digital quadrature mixer, and filtered by a quadrature decimation filter, to produce Nyquist-rate baseband data.

For a real (non-quadrature) system, signals beyond $f_s/2$ suffer from aliasing, whereas for a quadrature system the corresponding limits are $\pm f_s/2$. The total alias-free bandwidth is thus $f_s$. In order for $OSR = 1$ to correspond to no oversampling, the OSR of a quadrature system is defined as $OSR = f_s/f_B$. In other words, for a given signal bandwidth and sampling rate, a quadrature modulator has an OSR that is twice that of a real modulator. Lastly, since the minimum output data rate is $f_B$, decimation by a factor of $OSR$ is appropriate for a quadrature system.

A quadrature $\Delta\Sigma$ modulator can be either lowpass or bandpass. However, since a quadrature lowpass modulator is equivalent to a pair of regular lowpass modulators operating independently on the components of the quadrature signal, the advantages of a quadrature lowpass modulator over competing architectures are not as pronounced as they are for a quadrature bandpass modulator. In the bandpass case, a quadrature modulator is useful because it effectively doubles the OSR, and it does so without doubling the hardware. Section 5.6 will delve into this more deeply, but for the moment it suffices to note that a bandpass modulator having $n$ in-band zeros requires a loop filter of order $2n$ (containing $2n$ op-amps), whereas a

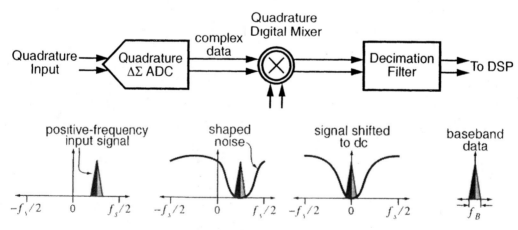

Figure 5.4: A quadrature $\Delta\Sigma$ ADC system.

quadrature modulator having $n$ in-band zeros requires a complex loop filter of order $n$, which also contains $2n$ op-amps.

ADCs employing bandpass modulation are less common than ADCs employing lowpass modulation, but have been reported in numerous publications [1-10], and are commercially available in a few devices [11-12]. Quadrature $\Delta\Sigma$ ADCs are more uncommon still, with only a few publications [13-14] and no commercial realization. Research into the utility of bandpass and quadrature $\Delta\Sigma$ DACs indicates some promise, but at this point demonstrating feasibility is more important than demonstrating significant performance advantages [15-16].

## 5.2 Bandpass NTF Selection

The NTF of a bandpass modulator is subject to the same realizability and stability constraints as the NTF of a lowpass modulator. In particular, Lee's rule for stability is as effective in the context of binary bandpass modulation as it is in the context of binary lowpass modulation. Zero optimization is likewise as useful for bandpass NTFs as it is for lowpass NTFs. However, one degree of freedom which is available to bandpass modulators but not to lowpass modulators is the choice of the center frequency $f_0$ relative to the sampling rate $f_s$.

If $f_0$ and $f_B$ are fixed by the application, choosing a high value of $f_s$ is desirable for two reasons: it maximizes the achievable SQNR by maximizing the OSR, and it eases the design of the anti-alias filter by minimizing the $f_0/f_s$ ratio. Of course, $f_s$ has a practical upper limit which is dictated by the available technology. If only $f_B$ is fixed, the choice of $f_s$ is still governed by the preceding considerations, but $f_0$ now becomes an independent parameter. To simplify the anti-alias filter, $f_0$ should be small. Furthermore, as depicted in Fig. 5.5, a low value of $\omega_0 = 2\pi f_0$ may improve the noise-shaping characteristics of the modulator. These considerations must be balanced against the corrupting influence of low-frequency noise as $f_0$ approaches zero. A further consideration in the choice of $f_0$ is that the digital quadrature mixer which precedes the decimation filter is simplest when $f_0/f_s$ is a simple rational such as $1/4$[†].

## 5 Bandpass and Quadrature Delta-Sigma Modulation

Once $f_B$, $f_0$ and $f_s$ have been selected, the NTF selection process for a bandpass modulator is essentially identical to that for a lowpass modulator. The designer chooses the modulator order, decides whether the NTF zeros should be optimally spread or not, and then uses a procedure such as that coded in the `synthesizeNTF` function of the $\Delta\Sigma$ Toolbox (see Chapter 8) to arrive at a candidate NTF. Simulation is then used to determine whether the NTF order and other parameter choices are acceptable. If not, iteration is necessary.

As an example, Fig. 5.6a shows the pole-zero distribution of a $6^{th}$-order bandpass NTF designed for a center frequency of $f_0 = f_s/6$, an oversampling ratio of $OSR = 32$ and an out-of-band gain of 2. Fig. 5.6b plots the frequency response of this NTF, as well as that of the STF which results when the modulator is realized with the CRFB structure and a single input feed-in. Since the mean depth of the NTF notch is 60 dB and the OSR is 32, if we assume that the quantization noise is white with power $\sigma_e^2$, then the in-band quantization noise power is $\sigma_e^2 \times 10^{-6}/OSR$, which is about 75 dB (12.5 bits) lower than $\sigma_e^2$. With a 4-bit quantizer, this NTF should therefore yield better than 16-bit performance.

To confirm this prediction, Fig. 5.7 plots the simulated output spectrum when the input is a $-1$ dBFS sine wave. The observed SQNR of 102 dB is in fact close to 17-

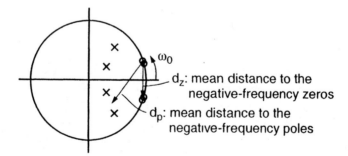

**Figure 5.5:** Lowering $\omega_0$ in a bandpass modulator will improve the suppression of quantization noise if $d_z < d_p$.

---

†. *Subsampling bandpass modulators*, in which $f_0/f_s = 3/4$, or $5/4$ etc. have equally simple digital mixers.

bit performance. Fig. 5.8 shows that this level of performance holds up well as the input amplitude is swept over a broad range.

Fig. 5.9 superimposes a portion of the output data on the input sine wave in order to demonstrate that the high resolution of a bandpass modulator, which is so readily apparent in the frequency domain, is nearly impossible to discern in the time domain. As Fig. 5.9 shows, the resemblance between the input and output

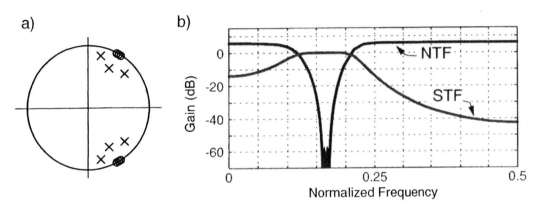

Figure 5.6: a) Pole-zero and b) NTF/STF magnitude plots for an $f_s/6$ bandpass modulator.

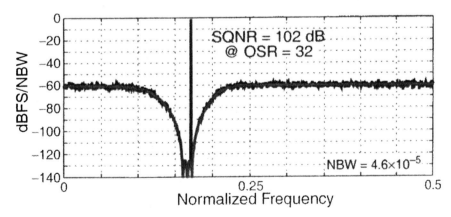

Figure 5.7: Example bandpass output spectrum.

## 5 Bandpass and Quadrature Delta-Sigma Modulation

waveforms is coarse at best, and gives no indication that in the signal band the accuracy of the modulation is in fact better than one part in $10^5$!

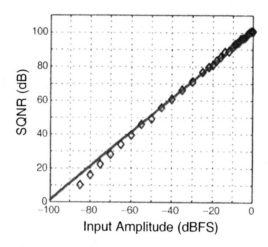

Figure 5.8: SNR vs. input amplitude for the example bandpass modulator.

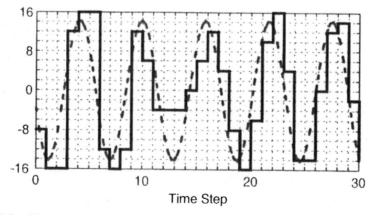

Figure 5.9: Simulated output data for the example bandpass modulator superimposed on the input waveform.

### 5.2.1 Pseudo N-path transformation

The substitution $z \rightarrow -z^2$ transforms a lowpass NTF (whose $n$ zeros are near $z = 1$) into a bandpass NTF with $n$ zeros near $z = j$ and $n$ zeros near $z = -j$.[†] This $(2n)^{\text{th}}$-order NTF has the same gain vs. frequency profile as the original NTF, but compressed by a factor of two and replicated, as illustrated in Fig. 5.10. One might therefore expect that a bandpass modulator derived in this way behaves in a manner similar to the original lowpass modulator. In fact, it is easy to show that the bandpass modulator is exactly equivalent to two copies of the original lowpass modulator, each operating on subsampled data with alternating polarities, as depicted in Fig. 5.11.

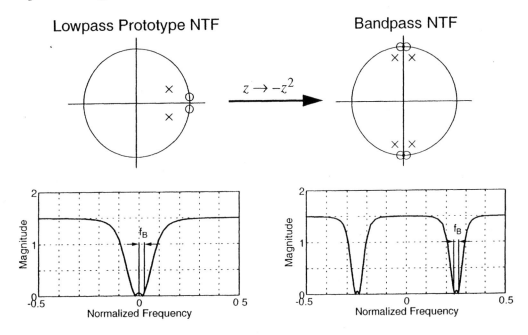

**Figure 5.10:** Pseudo 2-path transformation applied to a lowpass NTF.

---

†. A transformation of the form $z \rightarrow z^N$ is called an *N-path transformation* because it can be accomplished by replicating the original system $N$ times and processing the incoming signal with $N$ parallel paths operation in a time-interleaved fashion. Similarly, replicating all memory elements in the system also accomplishes this transformation. The transformation $z \rightarrow -z^N$ is called a *pseudo N-path transformation* because it is akin to an N-path transformation.

This equivalence shows that a $2n^{th}$-order bandpass modulator (derived from an $n^{th}$-order lowpass modulator via the $z \rightarrow -z^2$ transformation) has exactly the same stability properties and SNR curve as the lowpass modulator operated at the same OSR. Furthermore, the limit cycle characteristics of a bandpass modulator with an $f_s/4$ sine wave input correspond to the interleaved limit cycle characteristics of two lowpass modulators with dc inputs of $A\cos\phi$ and $A\sin\phi$, where $A$ is the amplitude of the sine wave and $\phi$ is its phase relative to the sampling clock.

The $z \rightarrow -z^2$ transformation can be applied to the NTF of a lowpass modulator to yield a bandpass NTF which can then be realized with any of the structures shown in Chapter 4. An alternative method is to apply the $z \rightarrow -z^2$ transformation directly to a lowpass modulator (at either the block-diagram level or the circuit level), simply by replacing each delay element with two delay elements and an inversion. As an illustration, Fig. 5.12 shows the $f_s/4$ bandpass analog of MOD1. The "integrator" now contains a pair of delays and an inversion in its feedback path, while the feedback path from the quantizer also contains two delays but loses the inversion that is normally present at the first summation. This structure is sim-

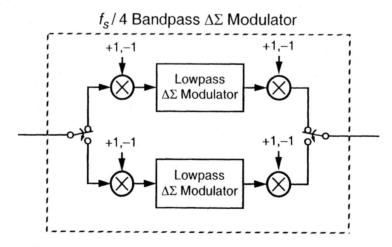

**Figure 5.11: Pseudo 2-path transformation applied to a lowpass modulator at the system level.**

pler than what would result from mapping the NTF onto any of the general-purpose structures presented in Chapter 4.

## 5.3 Architectures for Bandpass Delta-Sigma Modulators

### 5.3.1 Topology Choices

Bandpass modulators possess the same architectural variety as lowpass modulators, and the trade-offs between the different structures are also essentially the same. Bandpass modulators can be implemented in single-loop or cascade form, with a similar trade-off between improved stability and increased sensitivity to analog non-idealities such as parameter errors and finite op-amp gain. Likewise, the loop filter of a bandpass modulator can be constructed using any of the conventional forms found in lowpass modulators, including feedback, feedforward and hybrid topologies, with similar trade-offs between internal dynamic range and STF quality.

Fig. 5.13 contrasts the feedback and feedforward topology extremes for lowpass modulators. In the feedback topology, the quantizer output signal is fed back to the input of every integrator in the loop filter, whereas in the feedforward topology, the output signal of every integrator is fed forward to the input of the quantizer. The integrators may be any combination of continuous-time integrators (such as $G_m$-C or active-RC integrators) or discrete-time integrators (including delaying, non-delaying or half-cycle-delaying switched-capacitor integrators), provided the coef-

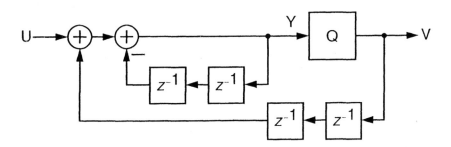

Figure 5.12: Pseudo 2-path transformation applied to MOD1 at the block level.

ficients and timing are chosen appropriately. To shift the loop-filter poles to non-zero frequencies, it suffices to add internal feedback paths such as the one shown with dashed lines in Fig. 5.13a. These loop filter topologies are readily used in the construction of bandpass modulators. For example, Fig. 5.14 shows the structure of the loop filter of a $4^{th}$-order bandpass modulator employing the feedback topology.

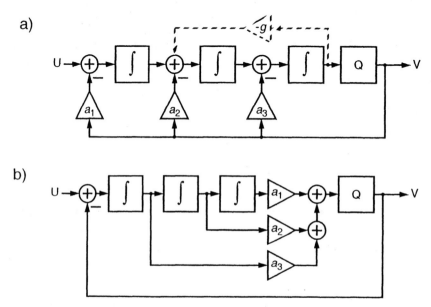

Figure 5.13: Basic loop topologies for lowpass modulators: a) feedback, b) feedforward.

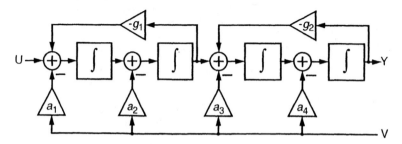

Figure 5.14: Loop filter of a $4^{th}$-order bandpass modulator employing the standard feedback topology.

When $f_0$ is a substantial fraction of the sampling rate, there is strong coupling between the two integrators that comprise a resonator, and thus the resonator output may be taken from the first integrator, as shown in Fig. 5.15. The integrators in this figure are shown as continuous-time blocks for convenience. Taking the resonator output from the first, rather than the second integrator output, changes the transfer function of the resonator from $\omega_0^2/(s^2+\omega_0^2)$, which is a lowpass response, to $s\omega_0/(s^2+\omega_0^2)$, which is a bandpass one. Since the bandpass response has a null at dc, it is clear that a lowpass modulator cannot make use of these bandpass resonators, whereas a bandpass modulator can. Since the $n/2$ resonators in a bandpass modulator may either be of the lowpass or bandpass variety, there are $2^{n/2}$ possible lowpass/bandpass resonator combinations for a given loop-filter category such as the feedback, feedforward or hybrid categories.

Fig. 5.16 illustrates how adding a feedforward path and thus connecting the output of one resonator to both of the integrators in the next resonator can eliminate one of the feedback coefficients (i.e., one of the feedback DACs) in a bandpass modulator. Since the transfer function from $V$ to $Y$ is the same in Fig. 5.16 as that of Fig. 5.15, the noise transfer function of a modulator employing the loop filter of Fig. 5.16 will be the same as that of a modulator employing the loop filter of Fig. 5.15. (Of course, the signal transfer functions may not be the same.) This transformation may be applied to each resonator section except the last one, thereby cutting the required number of DACs by nearly 50%. As will be seen in Section 5.4, this transformation is helpful in the construction of a bandpass modulators which employs an LC tank.

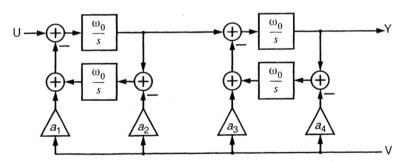

Figure 5.15: Loop filter of a 4$^{th}$-order bandpass modulator employing a feedback topology with bandpass resonators.

Fig. 5.17 shows a portion of a loop filter which encompasses all of the above variants. Each resonator section is coupled to the next through 4 arbitrary gains, so the choice of a lowpass vs. a bandpass section is simply a special case wherein all coefficients are zero except for one. The feedback DACs are not shown, and could be added to any or all of the integrator summing junctions, according to whether a feedback, feedforward or hybrid modulator topology is used.

### 5.3.2 Resonator Implementations

The primary difference between the realization of a lowpass modulator and a bandpass modulator is that a lowpass modulator requires good *integrators*, whereas a bandpass modulator needs good *resonators*. The degradation caused by a finite

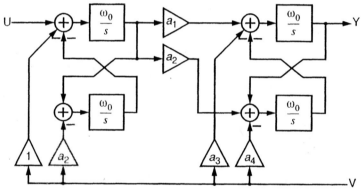

Figure 5.16: Eliminating a feedback DAC by adding a feedforward path.

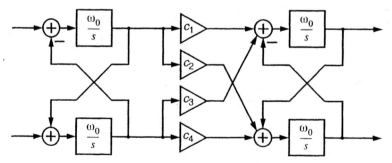

Figure 5.17: Internal structure of a more general loop filter for a bandpass modulator.

quality factor ($Q$) in the resonators of a bandpass modulator is analogous to the degradation caused by finite dc gain in the integrators of a lowpass modulator: both cause reduced SQNR and increased susceptibility to tonal behavior. The SQNR degradation is significant when $Q$ falls below $f_0/f_B$. Thus, in order to take full advantage of a high value of $OSR$, the $Q$ of each resonator should be high. Conversely, when the signal is not especially narrowband, i.e. when $f_0/f_B$ is not very high, the $Q$ requirements for nearly ideal operation are relaxed. The resonant frequency of the resonator must be accurate for similar reasons. A frequency error that is an appreciable fraction of $f_B$, say 20%, is usually close to the level of significance. Once again, a high value for $OSR$ dictates more stringent accuracy requirements. This section presents several resonator circuits which have been used in the construction of bandpass $\Delta\Sigma$ ADCs, and comments on the ability of each to achieve an accurate and high-$Q$ resonance.

Fig. 5.18a depicts the *lossless discrete integrator* (LDI) loop, which may be realized in switched-capacitor form as shown in Fig. 5.18b. The structure of this circuit is such that the poles are the roots of the characteristic equation

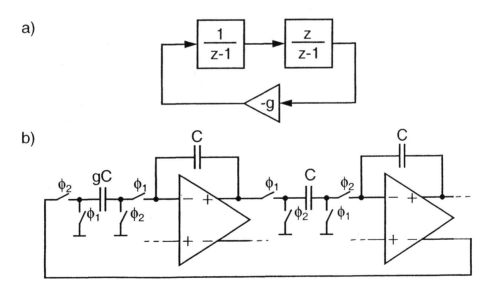

Figure 5.18: a) The LDI loop. b) A switched-capacitor implementation (half of a differential circuit).

$$1 + \frac{gz}{(z-1)^2} = 0. \tag{5.1}$$

The roots of (5.1) are $z_p = \sigma \pm j\sqrt{1-\sigma^2}$, where $\sigma = 1 - g/2$. Clearly, for $|\sigma| \leq 1$ ($0 \leq g \leq 4$), the poles of the LDI loop lie on the unit circle, and thus the Q of the resonator is ideally infinite. In practice, $Q$ is limited by finite op-amp gain, but since $Q > 100$ is readily achieved, finite resonator Q is usually not problematic.

The frequency of the resonance is given by $\omega_0 = \cos^{-1}(\sigma)$, which shows explicitly the dependence of $\omega_0$ on capacitor ratios. The sensitivity of $\omega_0$ to capacitor ratio errors is an increasing function of $\omega_0$, but even at the relatively high value of $\omega_0 = \pi/2$, a 1% shift in capacitor ratios translates into only a 0.6% shift in $\omega_0$. Since capacitor matching is typically much better than 1%, the $\omega_0$-accuracy of an LDI-based resonator is insufficient only if the signal bandwidth is a small fraction of the center frequency.

The LDI loop provides a good way to implement a switched-capacitor resonator possessing an arbitrary resonant frequency, but requires the use of two op amps. When the resonant frequency is $f_s/4$, the pseudo 2-path transformation described in Section 5.2.1 leads to circuits such as that depicted in Fig. 5.19 which are able to implement a resonator with a single op amp. Note that since such circuits imple-

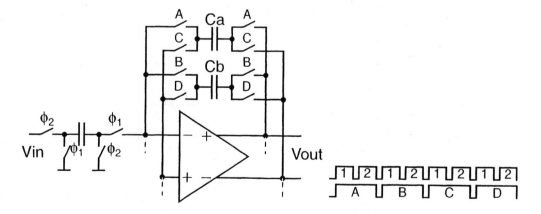

**Figure 5.19: A pseudo-2-path switched-capacitor $f_s/4$ resonator.**

ment the desired center frequency by virtue of their topology, rather than through the use a particular set of capacitor ratios, capacitor errors do not translate into center-frequency errors.

Although the center frequency of the circuit shown in Fig. 5.19 is insensitive to capacitor ratios, mismatch in the paths (in particular the Ca and Cb capacitors) causes the circuit to be *periodically time-varying*, instead of time-invariant. The time-varying nature of the circuit in turn causes mixing of the signal with $f_s/4$ and its harmonics, and it is the mixing of the signal with $f_s/2$ that results in the appearance of an *image signal*, a frequency-inverted copy of the signal centered on $f_0$. Another source of difficulty in this circuit stems from the use of clocks whose frequency is $f_s/4$. The large-amplitude clocks can leak into the signal band and appear as a tone at band-center.

The circuit depicted in Fig. 5.20 avoids these problem to a large degree. Since the Ca and Cb path capacitors are only used for charge storage, and since the conversion of charge to voltage is performed the (path-independent) Co capacitors, the time-varying nature of the circuit is essentially hidden. (In practice, the op-amp gain must be high enough to ensure adequate charge-transfer efficiency.) Also, since this circuit does not use $f_s/4$ clocks, this spur-generating mechanism is not an issue.

Fig. 5.21 shows the structure of a $G_m$-C resonator. Since the center frequency is given by $\omega_0 = g_m/C$, and since the value of $g_m/C$ implemented with on-chip capacitors and transconductors typically has 30% variability, the center frequency of a $G_m$-C resonator will be poorly controlled unless some means for tuning is provided. A common method for tuning a $G_m$-C filter is to adjust all the $G_m$ elements of the filter along with those of a simpler *reference filter* until the reference filter has the desired response. However, since the resonator can be converted into an oscillator with only a small amount of positive feedback, it suffices to measure the oscillation frequency of the resonator itself and adjust $G_m$ (or $C$) directly. Since this calibration must be done off-line, the designer must ensure that the drift of $G_m$ over temperature is sufficiently small. If the drift cannot be made sufficiently small, a continuous-tuning method involving a (scaled) copy of the resonator is the next best choice.

Once the problem of resonator tuning has been addressed, the next set of concerns revolve around the resonator's Q. Non-idealities such as finite output impedance and non-zero phase shift in the transconductors limit resonator Q. Techniques such as cascoding can boost output impedance, while the phase shift can be reduced by using a wide-band $G_m$, or compensated by adding a small resistor in series with the capacitors.

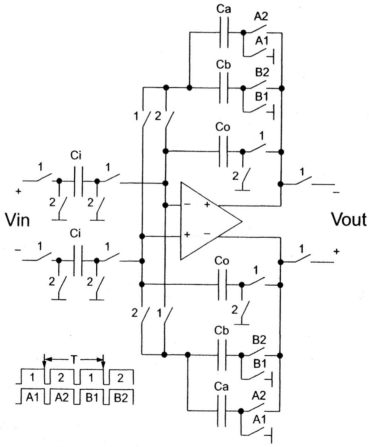

Figure 5.20: A 2-path resonator with reduced sensitivity to capacitor mismatch.

Fig. 5.22 shows the structure of an active-RC resonator. Here the center frequency is given by $\omega_0 = 1/(RC)$, and once again the highly variable nature of the RC product necessitates the use of tuning. Tuning may be accomplished by adjusting $R$ (continuously via MOS devices, or in discrete steps using a resistor array), by adjusting $C$ (here an array is most practical), or by a combination of the two approaches. Once again, configuring the resonator as an oscillator is straightfor-

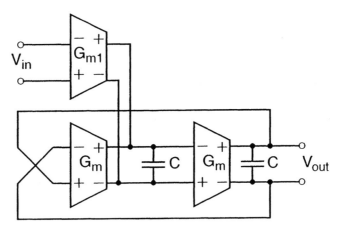

**Figure 5.21: A $G_m$-C resonator.**

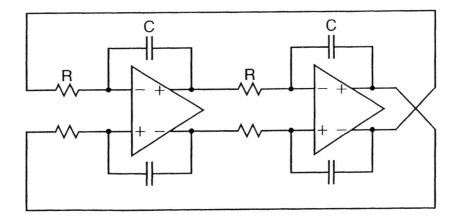

**Figure 5.22: An active-RC resonator.**

ward and eliminates the need for a replica block, but can only be done when the converter is off-line.

The last resonator we consider is the LC tank driven by a current source, shown in Fig. 5.23. From the viewpoint of complete integration, this topology represents somewhat of a backwards step. On-chip inductors possess only a few nanohenries of inductance, and so would only be useful if the center frequency is above 1 GHz or so. Since such high frequencies are beyond the reach of existing mainstream technologies, bandpass modulators which employ inductors have all relied on external components. As with the $G_m$-C and active-RC resonators, the accuracy of the LC tank's center frequency is determined by the accuracy of its components. Since discrete inductors with tolerances on the order 2% and $Q > 50$ are available, as are capacitors with even tighter tolerances and higher Q, it is possible to implement a high-Q LC resonator without incorporating means for tuning. Furthermore, since inductors and capacitors are ideally noiseless, a resonator based on an LC tank enjoys an enormous noise advantage over the preceding resonator circuits. The distortion of an LC tank is also quite small compared to what can be achieved with active circuitry. Lastly, since an LC tank implements a *physical resonance* (as opposed to the *synthesized resonance*), an LC resonator needs no bias power. Despite its resistance to integration, an LC tank possesses a number of important attributes (namely low noise, distortion and power!) that make its use in a bandpass converter highly advantageous. The following section will illustrate these advantages in the context of an example.

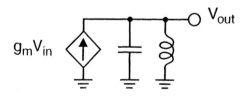

Figure 5.23: A resonator based on an LC tank.

## 5.4 Bandpass Modulator Example

As described in the introduction, bandpass modulators are most attractive in narrowband receiver systems. System specifications regarding portability, operating time, range and reliability translate into requirements for low power consumption and high dynamic range. When these requirements are extreme, as they are for the two-way radios used by military and safety personnel, a structure such as that shown in simplified form in Fig. 5.24 and more detailed form in Fig. 5.25 offers a number of important advantages.

The main advantages stem from the fact that the output of the active mixer (which is a current) is applied to a high-Q LC tank. The high impedance of the tank at resonance endows the LNA/mixer with high gain (an effective gain of 60 was reported in [17]), thereby reducing the noise of subsequent blocks in the ADC by the same factor when referred to the LNA input. This high gain can be supported without a large voltage swing on the tank since the tank is part of a bandpass $\Delta\Sigma$ converter. (In effect, IDAC, the current-mode feedback DAC, cancels the in-band portion of the mixer's output current and thus prevents unbounded voltage swing on the tank even if the Q of the tank is infinite.) Both IDAC and the mixer benefit by having an inductive connection to the supply, since this arrangement eliminates the dc drop associated with bias currents flowing through resistive loads and allows signals to swing above the supply, thereby giving plenty of headroom to both

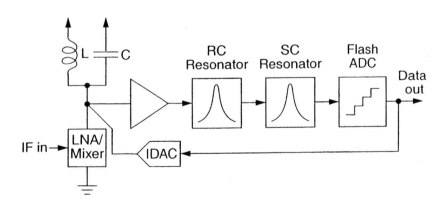

Figure 5.24: **Simplified block diagram of the IF-to-bits portion of a high-performance receiver.**

blocks. Finally, as mentioned in the preceding section, an ideal LC tank adds no noise or distortion, and consumes no power. The combination of increased headroom and increased gain without penalties in noise, distortion or power is a rare combination, one that makes using an LC tank as the first resonator in a bandpass $\Delta\Sigma$ ADC highly desirable.

For the user, the only disadvantage of an LC tank is reduced integration. With an IF in the 2-3 MHz range, the inductance range is roughly 5-10 µH. Since suitable inductors are fairly small (measuring $1.3 \times 2.0 \times 1.4$ mm) and cost only a few cents, this disadvantage is not particularly prohibitive.

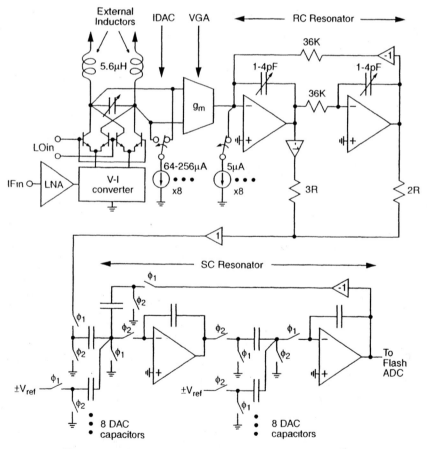

**Figure 5.25: A more detailed ADC schematic.**

For the ΔΣ designer, the disadvantage of an LC tank is that it complicates the design process. In a purely feedforward topology, a linear combination of all loop-filter state variables is fed to the quantizer, but since the inductor current is difficult to sense without adding noise or degrading the Q of the tank, it is impractical to incorporate the inductor current in the summing network which drives the quantizer. On the other hand, in a purely feedback topology the quantizer output must be fed to the summing node of each integrator. Since the integration performed by an inductor ($i = \int v \, dt / L$) requires a summation of voltages, incorporating a feedback term for controlling the inductor current requires a voltage-mode DAC, which is also impractical. These difficulties can be overcome as described in Section 5.3.1, namely by feeding the output voltage of the tank into both integrators of the subsequent resonator. Alternatively, as done in this modulator, the outputs of both integrators within the second resonator can be fed forward to the third resonator. This output-combining arrangement is implemented via the resistors labelled 3R and 2R in Fig. 5.25.

The second resonator in the modulator could also make use of an LC tank, but here the noise, power and distortion advantages are outweighed by the advantages of fewer external components and associated pins. Since an active-RC resonator offers a better noise/power trade-off than a switched-capacitor resonator, the second resonator is implemented with active-RC techniques. The first two resonators in the loop filter are thus continuous-time blocks, and sampling occurs after the signal has experienced two stages of continuous-time amplification. As a result, sampling errors (such as those associated with aliasing and jitter) are referred to the input through noise-shaping functions with double zeros. The alias suppression in this modulator is more than 80 dB, even at a relatively low-valued OSR of 48.

A switched-capacitor resonator is used as the final resonator in this modulator since its Q is high and its drift is low. Without such a stable resonator, the modulator would require a fourth resonator stage in order to guarantee sufficient attenuation of quantization noise in the face of finite Q and center-frequency drift in the resonators. Section 9.5 describes the design process for this modulator in greater detail.

Fig. 5.26 shows the PSD of the undecimated modulator output, together with the theoretical STF and NTF (scaled as described in Appendix A to align with the observed PSD). This figure shows the bandpass nature of the STF, and also that the NTF provides at least 50 dB of attenuation of quantization noise in the band of interest.

Fig. 5.27 shows the in-band portion of the modulator output spectrum when the input signal is a nearly full-scale tone at a 25 kHz offset from the 103.25 MHz center frequency of the system, while the LO and CK frequencies are 100 MHz and 26 MHz, respectively. This figure exhibits an SNR of 81 dB and a spurious-free dynamic range of 103 dB over a 270 kHz bandwidth ($OSR = 48$). Fig. 5.28 plots the measured SNR as a function of input power (in dBm) with LO and CK frequencies of 269 MHz and 32 MHz, respectively. As this figure shows, the system achieves a dynamic range of 90 dB for $OSR = 48$ and 105 dB for $OSR = 900$. This high level of performance is achieved with an analog power consumption of only 50 mW, a figure which includes the 30 mW power consumption of the LNA and mixer.

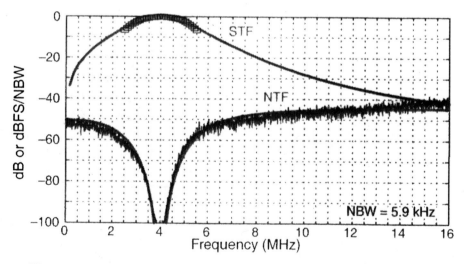

**Figure 5.26:** Measured STF and noise PSD. ($f_{CLK}$ = 32 MHz)

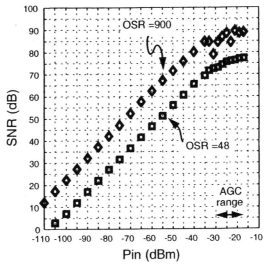

**Figure 5.28: SNR vs. input power.**
$f_{IF} = 273$ MHz, $f_{LO} = 269$ MHz, $f_{CK} = 32$ MHz.

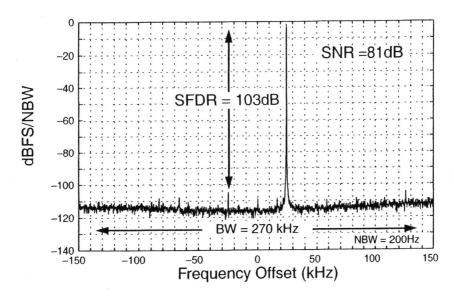

**Figure 5.27: In-band spectrum at OSR = 48.**
$f_{IF} = 103.25$ MHz, $f_{LO} = 100$ MHz, $f_{CK} = 26$ MHz

## 5.5 Quadrature Signals

This section reviews quadrature signal processing in preparation for the upcoming section on quadrature $\Delta\Sigma$ modulation. As will be explained in greater detail shortly, quadrature signals are (most often) produced by quadrature mixers, which are themselves useful because of their *image-rejection* properties.

A *quadrature signal* is an abstract signal composed of two real signals, $v_x$ and $v_y$, viewed as a single complex entity $v = v_x + jv_y$[†]. Since a quadrature signal has a non-zero imaginary part, its Fourier transform need not be symmetric about zero frequency. In other words, with quadrature signals, positive frequencies and negative frequencies contain independent information.

Quadrature analog signals are often created via *quadrature mixing*. In a quadrature downconversion mixer, a real (or quadrature) signal is multiplied by the quadrature signal $e^{-j\omega_{LO}t}$, which we will refer to as the LO (for local oscillator). The LO consists of two real signals, $\cos\omega_{LO}t$ and $-\sin\omega_{LO}t$, as illustrated in Fig. 5.29. Sup-

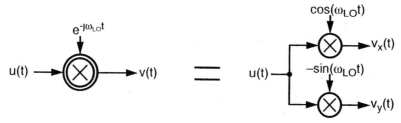

**Figure 5.29:** Quadrature mixing.

---

[†]. Since this text has a strong circuits emphasis, denoting the components of a quadrature signal by *I* and *Q* may lead to confusion since *I* is often used to represent current, while *Q* is often used to represent quality factor or charge. Similarly, denoting the components *re* and *im* leads to confusion when dealing with transforms. (The imaginary part of the Laplace transform of a quadrature signal is *not* the Laplace transform of the imaginary component of the quadrature signal.) Instead of using either of the two prevailing conventions, we adopt the notation (inspired by the Cartesian representation of a complex number) that a quadrature signal $v$ is decomposed as $v = v_x + jv_y$, and refer to the components of $v$ as the $x$ and $y$ components, respectively. The Laplace transform of $v$ is then $V = V_x + jV_y$, where $V_x$ and $V_y$ are the Laplace transforms of the $x$ and $y$ components of $v$.

pose that the input to such a mixer is the real signal $u(t) = A\cos((\omega_{LO} + \omega_{IF})t)$. Then the output of the mixer is

$$v(t) = A\cos(\omega_{LO} + \omega_{IF})t \times e^{-j\omega_{LO}t}$$

$$= A\left[\frac{e^{j(\omega_{LO} + \omega_{IF})t} + e^{-j(\omega_{LO} + \omega_{IF})t}}{2}\right]e^{-j\omega_{LO}t}$$

$$= \left(\frac{A}{2}e^{j\omega_{IF}t} + \frac{A}{2}e^{-j(2\omega_{LO} + \omega_{IF})t}\right) \quad (5.2)$$

Removing the second term in the last expression with a lowpass filter leaves a frequency-shifted version of the original signal, centered at the angular frequency $\omega_{IF}$.

A quadrature downconversion mixer is useful because it performs a frequency translation operation that distinguishes between signal frequencies above the LO and signal frequencies below the LO, whereas a conventional mixer does not make that distinction. To see this, suppose that the input to the quadrature mixer is a combination of two signals having frequencies above and below the LO by an amount equal to the IF: $x(t) = A_1\cos(\omega_{LO} + \omega_{IF})t + A_2\cos(\omega_{LO} - \omega_{IF})t$. Applying this signal to a quadrature mixer and lowpass filter yields the output

$$v(t) = \frac{A_1}{2}e^{j\omega_{IF}t} + \frac{A_2}{2}e^{-j\omega_{IF}t}, \quad (5.3)$$

The signal that was located above the LO frequency could be recovered with a quadrature filter whose passband is located at $\omega_{IF}$, while the signal below the LO frequency could be recovered with a quadrature filter whose passband is located at $-\omega_{IF}$. However, since the output of a conventional mixer is simply the real part of (5.3), namely

$$v_x(t) = \frac{A_1 + A_2}{2}\cos\omega_{IF}t. \quad (5.4)$$

It is thus clear that $A_1$ and $A_2$ cannot be computed from knowledge of $v_x(t)$ alone.

In practice, the ability of a quadrature mixer to distinguish between frequencies offset from the LO by equal positive and negative amounts is limited by the degree to which the two real mixers match and the degree to which the two components of the LO are in quadrature. The *image rejection ratio* (IRR) specifies the signal power appearing at $-\omega_{IF}$ relative to the signal power appearing at $\omega_{IF}$ as a result of an input at $\omega_{LO} - \omega_{IF}$. For small errors, IRR is approximately [18]

$$IRR = 6 - 10\log_{10}\left[\left(\frac{\Delta A}{A}\right)^2 + (\Delta\phi)^2\right], \tag{5.5}$$

where $(\Delta A)/A$ is the relative amplitude imbalance and $\Delta\phi$ is the phase error (in radians). Fig. 5.30 indicates that an amplitude imbalance of 2% (0.17 dB), or a phase error of 0.02 rad (1.1 degree), is sufficient to limit IRR to 40 dB. Higher image suppression requires proportionally greater amplitude and phase accuracy.

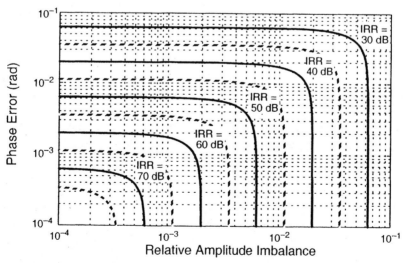

Figure 5.30: **Image-rejection ratio as a function of amplitude imbalance and phase error.**

Quadrature signals may be filtered using a *quadrature filter*. The transfer function, $H$, of a quadrature filter differs from that of a real filter in that the poles and zeros of $H$ need not come in complex-conjugate pairs, i.e. $H$ may have an asymmetric frequency response. Formal manipulation of such transfer functions in symbolic form is straightforward; realizing a quadrature filter is more cumbersome. One way to implement a quadrature filter $H$ starts with decomposing $H$ into $H = H_x + jH_y$, where $H_x$ and $H_y$ are real transfer functions.[†] The output of the filter is

$$\begin{aligned} V &= HU \\ &= (H_x + jH_y)(U_x + jU_y) \\ &= (H_x U_x - H_y U_y) + j(H_x U_y + H_y U_x) \\ &= V_x + jV_y \end{aligned} \qquad (5.6)$$

which indicates that a quadrature filter may be implemented with the lattice structure shown in Fig. 5.31. This figure depicts a two-input/two-output linear system in which the transfer function from the input $U_x$ to the output $V_x$ is equal to that from $U_y$ to $V_y$, while the transfer function from $U_x$ to $V_y$ is the negative of that from $U_y$ to $V_x$. In practice, these symmetries are not exact, and the reader may well wonder what impact such an imperfection has.

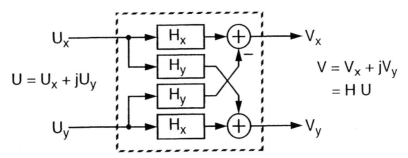

Figure 5.31: A quadrature filter.

---

†. This decomposition may be performed by multiplying the numerator and denominator of $H$ by the complex conjugate of its denominator polynomial, and then gathering numerator terms.

To address this question, Fig. 5.32a shows an arbitrary two-input/two-output real linear system, whose inputs and outputs are to be interpreted as quadrature signals. As depicted in Fig. 5.32b, this system can be represented with two complex filters: one ($H$) operating on the unaltered signal, $U$, and the other ($H_i$) operating on its conjugate, $U^*$. To derive the equivalence, simply write the output of the second system in expanded form:

$$\begin{aligned} V &= HU + H_i U^* \\ &= (H_x U_x - H_y U_y) + j(H_x U_y + H_y U_x) \\ &\quad + (H_{i,x} U_x + H_{i,y} U_y) + j(-H_{i,x} U_y + H_{i,y} U_x) \\ &= (H_x + H_{i,x}) U_x + (H_{i,y} - H_y) U_y \\ &\quad + j((H_y + H_{i,y}) U_x + (H_x - H_{i,x}) U_y) \end{aligned} \qquad (5.7)$$

Thus

$$\begin{bmatrix} H_{11} & H_{12} \\ H_{21} & H_{22} \end{bmatrix} = \begin{bmatrix} H_x + H_{i,x} & H_{i,y} - H_y \\ H_y + H_{i,y} & H_x - H_{i,x} \end{bmatrix}, \qquad (5.8)$$

or in inversely

$$\begin{bmatrix} H_x & H_y \\ H_{i,x} & H_{i,y} \end{bmatrix} = \frac{1}{2} \begin{bmatrix} H_{11} + H_{22} & H_{21} - H_{12} \\ H_{11} - H_{22} & H_{21} + H_{12} \end{bmatrix}. \qquad (5.9)$$

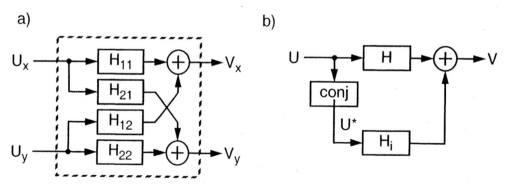

Figure 5.32: Mismatch in a quadrature filter creates an image response.

The upshot of (5.7-5.9) is that path mismatch ($H_{11} \neq H_{22}$ and/or $H_{12} \neq -H_{21}$) causes the output of a quadrature filter to contain the conjugate of the input, multiplied by the *image transfer function* $H_i = H_{i,x} + jH_{i,y}$, where $H_{i,x} = H_{11} - H_{22}$ and $H_{i,y} = H_{21} + H_{12}$. Since taking the conjugate of the input reflects its Fourier transform about $f = 0$, i.e. $(x(t) \leftrightarrow X(f)) \Rightarrow (x^*(t) \leftrightarrow X^*(-f))$, the image transfer function is responsible for transferring signal energy from positive frequencies to negative frequencies, and vice versa. We shall see shortly that in a quadrature $\Delta\Sigma$ modulator, this mirroring action can be highly detrimental to the performance of a quadrature modulator.

At this point, it is helpful to consider a more concrete example. Suppose we want to implement a quadrature filter with transfer function

$$H(s) = \frac{\omega_0}{s - j\omega_0}. \qquad (5.10)$$

Since this is a first-order transfer function with a single pole at $s = j\omega_0$, the resulting filter will therefore be a positive-frequency resonator. The $H_x$ and $H_y$ components of $H$ are found by multiplying the numerator and denominator by the complex-conjugate of the denominator, as follows:

$$H(s) = \frac{\omega_0}{s - j\omega_0}\left(\frac{s + j\omega_0}{s + j\omega_0}\right) = \frac{\omega_0 s + j\omega_0^2}{s^2 + \omega_0^2}. \qquad (5.11)$$

Thus the required filters are

$$H_x(s) = \frac{\omega_0 s}{s^2 + \omega_0^2} \text{ and } H_y(s) = \frac{\omega_0^2}{s^2 + \omega_0^2}. \qquad (5.12)$$

These two second-order filters, as well as the computations of (5.6) can be implemented with only two real integrators, configured as shown in Fig. 5.33.

## 5.6 Quadrature Modulation

As with other modulator types, the starting point in the design of a quadrature modulator is the NTF. The causality constraint ($h(0) = 1$) is the same as that of a real modulator, and optimized zeros are as useful in quadrature systems as they are in real systems. The stability-imposed constraints on the out-of-band NTF gain appear to be similar between real and quadrature systems. The only real difference (pun intended) is that the pole/zero distribution of a quadrature NTF need not be symmetric about the real axis. Fig. 5.34a shows the pole-zero plot of a quadrature NTF intended for an $f_0 = f_s/4$, $OSR = 32$ application. Observe that the NTF

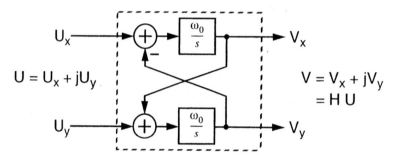

Figure 5.33: A quadrature resonator, $H(s) = \omega_0/(s - j\omega_0)$.

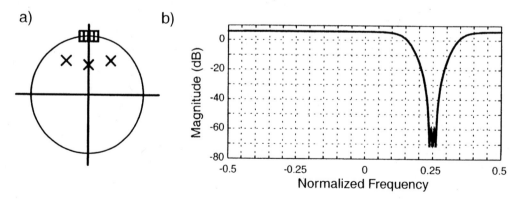

Figure 5.34: a) Pole-zero and b) magnitude plots for a quadrature NTF with $OSR = 32$.

zeros are located only in the positive-frequency passband. Fig. 5.34b plots the associated NTF magnitude. Since the rms attenuation of quantization noise is approximately 60 dB, this NTF provides 10 bits of resolution in addition to the 2.5 bits resulting from OSR = 32. The expected SNR of a modulator employing this NTF in conjunction with a pair of 4-bit quantizers is thus about 100 dB (16.5 bits). (This SNR estimate is identical to that given on page 146 for a bandpass modulator having similar NTF attenuation, OSR and number of quantizer levels.)

Fig. 5.35 shows the simulated quadrature output data of such a modulator when the input is a −3 dBFS quadrature sine wave. As with the bandpass example, the correspondence between the input and output waveforms appears very coarse in the time domain, but a much clearer picture emerges in the frequency-domain plot of Fig. 5.36, from which an SQNR of nearly 100 dB is calculated.

As a final demonstration that the quadrature modulator is indeed operating in accordance with expectations, Fig. 5.37 plots the simulated SQNR as well as the

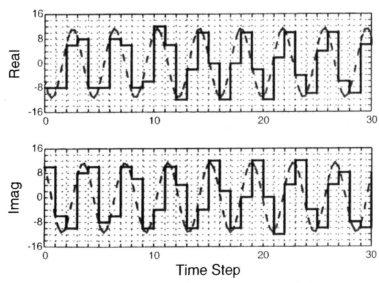

**Figure 5.35:** Real (I) and imaginary (Q) components of the output data from a quadrature modulator containing 16-step quantizers.

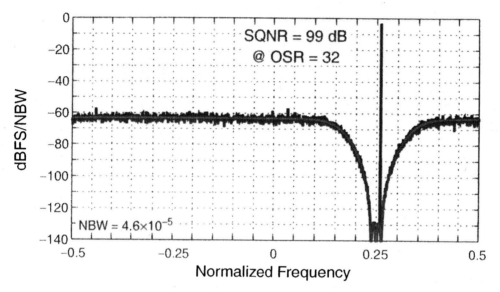

Figure 5.36: Output spectrum for a quadrature modulator.

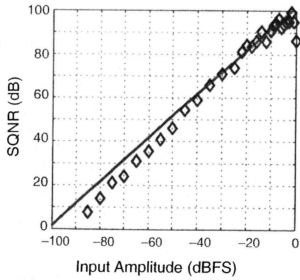

Figure 5.37: SQNR vs. input amplitude for the quadrature modulator.

expected SQNR vs. input amplitude curve. For moderately strong inputs (i.e. input levels above −40 dBFS), the observed SQNR follows the theoretical curve well. A peak SNR of about 100 dB is observed, and the modulator appears to be stable for inputs up to −2 dBFS.

Under ideal circumstances, the performance of a quadrature modulator is essentially as predicted. Non-idealities such as finite resonator Q and shift in resonator frequency have deleterious effects that are similar in magnitude to those found in real systems, and so are usually not problematic. However, quadrature errors (caused by mismatch in the two channels) can be a serious source of degradation. To see this, observe that in the spectrum of Fig. 5.36 the level of the quantization noise in the passband (around $f_s/4$) is nearly 65 dB below the level of the quantization noise in the image band (around $-f_s/4$). Path mismatch on the order of 0.1% (caused, for example, by mismatch in the full-scale outputs of the DACs which feed back to the first quadrature resonator) is sufficient to reflect enough image band noise to degrade the SQNR by more than 6 dB. Since much more stringent matching would be needed to ensure negligible performance degradation, path mismatch can easily be the dominant error source in a quadrature modulator.

Two methods for countering path mismatch have been described in the literature. The first involves adding one or more image zeros (and corresponding image poles) to the NTF so that the noise present in the image band is reduced. The depth of the image notch is adjusted to achieve the desired immunity to path mismatch. In addition to the increased hardware complexity, reducing the mismatch sensitivity in this way comes at the price of a reduction in the suppression of quantization noise and possibly a reduction in the stable input range. The second method for combatting mismatch applies only to DAC mismatch, and involves the use of *quadrature mismatch-shaping* [20].

The structure of a single-loop quadrature modulator follows that of real modulators, namely a loop filter attached to a quantizer. However, the loop filter and the quantizer are now quadrature systems. A quadrature quantizer is simply a pair of real quantizers, while the quadrature loop filter is a cascade of quadrature resonator sections such as that shown in Fig. 5.33. The usual variety of feedback and feedforward topologies, as well as single-loop and multi-loop architectures, are all

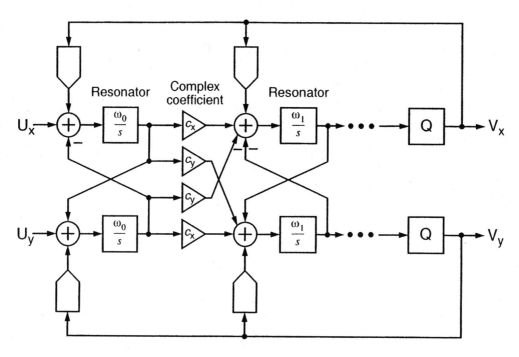

**Figure 5.38: A quadrature modulator employing the feedback topology.**

applicable to quadrature modulators. For example, a feedback topology (using continuous-time resonators) is depicted in Fig. 5.38, while [21] describes the use of a feedforward topology.

## 5.7 Conclusions

Delta-sigma modulation is applicable to narrowband bandpass and narrowband quadrature signals, and in this capacity delta-sigma converters are unique in the converter world. Efficient realizations of high-performance ADCs are possible, especially if a physical resonance, such as that of an LC tank, is exploited. Bandpass modulators are currently commercially available as stand-alone ICs and as sub-blocks in larger ASICs. Quadrature modulators are a more recent development and appear to be on the verge of achieving similar levels of commercial success.

# References

[1] T. H. Pearce and A. C. Baker, "Analogue to digital conversion requirements for HF radio receivers," *Proceedings of the IEE Colloquium on system aspects and applications of ADCs for radar, sonar and communications,* London, Nov. 1987, Digest No. 1987/92.

[2] S. A. Jantzi, M. Snelgrove and P. F. Ferguson, Jr., "A $4^{th}$-order bandpass sigma-delta modulator," *Proceedings of the IEEE 1992 Custom Integrated Circuits Conference*, Boston, MA, pp. 16.5.1 - 16.5.4, May 3-6 1992.

[3] G. Tröster, H.-J. Dreßler, H.-J. Golberg, W. Schardein, E. Zocher, A. Wedel, J. Arndt, "An interpolative bandpass converter on a 1.2µm BiCMOS analog/digital array," *IEEE Journal of Solid-State Circuits*, vol. 28, no. 4 pp. 471-477, April 1993.

[4] A. Hairapetian, "An 81MHz IF receiver in CMOS," *IEEE Journal of Solid-State Circuits,* vol. 31, no. 12, pp. 1981-1986, December 1996.

[5] O. Norman, "A bandpass ΣΔ modulator for ultrasound imaging at 160 MHz clock rate," *IEEE Journal of Solid-State Circuits*, vol. 31, no. 12, pp. 2036-2041, December 1996.

[6] S. Bazarjani and W. M. Snelgrove, "A160-MHz fourth-order double-sampled SC bandpass sigma-delta modulator," *IEEE Transactions on Circuits and Systems II*, vol. 45, no. 5, pp. 547-555, May 1998.

[7] J. A. E. P. Van Engelen, R. J. Van De Plassche, E. Stikvoort and A. G. Venes, "A sixth-order continuous-time bandpass sigma-delta modulator for digital radio IF," *IEEE Journal of Solid-State Circuits*, vol. 34, no. 12, pp. 1753 -1764, Dec. 1999.

[8] G. Raghavan, J. F. Jensen, J. Laskowski, M. Kardos, M. G. Case, M. Sokolich and S. Thomas III, "Architecture, design, and test of continuous-time tunable intermediate-frequency bandpass delta-sigma modulators," *IEEE Journal of Solid-State Circuits*, vol. 36, no. 1, pp. 5-13, Jan. 2001.

[9] P. Cusinato, D. Tonietto, F. Stefani and A. Baschirotto, "A 3.3-V CMOS 10.7-MHz sixth-order bandpass ΣΔ modulator with 74-dB dynamic range," *IEEE Journal of Solid-State Circuits*, vol. 36, no. 4, pp. 629-638, April 2001.

[10] T. Salo, S. Lindfors and K.A.I Halonen, "A 80-MHz bandpass ΔΣ modulator for a 100-MHz IF receiver," *IEEE Journal of Solid-State Circuits,* vol. 37, no. 7, pp. 1798 -808, July 2002.

[11] Analog Devices, Inc., *AD9870 datasheet*, Norwood, MA: Analog Devices, 1999.

[12] Analog Devices, Inc., *AD9874 datasheet*, Norwood, MA: Analog Devices, 2002.

[13] S. A. Jantzi, K. W. Martin and A. S. Sedra, "Quadrature bandpass ΔΣ modulation for digital radio," *IEEE Journal of Solid-State Circuits*, vol. 32, no. 12, pp. 1935-1950, Dec. 1997.

[14] T. Paulus, S. S. Somayajula, T. A. Miller, B. Trotter, C. Kyong and D. A. Kerth, "A CMOS IF transceiver with reduced analog complexity," *IEEE Journal of Solid-State Circuits*, vol. 33, no. 12, pp. 2154 -2159, Dec. 1998.

[15] C. H. Leong and G. W. Roberts, "High-order bandpass sigma-delta modulators for high-speed D/A applications," *Electronics Letters*, vol. 33, no. 6, pp. 454-455, March 1997.

[16] J. Keyzer, R. Uang, Y. Sugiyama, M. Iwamoto, I. Galton and P. M. Asbeck, "Generation of RF pulsewidth modulated microwave signals using delta-sigma modulation," *IEEE MTT-S International Microwave Symposium Digest*, vol. 1, pp. 397 -400, Sept. 2002.

[17] R. Schreier, J. Lloyd, L. Singer, D. Paterson, M. Timko, M. Hensley, G. Patterson, K. Behel and J. Zhou, "A 10-300 MHz IF-digitizing IC with 90-105 dB dynamic range and 15-333 kHz bandwidth," *IEEE Journal of Solid-State Circuits*, vol. 37, no. 12, pp. 1636-1644, Dec. 2002.

[18] B. Razavi, *RF Microelectronics*, Englewood Cliffs, NJ: Prentice-Hall, 1997.
[19] J. Van Engelen and R. Van De Plassche, *Bandpass sigma delta modulators— stability analysis, performance and design aspects*, Norwell, MA: Kluwer Academic Publishers 1999.
[20] R. Schreier, "Quadrature mismatch-shaping," *Proceedings, IEEE International Symposium on Circuits and Systems*, Vol. 4, pp. 675-678, May 2002.
[21] K. Philips, "A 4.4mW 76dB complex $\Sigma\Delta$ ADC for bluetooth receivers," *IEEE International Solid-State Circuits Conference Digest of Technical Papers*, pp. 64-65, Feb. 2003.

# CHAPTER 6
# *Implementation Considerations For $\Delta\Sigma$ ADCs*

In this chapter, some of the implementation issues arising in the design of $\Delta\Sigma$ A/D converters will be discussed. These include the choice between single-bit and multi-bit quantizers, and algorithms which can effectively linearize the internal DAC in multi-bit quantizers. Also, the relative advantages of sampled-signal and continuous-time loop filters will be analyzed, and the design of modulators with continuous-time loop filters will be addressed.

## 6.1 Modulators with Multi-Bit Internal Quantizers

As already discussed in Chapter 4, the use of multi-bit quantizers offers many advantages, including the following:

1. For a fixed full-scale, the quantization error is reduced by 6 dB for every bit added to the resolution of the quantizer. Hence, the SQNR is increased by this amount, and the stopband performance of the decimation or reconstruction filter used to suppress the out-of-band noise can be correspondingly relaxed.

2. The feedback loop becomes more linear, since the variations of the effective gain of the quantizer with its input signal are reduced. Hence, the stability is more robust, and the performance is closer to what a linear analysis predicts. Thus, the simulation effort needed to predict the expected performance can be reduced.

3. Due to the improved loop stability, the NTF can be chosen more aggressively. Hence, better SQNR can be obtained, and/or larger input signals can be allowed. For example, for a $5^{th}$-order single-bit modulator with $OSR = 16$, stability considerations limit the SQNR to about 60 dB; using a 4-bit quantizer, the reduced quantization noise and improved stability combine to increase the theoretical peak SQNR to about 120 dB!

4. Since the DAC input to the loop filter changes less from sample to sample, the required slew rate of the input op amp of the loop filter is reduced. This may allow lower power dissipation.

5. Since the difference between the input and the DAC signals entering the loop filter is smaller, the linearity requirements on the input stage of the filter can be eased.

6. Finally, for DACs and for ADCs with continuous-time loop filters, the smaller steps in the DAC waveform make the operation less sensitive to clock jitter, as will be discussed in Section 6.6.

Regarding item 2 and referring to Fig. 6.1, the "instantaneous gain" of the multi-bit quantizer is

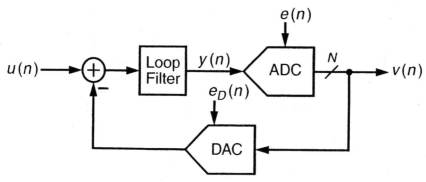

Figure 6.1: Signals in a common $\Delta\Sigma$ ADC topology.

$$k(n) = \frac{v(n)}{y(n)} = \frac{y(n) + e(n)}{y(n)} = 1 + \frac{e(n)}{y(n)} \tag{6.1}$$

Hence, the gain deviation from the nominal gain 1 is at most $\Delta/|2 \cdot y(n)|$, where $\Delta$ is the LSB voltage of the quantizer. Note that the deviation is large for small $y(n)$ which cannot overload the quantizer, and smallest for large quantizer inputs, where instability is most likely. By contrast, for a 1-bit quantizer, the instantaneous gain (with the normalization used in Fig. 2.4) is $1/y(n)$, which can take on any value as $y(n)$ changes, and tends to zero as $y(n)$ becomes large. As the discussions of Chapter 4 showed, small values of $k(n)$ are likely to result in instability for higher-order loops.

Against all these advantages stands a single large disadvantage: as can be seen from the location of the DAC in the feedback path between the output and the input of the loop (Fig. 6.1), any error $e_D(n)$ in the DAC response will be directly subtracted from the input signal $u(n)$ and hence it appears at the output without the benefit of noise shaping.

Especially damaging is the fact that any nonlinearity of the DAC will introduce a corresponding nonlinear signal distortion into the overall ADC response. To see this, assume that the nonlinear input/output characteristic of the DAC is $w(n) = F[v(n)]$. Since, due to the operation of the feedback loop, to a very good approximation $w = u$ holds within the signal band, for a nonlinear characteristics $F[\bullet]$ the output $v(n) = F^{-1}[w(n)] = F^{-1}[u(n)]$ will contain the nonlinearly distorted version of $u(n)$.

Clearly, for an $N$-bit ADC linearity, the DAC needs to be linear to at least $N$ bits. Since it is hard to realize a DAC with a linearity much better than $10 \sim 12$ bits without trimming, other methods must be used to achieve the 16- to 19-bit ADC linearity typically needed. The most useful ones of these methods can be classified into dual-quantization, mismatch-shaping and digital correction schemes. They will be discussed in the next sections.

## 6.2 Dual-Quantizer Modulators

The fundamental idea behind the dual-quantization architectures is to use a single-bit quantizer where the DAC nonlinearity is most critical, i.e., where the DAC output is directly subtracted from the input, and use one or more multi-bit quantizers elsewhere in the system. The circuit will have two or more digital outputs which are combined in such a way that the large error introduced by the single-bit quantizer is cancelled. The simplest example of this technique is the Leslie-Singh architecture, already described in Section 4.5.1 and depicted in Fig. 4.20. The single-bit quantizer $Q$ makes it possible to achieve high linearity in the first stage which processes the input signal $u$. The large quantization error $e_1$ of $Q$ is then converted by the multi-bit ADC into digital form, and cancelled in the output using the noise-cancellation logic implemented by the digital blocks $H_1$ and $H_2$. The final output $v$ ideally contains only the input signal $u$ and the filtered and inherently small quantization error $e_2$ of the second-stage ADC.

Next, some more elaborate schemes will be shown for implementing the dual-quantization principle.

### 6.2.1 Dual-Quantization MASH Structure

As an extension of the Leslie-Singh scheme, a two-stage MASH can be implemented so that the second stage contains a multi-bit quantizer (Fig. 6.2) [1]. The nonlinearity of the multi-bit DAC (here, a 3-bit DAC) in the second stage will not introduce signal distortion, since the input to the second stage (at least ideally) contains only the quantization error $e_1$ of the first stage and not the input signal $u$. Also, the nonlinearity error $e_D$ of this DAC will be filtered along with $e_1$ by the digital highpass filter $H_2$, and hence will be suppressed in the signal band. The overall output is still given by Eq. (4.52), where $E_2$ is now the small error due to the multi-bit second-stage quantizer. The improved stability of the second stage makes it possible to use a more aggressive noise shaping function $NTF_2$ here.

The principle can be extended to 3-stage MASH structures, with a binary first stage and multi-bit second and third stages.

As discussed in Section 4.5.3, the imperfections of the analog elements will affect the cancellation of the first-stage quantization error. This is a key limitation on the performance of all multiple-quantizer schemes.

### 6.2.2 Dual-Quantization Single-Stage Structure

It is also possible to use dual (both single-bit and multi-bit) quantizers in a single delta-sigma loop [2], as shown in Fig. 6.3 for a third-order structure. Here, the DAC signals fed back to the inputs of the first and second integrators are single-bit ones, and hence it is possible to achieve high linearity for these signals. A multi-bit signal is fed back to the third integrator. The nonlinear distortion introduced by the $N$-bit DAC generating this signal is divided by the product of the transfer functions of the first two integrators, and hence can be negligible for high OSR. By combining the outputs of the single-bit and multi-bit ADCs, $e_1$ can be cancelled and replaced in the output by the quantization error $e_2$ of the $N$-bit ADC. The overall output is then given by

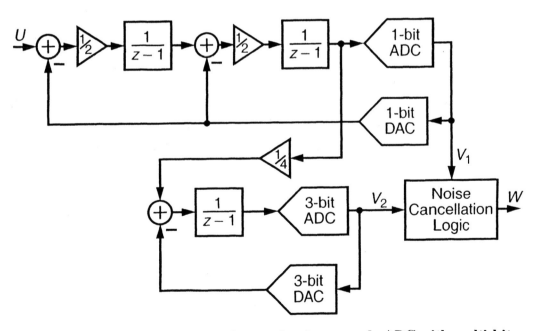

**Figure 6.2:** A third-order dual-quantization cascade ADC with multi-bit second stage.

$$V(z) = z^{-1}U(z) + 2(1-z^{-1})^3 \cdot E_2(z) - 2z^{-1}(1-z^{-1})^2 \cdot E_D(z) \qquad (6.2)$$

Since the use of multi-bit quantizers reduces the input signals of both quantizers, the loop stability is improved compared to that of a simple binary loop. The effect of nonlinearity in the multi-bit DACs can be further reduced by using one of the low-distortion structures discussed in Section 4.4 for the loop filter.

Since the cancellation of $e_1$ is contingent upon the equality of analog and digital transfer functions (just as in the MASH structure), the imperfections of analog components (finite op-amp gains, capacitor mismatches) will introduce single-bit quantization noise leakage in this structure also.

## 6.3 Dynamic Element Randomization

Element mismatches in the multi-bit DAC introduce an output error, which consists of the harmonics of the input signal as well as an increased noise floor due to the folding of high-frequency quantization noise into the signal band. In some cases, the increased noise floor (and the resulting reduction of the SQNR) is acceptable, but the harmonic signal distortion is not. In this case, a randomization of the static nonlinearity of the DAC may provide a simple solution to the problem, by converting the energy of the harmonic spurs into pseudo-random noise [3]. The process can be carried out conveniently if the input of the DAC is a thermometer-

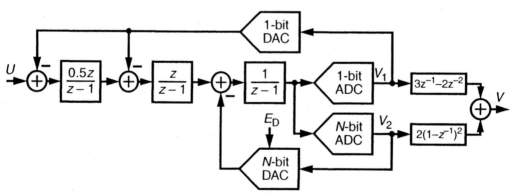

Figure 6.3: A third-order, dual-quantization, single-path ADC.

coded digital signal, and the DAC is built from unit elements (equal-valued switched capacitors or current sources). The data conversion is performed by activating $K$ unit elements if the value of the input code is $K$. The error randomization is achieved by choosing these $K$ unit elements randomly each time. Then the DAC error $e_D(n)$ at time $n$ will not be correlated with the value of its input $v(n)$, and hence the signal distortion is replaced by random noise in the DAC output. A schematic block diagram of the system implementing the process is shown in Fig. 6.4 [4].

To illustrate the effect of error randomization, Fig. 6.5 [4] compares the output error spectra of a third-order ADC with a 3-bit quantizer, without and with randomization. It is assumed that the DAC is built from unit elements with a [–0.1%, +0.1%] linear-gradient mismatch. Fig. 6.5 shows that applying randomization eliminates the large second harmonic spur caused by this static DAC nonlinearity, but unfortunately also increases the noise floor.

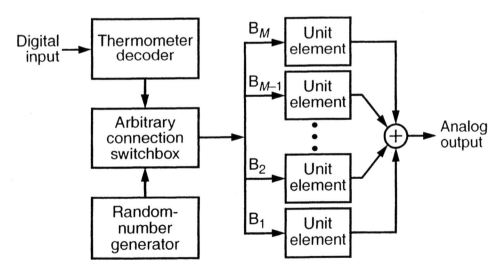

Figure 6.4: Parallel-unit-element DAC with randomized element selection.

## 6.4 Mismatch Error Shaping

The basic operating principle of delta-sigma data conversion, namely noise shaping, can also be utilized to reduce the effects of the nonlinearity in a multi-bit DAC. Here, the "noise" is the mismatch error introduced by the uneven spacings of the DAC levels, rather than the quantization error, but the procedure used is the same: use filtering to suppress the noise spectrum in the signal band and to shift its power to out-of-band frequencies. There are a number of strategies which can achieve this [4]. The most commonly used ones are *element rotation* (also called *data-weighted averaging*), *individual level averaging*, *vector-based mismatch shaping* and *tree-structure element selection*. All these techniques utilize the unit-element DAC structure shown in Fig. 6.4, but each uses a different strategy for the selection of the unit elements for a given digital input code. These strategies will be discussed in the next subsections.

A common underlying assumption of these element-selection algorithms is that the offset and gain errors of the DAC (which introduce corresponding offset and gain errors into the transfer characteristics of the overall $\Delta\Sigma$ ADC) are acceptable, and only the DAC nonlinearity error is of concern. This assumption is valid in many

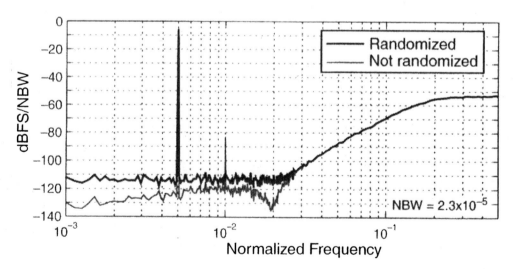

Figure 6.5: Output spectra of a $3^{rd}$-order 3-bit modulator, with and without element randomization.

practical applications, and it is necessary to make, since no element selection method can reduce the offset $OS$ (i.e., the DAC output when the input code is zero, so no unit element is used) or affect the full-scale value $FS$ when all elements are used (Fig. 6.6). Hence, the effective gain $(FS - OS)/M$ (where $M$ is the full scale digital input to the DAC) cannot be changed by any element selection logic either. Therefore, we must assume that these values are the correct ones, and hence that the "ideal" $M$-element DAC input-output characteristics is described by the linear function

$$w(k) = OS + \frac{FS - OS}{M} \cdot k, \qquad k = 0, 1, 2, ..., M \qquad (6.3)$$

Next, let each unit element have the actual value $u_i = U + d_i$ ($i = 1, 2, ..., M$). Here, $U$ is the average element value

$$U = \sum_{i=1}^{M} u_i / M \qquad (6.4)$$

and the $d_i$ are the deviations of the $u_i$ from the average. From (6.4), it follows that

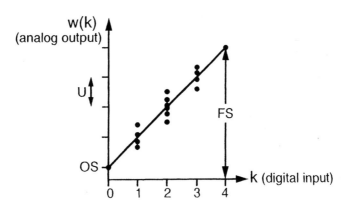

**Figure 6.6: Outputs of a 4-element DAC.**

$$\sum_{i=1}^{M} d_i = 0 \tag{6.5}$$

The full-scale output is therefore

$$FS = OS + \sum_{i=1}^{M} u_i = OS + M \cdot U \tag{6.6}$$

and the ideal output for an input $k$ is

$$w(k) = OS + k \cdot U, \tag{6.7}$$

An interesting and important property of unit-element DACs concerns the symmetry properties of the possible DAC outputs shown in Fig. 6.6, which shows the possible characteristics of a DAC with $M = 4$ inaccurate unit elements. As explained earlier, for input codes $k = 0$ and $k = M = 4$ the analog output is unique. For all other inputs, however, due to the matching errors $d_i$, different outputs result from different selections of the unit elements used to produce the output. For example, for $k = 2$, the first unit element used may be chosen four different ways, while the other one three different ways. Since the order of selection is immaterial, we have $4 \times 3/2 = 6$ possible outputs for the same input code. Three of these selections will include $u_1$, three will include $u_2$, etc. Hence, when we calculate the average value of all 6 possible outputs for $k = 2$, we find

$$\overline{w}(k) = OS + 3 \cdot \sum_{i=1}^{4} u_i/6 = OS + 2U + 0.5 \cdot \sum_{i=1}^{4} d_i = OS + 2U \tag{6.8}$$

which is the same as the ideal value calculated from (6.7). Thus, the average value of all possible outputs lies on the ideal DAC characteristic line connecting the $k = 0$ and $k = M$ points. This statement is generally true, and is utilized in some of the element selection algorithms described below.

### 6.4.1 Element Rotation or Data-Weighted Averaging

The element rotation [5] or data-weighted-averaging (DWA) [6] algorithm aims to make the long-term average use of each unit element in the DAC the same, by rotating the pattern of unit elements. Consider a DAC with 8 unit elements, and visualize them arranged in a circle (Fig. 6.7). Let the first input code be 3; then elements 1, 2 and 3 will be used. Let the next code be 4; then the next 4 elements (4, 5, 6 and 7) will be called to duty. If the next code is a 6, then we wrap around the circle, and use elements 8, 1, 2, 3, 4 and 5. Thus, we use all components sequentially, and as often as possible.

To understand the effect of DWA on the time response and spectrum of the DAC error induced by the mismatches $d_i$, next we shall examine the sequence of DAC output errors $e(n)$ ("mismatch noise") generated during the conversion. A limit can be found for the sum of all errors

$$s(N) = \sum_{n=1}^{N} e(n) \qquad (6.9)$$

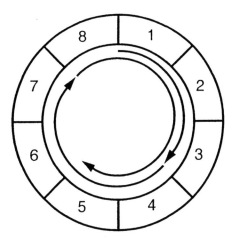

**Figure 6.7: Element rotation.**

accumulated during the operation over $N$ clock periods. Specifically, as the conversion progresses and cycles around the dial of unit elements, every time it reaches element 8, $s(N)$ is reset to 0, since the sum of errors $d_i$ for all 8 elements is zero. Hence, for any $N$, $|s(N)|$ is at most equal to $P/2$, where $P$ is the sum of the absolute values of the errors in the 8 unit elements. $P$ is a finite (in fact, relatively small) number. The average value of the error samples $e(n)$ after $N$ clock periods thus satisfies

$$|e(N)|_{ave} = |s(N)|/N \leq P/(2N) \to 0 \text{ as } N \to \infty. \tag{6.10}$$

Hence, the long-term average value of $e(n)$ is zero, a very desirable result.

Next, let the power spectrum of $s(n)$ be $S(\omega)$. Since $s(n)$ is bounded by $P/2$, it is reasonable to assume that (for a random and busy input signal) $s(n)$ will behave like a bounded white noise so that $S(\omega)$ will be a uniform noise spectrum. Since, by the definition of $s(n)$, the relation $e(n) = s(n) - s(n-1)$ holds, the power spectrum of $e(n)$ satisfies

$$E(\omega) = |1 - e^{-j\omega}|^2 \cdot S(\omega). \tag{6.11}$$

Hence, the mismatch noise is shaped by a first-order highpass filter function. The noise shaping is thus similar to that encountered by the quantization noise in MOD1.

As in the case of MOD1, however, tone generation is a major problem with the DWA algorithm, if the DAC input is not a busy random signal but a dc or a low-frequency periodic one. Assume, e.g., that the input to the 8-element DAC of Fig. 6.7 is a dc digital signal with a value of 6. Then the element sets used will be $S_1 = (1, 2, 3, 4, 5, 6)$; $S_2 = (7, 8, 1, 2, 3, 4)$; $S_3 = (5, 6, 7, 8, 1, 2)$; and $S_4 = (3, 4, 5, 6, 7, 8)$. After that $S_5 = S_1$, $S_6 = S_2$, etc. Thus, $e(n)$ will be a periodic signal with a line spectrum, not a highpass-filtered noise. The tonal character of $e(n)$ can be changed by introducing a random or pseudo-random effect (e.g., random element skipping, or bidirectional rotation) into the process. In the latter scheme (named bi-DWA in [7]), the direction of the rotation is inverted in every cycle. Each directional rotation proceeds independently from the other. As the simulation results of Fig. 6.8 illustrate, this increases the spurious-free dynamic range, but also increases the noise floor by approximately 9 dB.

### 6.4.2 Individual Level Averaging

Individual level averaging (ILA) [8] is another technique for introducing a first-order highpass filtering of the mismatch noise. Instead of attempting to equalize the use of all elements for all codes over a long time period, ILA tries to equalize their usage separately for each code over time. Since the average error of all selections for a given code is zero, as shown in the derivations of (6.8), cycling through all possible selections sequentially will result in the elimination of the average error (dc error) of the DAC. An alternative is to cycle through all elements for a given code $k$, making the probabilities of element usages the same. The simplest way to carry this out is to use the DWA algorithm for each code, i.e., for a given code $k$ rotate the elements used by $k$ places each time the code occurs, and wrap around after the $M^{th}$ element has been used.

Compared to the DWA method, ILA converges more slowly to the zero-average condition, which reduces the effectiveness of the technique, especially for large $M$.

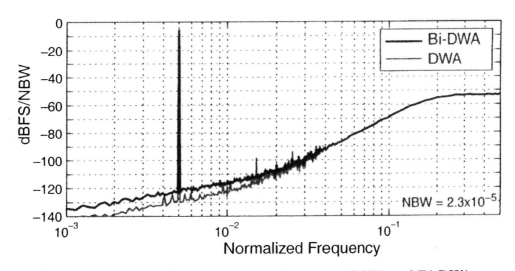

Figure 6.8: Simulated spectra for 8-element DWA and Bi-DWA with 1% element mismatch.

On the other hand, it is less affected by any correlation between the mismatch shaping and the DAC input signal, and is less likely to generate tones even for dc or periodic inputs.

Fig. 6.9 illustrates the output spectrum of a third-order $\Delta\Sigma$ loop with a 3-bit DAC, when ILA was used with DWA rotation for each code. No harmonics or spurious tones are visible in the spectrum, but the noise floor is higher than when element rotation is used.

### 6.4.3 Vector-Based Mismatch Shaping

The mismatch-error-shaping techniques described in the preceding sections are specifically tailored to achieve first-order error filtering, since they are based on the averaging of errors. A more general strategy, which can achieve higher-order spectral shaping [9] is described next. It actually involves the implementation of $M$ digital noise-shaping loops, one for each unit element. Fig. 6.10 shows an error-feedback loop which generates the bit $x_i$ turning on or off the $i^{th}$ unit element. Here, $x_i(n)$ is produced by a digital comparator whose output is 1 (0) if its input

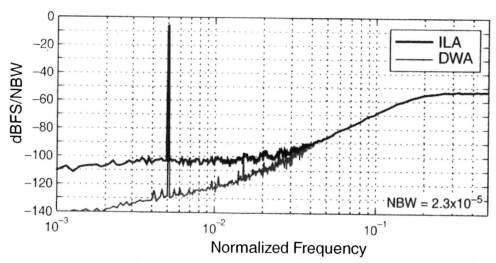

Figure 6.9: Spectrum for 8-element ILA with 1% element mismatch.

$w_i(n)$ is greater (smaller) than a time-varying reference $r(n)$. Its quantization error is $e_i(n)$. Note that the input $f(n)$ and the reference $r(n)$ are the same for all $M$ loops, as is the NTF $H(z)$.

The digital loop needed is very simple. For first-order mismatch shaping by $1 - z^{-1}$, the loop filter contains only a delay $z^{-1}$; for second-order shaping with $(1 - z^{-1})^2$ it contains two adders, two delays and a binary-point shift, realizing $z^{-1}(2 - z^{-1})$.

Linear analysis shows that

$$X_i(z) = F(z) + H(z)E_i(z) \qquad (6.12)$$

and in the time domain

$$x_i(n) = f(n) + [h*e_i](n) \qquad (6.13)$$

where the asterisk denotes discrete convolution.

The output of the DAC with input bits $x_i(n)$ will be

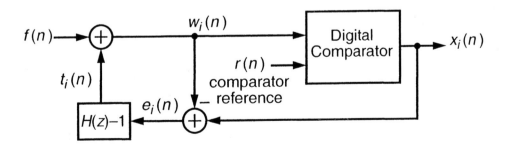

**Figure 6.10: One of $M$ error-feedback loops in vector mismatch-shaping.**

$$y(n) = \sum_{i=1}^{M} u_i(n) = \sum_{i=1}^{M} x_i(n)[U + d_i]$$

$$= U \sum_{i=1}^{M} x_i(n) + \sum_{i=1}^{M} \{f(n) + [h*e_i](n)\} \cdot d_i \qquad (6.14)$$

Here, the first term on the RHS is the ideal DAC output, for all unit-element errors $d_i = 0$. The second term gives the DAC error $e_D(n)$. It can be rewritten as

$$e_D(n) = f(n) \cdot \sum_{i=1}^{M} d_i + h(n) * \sum_{i=1}^{M} [d_i \cdot e_i(n)] \qquad (6.15)$$

The first term is zero, as (6.5) shows. The second term is the filtered and weighted sum of the output noises of all $M$ digital mismatch-shaping loops. Hence, $e_D(n)$ itself will be a noise-like signal shaped by $H(z)$. The order of the noise shaping is determined by that of $H(z)$, and (unless instability occurs) may be of arbitrarily high order. To prevent tone generation in the digital loops, dither signals may be injected at the inputs of the comparators. MASH and/or multi-bit loops may be utilized to improve the stability conditions, provided the elements can be used multiple times in a clock period. Finally, the system can be generalized to the use of elements with non-equal weights.

So far, the two inputs $f(n)$ and $r(n)$, which are common to all loops, have not been determined. In fact, $r(n)$ and the digital comparators need not be implemented; their presence in the loop shown in Fig. 6.10 is symbolic. The actual operation is to assign to a total of $K$ outputs $x_i(n)$ the value 1, and to the rest the value 0. Here, $K$ is the value of the input code $v(n)$ to the DAC. As (6.15) indicates, the DAC error can be reduced by making the $e_i(n)$ as small as possible; this will be the case if the $K$ largest $w_i(n)$ are quantized to $x_i(n) = 1$, and the rest to $x_i(n) = 0$.

# Mismatch Error Shaping

The choice of $f(n)$ can be based on making all data in the loop positive, and at the same time making their magnitudes as small as possible, to keep them bounded and the loop stable. This will be achieved if $f(n) = -\min_i [t_i(n)]$ is used.

Fig. 6.11 shows the output spectra of a third-order 16-level ADC for several mismatch scenarios when the input is a half-scale sine-wave and the oversampling ratio is 50. With an ideal DAC, $SNR = 108$ dB can be achieved. When 1% mismatch occurs between the DAC elements, the noise floor rises and large harmonics become apparent. The SNR deteriorates catastrophically, to about 50 dB. For second-order mismatch shaping, the SNR recovers to about 105 dB; for first-order shaping (which is equivalent to element rotation), the SNR is about 97 dB. Note that the slope of the noise-floor for the ideal ADC is 60 dB/decade, due to the third-order noise shaping of the loop; for the ADCs with nonideal DACs, the order of mismatch shaping determines the slope of the noise floor (40 dB/decade for second-order shaping, 20 dB/decade for first-order shaping).

The system described above can be regarded as a single delta-sigma loop operating on vectors containing the $x_i$, $e_i$, $w_i$ and $t_i$ utilizing a quantizer which truncates

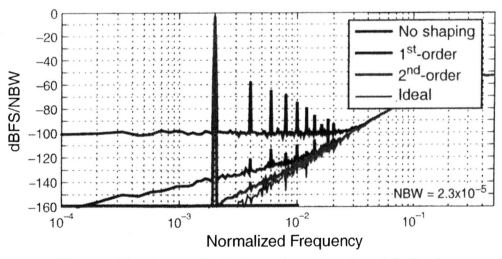

**Figure 6.11:** Spectra for 8-element vector mismatch shaping.

the vector of the $w_i$. Hence, it sometimes called a *vector quantizer* mismatch shaper.

### 6.4.4 Element Selection Using a Tree Structure

Another mismatch-shaping strategy using internal delta-sigma modulation, which can also provide higher-order (in practice, second-order) mismatch noise filtering, is based on the tree structure architecture [10] shown in Fig. 6.12 for $M = 8$ unit elements. The structure consists of $\log_2(M)$ layers of *switching blocks* (SBs), each of which splits its input data into two output signals whose sum equals the input.

The input to Layer 1 is slightly over 3 bits, corresponding to 9 possible values $(0, 1, 2, ..., 8)$. Hence, as illustrated, the input $x(n)$ must have a 4-bit wordlength. Alternatively, the signals shown in Fig. 6.12 can be encoded in unary (thermometer-code) form, in which case the total number of bits entering and leaving each layer is 8. Thermometer-coding is advantageous because it avoids the thermometer-to-binary conversion of the multi-bit flash converter's output.

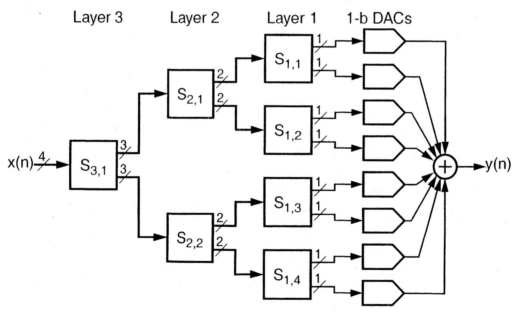

**Figure 6.12: Tree-structured mismatch-shaper.**

For switching block $S_{k,r}$, the first subscript $k$ denotes its layer, which also indicates the number of bits in its output. The second subscript $r$ denotes the SB's location within its layer. The last layer contains 1-bit outputs, each of which then activates a 1-bit DAC, i.e., a unit element. Since the sum of the codes in each layer remains the same, the tree simply rearranges the location of the 1s as a code ripples down from the single $M$-bit input to the final $M$ 1-bit outputs. The function of the structure is to select the final location of the 1s, and thus choose the unit elements activated, in such a way that the resulting DAC mismatch noise is highpass filtered.

The structure of switching block $S_{k,r}$ is shown in Fig. 6.13. It contains a sequence generator, which generates a 1-bit sequence $s_{k,r}(n)$ of special properties. It can be shown [10] that the final DAC mismatch noise will be a weighted sum of the $s_{k,r}$ sequences, and hence it will be shaped by an $L^{\text{th}}$-order highpass filter function, if each $s_{k,r}(n)$ sequence has an $L^{\text{th}}$-order spectral shaping.

A simple method which can generate $s_{k,r}(n)$ for first-order ($L = 1$) shaping is to use the rule

$$1, 0, 0, \ldots, 0, -1, 0, 0, \ldots, 0, 1, 0, 0, \ldots \tag{6.16}$$

Here, the sequence of zeros is interrupted by a 1 or $-1$ (alternating) when an odd $x_{k,r}(n)$ enters the SB. Clearly, the sum of the $s_{k,r}(n)$ from $n = 0$ to infinity is

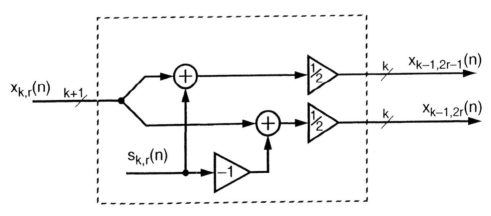

**Figure 6.13: Switching block.**

between −1 and 1, and hence its power spectral density has a zero at dc, indicating first-order noise shaping. A similar sequence can also be obtained by using a first-order digital $\Delta\Sigma$ modulator loop as the sequence generator; an $s_{k,r}(n)$ sequence with second-order shaping can also be obtained, by using a second-order $\Delta\Sigma$ loop. Figs. 6.14a and 6.14b show, respectively, a first- and second-order loop capable of generating shaped $s_{k,r}(n)$ sequences. In Fig. 6.14b, the quantizer is mid-tread when $x_{k,r}(n)$ is odd and mid-rise when $x_{k,r}(n)$ is even. As discussed above, using such a loop will result in first- or second-order mismatch shaping for the overall DAC.

Fig. 6.15 illustrates the performance of an 8-element DAC with ideal unit elements, as compared with 1%-mismatched unit elements selected by scrambling, as well as by first- and second-order mismatch shaping. Clearly, the low-frequency

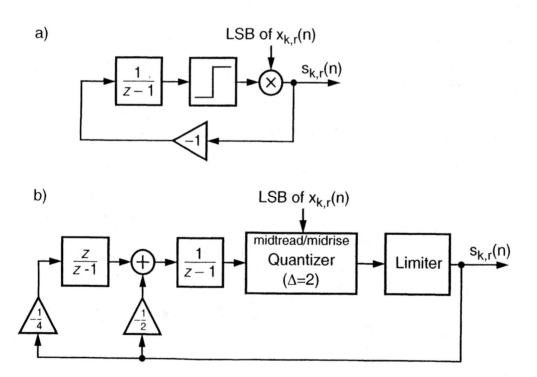

Figure 6.14: Switching sequence generators: a) first-order; b) second-order.

noise shaping is much improved by mismatch shaping. Also, note that the improvement offered by second-order shaping is significant only if the OSR is large (say, greater than 64).

## 6.5 Digital Correction of DAC Nonlinearity

The multi-bit modulators discussed in Section 6.2 reduce the effects of multi-bit DAC nonlinearities by using multi-bit DACs only in branches where only a filtered version of the resulting nonlinearity error appears in the output signal. The schemes described in Sections 6.3 and 6.4 convert the error signal introduced by the nonlinearity of the multi-bit DAC into a pseudo-random noise, and (in most systems) also shape the power spectral density of this noise such that the in-band noise power is reduced. All these schemes rely on oversampling and noise-shaping, and hence become ineffective for very low values (4 ~ 8) of the OSR, which are sometimes needed in broadband data converters.

In this section, different strategies will be discussed. These are based on acquiring the DAC errors in a digital form, and then cancelling their effects in the digital domain. The acquisition can be performed at power-up, or as a background process during normal operation.

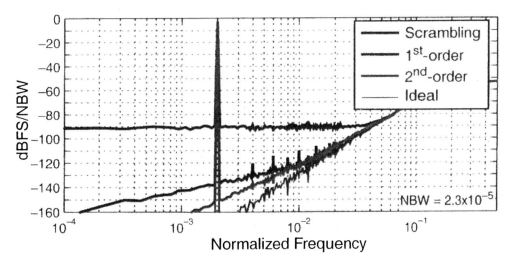

Figure 6.15: Spectra for 8-element tree-structured mismatch-shaping.

### 6.5.1 Digitally-Corrected Multi-Bit ΔΣ Modulator with Power-Up Calibration

Fig. 6.16 illustrates the basic correction scheme of a digitally-corrected ΔΣ ADC [11]. It will next be shown that the digital correction stage following the modulator loop can simply be a random-access memory (RAM), storing the accurate digital equivalents of the actual output values of the N-bit DAC for all possible input codes, and addressed by $v(n)$. Thus, for any loop output signal sample $v(n)$, the analog output $v'(n)$ of the DAC and the digital output $w(n)$ of the correction stage are the same. Since for sufficiently high inband loop gain the inband spectrum of $v'(n)$ is very close to that of the analog input $u(n)$, and since $v'(n) = w(n)$, it follows that the inband spectrum of $w(n)$ is very close to the input spectrum.

The calibration, i.e., the acquisition and storing of the digital versions of the actual DAC output values, can be performed at power-up. In the process, an N-bit digital counter (not shown) is used to produce all input codes of the DAC (Fig. 6.17). Each code is held for a duration of at least $2^M$ clock periods, where M is the required linearity of the overall ADC, in bits. During this time, the DAC output is fed to a single-bit (and hence very linear) ΔΣ ADC with a digitally filtered output. The resulting digital code is the desired DAC output equivalent, and is stored in the RAM.

The system can be made much more efficient and economical if the stored digital values represent the errors, rather than the actual values, of the DAC output levels. The reader is referred to [11] for the details.

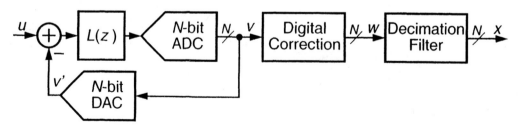

**Figure 6.16:** A digitally-corrected ΔΣ ADC.

Fig. 6.18 [11] verifies the effectiveness of the correction process. It shows the measured signal spectra at the loop output and at the output of the RAM for a sinewave input signal. The DAC linearity was about 10.5 bits, resulting in a spur-free dynamic range SFDR of about 63 dB and an elevated noise floor in the uncorrected output $v(n)$. The corrected signal $w(n)$ at the RAM output had a noise floor lowered by nearly 20 dB (achieving a peak SNR of 95 dB at $OSR = 128$), and a much improved SFDR of approximately 105 dB.

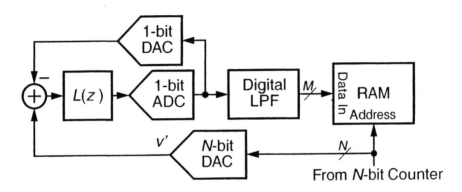

**Figure 6.17: Calibration scheme for digital correction.**

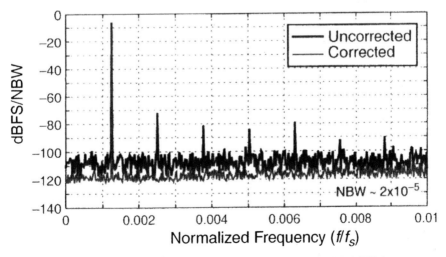

**Figure 6.18: Measured spectra ($f_s = 5.25$ MHz).**

The principle of Fig. 6.16 is easily extended to ΔΣ DACs. Fig. 6.19 illustrates a digitally corrected N-bit DAC. As before, the RAM should contain the accurate digital replicas of the actual DAC output levels. An argument very similar to that presented for the ADC of Fig. 6.16 proves that the inband spectrum of the analog output will be very close to the spectrum of the digital input signal $x(n)$ even if the DAC has static nonlinearities.

### 6.5.2 Digitally-Corrected Multi-Bit ΔΣ ADC with Background Calibration

While power-up calibration is feasible in most cases, it cannot track and correct the changes due to thermal drift during operation. Hence, background calibration methods, which continuously update the DAC error correction, are usually preferable. Such a technique, based on a correlation algorithm, has been reported in [12]. The basic principle is illustrated in Fig. 6.20. The ADC output $v(n)$ is represented in a thermometer code, and the resulting bits are permuted (scrambled) in a pseudorandom way. The objective of the scrambling is to reduce the correlation between the individual bit sequences $b_i(n)$ controlling the unit elements of the DAC and all other signals in the system, especially the input signal $u(n)$ and the quantization noise $q(n)$.

The DAC errors are estimated from the correlation between the scrambled bits $b_i(n)$ and the ADC output $v(n)$. The latter is given by

$$v(n) = [u*stf](n) + [q*ntf](n) + [e*etf](n) \qquad (6.17)$$

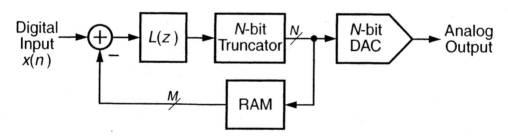

Figure 6.19: A digitally-corrected N-bit loop.

where $stf(n)$, $ntf(n)$ and $etf(n)$ are the impulse responses (i.e., inverse z-transforms) of the *STF*, *NTF* and *ETF*, respectively, and the asterisk ($*$) denotes the convolution operation. Here, *ETF* is the error transfer function, from the DAC error to the output. As before, we are only concerned with the deviations $d_i$ of the unit element values $u_i$ from the average element value $U$ given by (6.4). Hence, the DAC errors are given by

$$e(n) = \sum_{i=1}^{M} d_i \cdot b_i(n) \qquad (6.18)$$

An estimate of the $i^{th}$ DAC error $d_i$ can be obtained by finding the expected value of the product $v(n) \cdot b_i'(n)$, where $b_i'(n)$ denotes $[b_i * etf](n)$. From (6.17), denoting the first term in (6.17) by $u'(n)$, the second by $q'(n)$ and the third by $e'(n)$, we obtain

$$E\{v(n) \cdot b_i'(n)\} = E\{u'(n) \cdot b_i'(n)\} + E\{q'(n) \cdot b_i'(n)\} + E\{e'(n) \cdot b_i'(n)\} \quad (6.19)$$

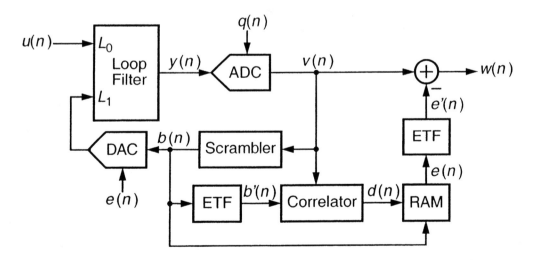

**Figure 6.20: Background calibration scheme for a digitally-corrected ΔΣ ADC**

Assuming that sequences $u'(n)$, $q'(n)$ and $e'(n)$ have zero mean values, and that $b_i(n)$ is uncorrelated with $u(n)$ and $q(n)$, the first two terms on the RHS of (6.19) will be zero. Assuming also that $b_i(n)$ is uncorrelated with all other $b_j(n)$ sequences ($j \neq i$), the last term also simplifies, by (6.18), to

$$E\{[d_i b_i(n) * etf(n)] \cdot [b_i(n) * etf(n)]\} = d_i E\{[b_i'(n)]^2\}. \qquad (6.20)$$

Hence, an estimate of the $i^{th}$ unit element error can be obtained from

$$d_i \approx \frac{E\{v(n) \cdot b_i'(n)\}}{E\{[b_i'(n)]^2\}} \qquad (6.21)$$

The ratio of expected values $E\{\bullet\}$ can be obtained by accumulating the products $v(n) \cdot b_i'(n)$ and $[b_i'(n)]^2$, and finding the ratio of the two sums. This operation is performed by the correlator block in Fig. 6.20 either in every clock period, or after a predetermined number of clock periods. The $d_i$ values can then be stored in a RAM (Fig. 6.20), updated periodically, and used to correct the output $v(n)$ by subtracting the expected output error $e'(n)$ due to the DAC, as found from (6.18).

Unfortunately, the crucial assumption about the $b_i(n)$ being completely uncorrelated to all other sequences involved in the derivation, and the zero-mean property assumed for various signal and error sequences does not hold. Hence, in spite of the scrambling of bits at the input of the DAC, the remaining correlations cause slow convergence and a large residual error in the adaptive correction. However, there are several simple steps that can be taken to reduce the harmful correlations [12]. The resulting process is quite accurate, as illustrated in Fig. 6.21 for a 2-0 MASH ADC with a very low oversampling ratio ($OSR = 4$) and a 32-element DAC [12]. Fig. 6.21 compares the output spectra of the modulator for ideal DAC elements and for a DAC with a 0.1% random element errors, before and after correction. The inband SNDR is 102.6 dB for the ideal case, and only 76.2 dB for the modulator with the inaccurate DAC. Using the adaptive correction scheme, after $2^{17} = 131072$ clock periods the inband SNDR rose to 101.5 dB.

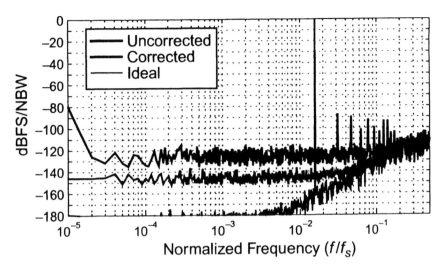

**Figure 6.21:** Simulated spectra for a 2-0 MASH modulator with adaptive correction of errors in the multi-bit DAC.

## 6.6 Continuous-Time Implementations

The earliest implementations of $\Delta\Sigma$ ADCs used continuous-time (CT) loop filters, but since the advent of switched-capacitor (SC) circuits in the 1980s the majority of $\Delta\Sigma$ ADCs have been constructed with SC loop filters. SC filters are attractive because they exhibit both good accuracy and good linearity. In addition, the difference equations which describe an SC circuit are independent of the clock rate, and hence the transfer function of an SC circuit scales naturally with the clock frequency. In contrast, CT filters generally have inferior linearity and accuracy. The time-constants of a CT filter are subject to large variation and furthermore do not track the clock rate. Consequently, the time-constants of a CT loop filter typically require calibration and that calibration is only valid for a single clock frequency.

Despite the above disadvantages, $\Delta\Sigma$ ADCs employing CT elements in their loop filters are appearing increasingly often in both the literature and the marketplace. There are two important reasons for this trend:

1. CT modulators possess *inherent anti-aliasing*. As will be demonstrated shortly, the degree of alias suppression is equivalent to the degree of quantization-noise suppression. Inherent anti-aliasing simplifies system design by eliminating the anti-alias filter, which typically must precede other ADCs. Inherent anti-aliasing also allows improved system performance, since it also eliminates the noise-folding associated with sampling the incoming signal.

   Effectively, the use of a CT filter postpones the inevitable sampling of the signal, which now takes place at the output of the loop filter. Thus, imperfections of the sampling process, and the folding of the wideband noise, both take place at a much less sensitive point in the loop.

2. The theoretical limit on the clock rate of a CT modulator is determined by the regeneration time of the quantizer and the update rate of the feedback DAC, whereas in an SC modulator the clock rate is limited by the op-amp settling requirements, to about 20% of the unity-gain frequency $f_u$ of the amplifiers within it. In practice, a CT modulator can operate with a clock frequency (and thus achieve a signal bandwidth) which is 2-4 times greater than that which can be achieved with SC techniques, albeit with a lower linearity and accuracy.

   This difference is due to the fact that the settling error in an SC stage depends exponentially on the $f_u/f_s$ ratio, and hence the error grows rapidly when this ratio drops below about 4. By contrast, the error introduced by op-amp bandwidth in a CT stage increases only gradually when $f_u/f_s$ is reduced.

The above two advantages are driving the development of CT ADCs with bandwidths in the multi-MHz range. These ADCs are geared towards wired and wireless communications systems, and will enable digital signal processing to replace analog signal processing to a greater extent than is feasible today. The purpose of this section is to demystify the operation of CT ΔΣ modulators, while providing deeper justification for the above advantages along the way.

### 6.6.1 A Continuous-Time Implementation of MOD2

This section examines a CT implementation of MOD2. The mathematical equivalence between a CT modulator and a discrete-time (DT) modulator with pre-filters is established and used to demonstrate the inherent anti-aliasing property of a CT modulator. In Section 9.4, we provide a more in-depth discussion of a practical implementation of CTMOD2.

Consider the second-order modulator depicted in Fig. 6.22. There are several ways to analyze this system, but perhaps the most direct method starts with simply writing the time-domain equations which describe the system's operation. As usual, we assume that the sampling rate is 1 Hz. We will also assume that the quantizer operates instantaneously, so that $v(n) = Q(x_2(n))$, and we will also assume that the DAC acts as a delay-free sample-and-hold stage, so that $v_c(t) = v(n)$ for $t \in [n, n+1]$. (Note that we denote a CT signal by attaching a subscript "c" to it.)

Next, we shall derive a sample-data model of the CT loop filter, which provides the correct sampled input for the clocked quantizer Q.

Using the above assumptions, the output of the first integrator between sample times $n$ and $n+1$ may be expressed in terms of its value at time $n$ and the value of the feedback signal at time $n$:

$$x_{1c}(t) = x_{1c}(n) + \int_n^t [u_c(\tau) - v(n)]d\tau, \text{ for } n \leq t \leq n+1. \quad (6.22)$$

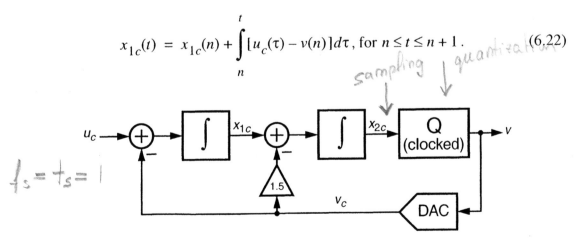

Figure 6.22: MOD2-CT: A continuous-time implementation of MOD2.

Thus the sampled values of this waveform obey the difference equation

$$x_1(n+1) = x_1(n) - v(n) + u_1(n) \qquad (6.23)$$

where

$$u_1(n) = \int_n^{n+1} u_c(\tau)d\tau = (g_{p1c} * u_c)(n). \qquad (6.24)$$

Hence, we can define a hypothetical pre-filter with a (non-causal) unit-rectangle impulse response $g_{p1c}$ †

$$g_{p1c}(t) = \begin{cases} 1, & -1 \le t < 0 \\ 0, & \text{otherwise} \end{cases}. \qquad (6.25)$$

The pre-filter's input is $u_c(t)$ and the pre-filter's output is $u_1(n)$ (Fig. 6.23).

Similarly, the output of the second integrator is given by

$$x_{2c}(n+1) = x_{2c}(n) + \int_n^{n+1} [x_{1c}(\tau) - 1.5v(n)]d\tau$$

$$= x_{2c}(n) + \int_n^{n+1} \left( x_{1c}(n) + \int_n^t u_c(\tau) - v(n)d\tau \right) dt - 1.5v(n)$$

$$= x_{2c}(n) + x_{1c}(n) + \int_n^{n+1} \left( \int_n^t u_c(\tau)d\tau \right) dt - 2v(n)$$

*trapezoidal* $\frac{u_c(t) + u_c(n)}{2} \cdot (t-$

*approximation*

*rectangular* $\approx u_c(t) \cdot (t-n)$ !

---

†. The pre-filter is non-causal by virtue of our definition of $u_1$. It is possible to alter the definition of $u_1$ in such a way that the pre-filter becomes causal, but the analysis and diagrams are tidier if we (temporarily) introduce non-causal filtering.

$$= x_{2c}(n) + x_{1c}(n) + \int_{-1}^{0} (\tau + 1)u_c(n-\tau)d\tau - 2v(n). \tag{6.26}$$

Thus, the sampled values of this waveform obey

$$x_2(n+1) = x_2(n) + x_1(n) + u_2(n) - 2v(n) \tag{6.27}$$

where

$$u_2(n) = (g_{p2c} * u_c)(n), \tag{6.28}$$

and $g_{p2c}$ is a (non-causal) unit-ramp pre-filter impulse response

$$g_{p2c}(t) = \begin{cases} (t+1), & -1 \leq t < 0 \\ 0, & \text{otherwise} \end{cases} \tag{6.29}$$

Fig. 6.23 depicts a system which embodies the above difference equations. Note that by virtue of (6.27), the feedback coefficient for the second integrator is 2 in the DT system, whereas in the CT system this coefficient was 1.5. Clearly, constructing the DT equivalent of a CT system is not simply a matter of replacing the CT integrators with DT integrators!

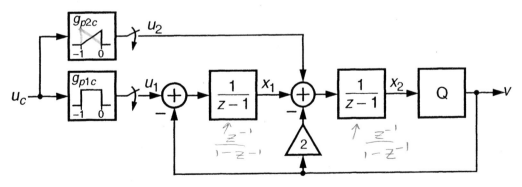

Figure 6.23: Discrete-time equivalent of MOD2-CT.

Fig. 6.23 also shows that in order for the DT system to follow the behavior of the CT system exactly, the CT input $u_c(t)$ must be processed by the two pre-filters, $g_{p1c}$ and $g_{p2c}$, the outputs of which are sampled and then injected into the loop filter. The rest of the system operates in discrete-time. A brief analysis shows that

$$V(z) = z^{-2}U_1(z) + z^{-1}(1-z^{-1})U_2(z) + (1-z^{-1})^2 E(z), \quad (6.30)$$

where $E(z)$ is the z-transform of the quantization error. Since the noise transfer function is identical to that of MOD2, all previously obtained results regarding the quantization noise and stability properties of MOD2 carry over to MOD2-CT. However, the signal transfer function of MOD2-CT must be obtained separately. According to (6.30), the impulse response of the STF is a twice-delayed version of $g_{p1c}$ plus a delayed version of $g_{p2c}$ minus a twice-delayed version of $g_{p2c}$. These operations (depicted in Fig. 6.24) result in a causal triangular impulse response

$$g_{pc}(t) = \begin{cases} t, & 0 \le t < 1 \\ t-2, & 1 \le t < 2 \\ 0, & \text{otherwise} \end{cases} \quad (6.31)$$

If this pre-filter is cascaded with MOD2, as shown in Fig. 6.25, the resulting system will have the same input-output behavior as MOD2-CT.

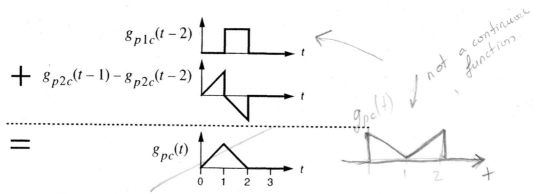

Figure 6.24: Impulse response of the STF of MOD2-CT.

The Laplace transform of $g_{pc}(t)$ is

$$G_{pc}(s) = \left(\frac{1-e^{-s}}{s}\right)^2. \tag{6.32}$$

Evaluating $G_{pc}(s)$ for physical frequencies $s = j2\pi f$ yields the frequency response of the STF

$$|STF(f)| = \left(\frac{\sin \pi f}{\pi f}\right)^2, \tag{6.33}$$

which is plotted in Fig. 6.26. This function has second-order zeros at all non-zero multiples of the sampling frequency, and thereby provides alias protection commensurate with the order of the modulator.

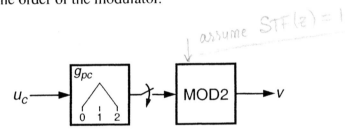

Figure 6.25: MOD2-CT is equivalent to MOD2 preceded by a pre-filter.

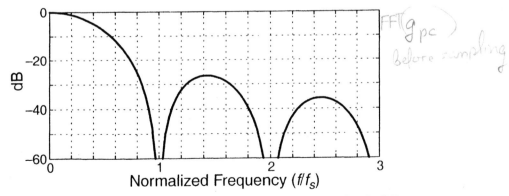

Figure 6.26: Frequency response of the STF of MOD2-CT.

## 6 Implementation Considerations For ΔΣ ADCs

In general, the alias protection of a high-order CT modulator is at least as good as its quantization noise suppression. The reason is simply that both sampling and quantization occur at the same point in the loop. We regularly model quantization as the addition of noise, so if we similarly regard sampling as the addition of aliases, then the quantization noise suppression and the alias suppression should be the same. This argument will be made more precise shortly.

### 6.6.2 Inherent Anti-Aliasing in CT ΔΣ ADCs

The above example demonstrated that MOD2-CT is mathematically equivalent to MOD2 with a pre-filter. In a more general case, the samples of the signal entering the quantizer are given by

$$y(n) = [l_{0c} * u_c](n) + [l_1 * v](n) \qquad (6.34)$$

where $l_{0c}$ is the continuous-time impulse response from the input $u_c(t)$ to the quantizer, while $l_1$ is the discrete-time impulse response from the quantizer's output, through the feedback DAC(s) and the loop filter, back to the quantizer's input [13]. As usual, modeling the discrete-time quantization process by the addition of an error signal yields the noise transfer function $NTF(z) = 1/(1 - L_1(z))$. However, instead of the usual $STF(z) = L_0(z)NTF(z)$, formal manipulation of the equations yields

$$STF = L_{0c}(s)NTF(z). \qquad (6.35)$$

Note that the STF is a mixture of continuous-time ($s$-domain) and discrete-time ($z$-domain) transfer functions, and that we have side-stepped the issue of which argument is appropriate for the STF. For physical frequencies, $s = j2\pi f$ and $z = e^{j2\pi f}$, we may write the less ambiguous relation

$$STF(f) = L_{0c}(j2\pi f)NTF(e^{j2\pi f}). \qquad (6.36)$$

At a pole frequency of $L_{0c}$ (which is assumed to lie in the modulator's passband), NTF is zero and a pole-zero cancellation results in a finite (often unity) signal transmission. For example, in MOD2-CT, $L_{0c}(s) = 1/s^2$ and $NTF(z) = (1-z^{-1})^2$. According to (6.35),

$$STF = \frac{(1-z^{-1})^2}{s^2} = \left(\frac{1-z^{-1}}{s}\right)^2, \qquad (6.37)$$

and we see that the double pole at dc ($s = 0$) is cancelled by a double zero at dc ($z = 1$). Substituting $s = j2\pi f$ and $z = e^{j2\pi f}$ yields

$$|STF(f)| = \left|\frac{1-e^{-j2\pi f}}{2\pi f}\right|^2 = \left(\frac{\sin \pi f}{\pi f}\right)^2 \qquad (6.38)$$

as before in (6.33), from which it is clear that $|STF(0)| = 1$.

Next, consider a frequency $f$ that aliases to an NTF zero. Since this frequency is outside the passband, $L_{0c}(j2\pi f)$ is finite by our first assumption. Since $f$ aliases to an NTF zero, $NTF(e^{j2\pi f}) = 0$ and thus $STF(f) = 0$. More generally, (6.36) shows how the alias suppression is intimately linked with the NTF. Any frequency which aliases to the passband is attenuated by the $NTF(e^{j2\pi f})$ factor, and the $L_{0c}(j2\pi f)$ factor cannot undo the large amount of attenuation provided by the $NTF(e^{j2\pi f})$ factor. In fact, for frequencies that are outside the passband, $L_{0c}$ is usually less than one and thus the alias attenuation is usually greater than the quantization noise attenuation.

### 6.6.3 Design Issues for Continuous-Time Modulators

The key theoretical result regarding a CT modulator is that a CT modulator is functionally equivalent to a DT modulator preceded by a pre-filter. This fact suggests that one could first select a DT prototype and then transform it into a CT system which has the same NTF. In order for this procedure to work, the designer needs to ensure that the path from the quantizer output back to its input has the same impulse response in the two structures. In the DT case, this impulse response is $l_1(n)$. In the CT case, the impulse response is obtained by tracing the signal path from the (discrete-time) quantizer output, through the DT-CT transformation of the DAC(s) and then through the loop filter, whose output is sampled to yield a discrete-time response. In more concise (and only slightly less precise) terms, *the* sampled pulse response of the CT loop filter must match the impulse response of the DT loop filter. If the designer is successful in achieving this correspondence, the CT system and the DT system will implement the same NTF.

A number of practical concerns act as impediments to the above goal, the most important of which derives from the assumption that the quantizer operates instantaneously and provides feedback immediately after sampling its input. If one were to try to approximate this ideal behavior by using a fast comparator, then metastability, or more generally, input-dependent comparator delay, would affect the feedback timing and thereby degrade the performance of the system. The following approximate calculation illustrates the severity of this problem.

Suppose that the modulator is a single-bit system clocked at 100 MHz, and that once every 100 cycles or so the comparator output (and thus the feedback) is delayed from its nominal timing by $\Delta t = 0.1$ ns, as illustrated in Fig. 6.27. The equivalent relative error signal therefore has an average power of

$$\frac{\left(2\frac{\Delta t}{T}\right)^2}{100} = \frac{\left(2\frac{0.1 \text{ ns}}{10 \text{ ns}}\right)^2}{100} = 4\times 10^{-6}, \quad (6.39)$$

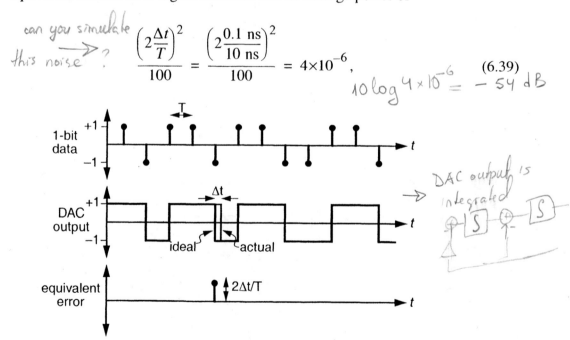

Figure 6.27: Timing variation in a single-bit CT modulator.

*cannot be white noise (digital errors in between = 0)*  $10 \log \frac{1^2}{2} = -3 dB$

which is only 51 dB below the power of a full-scale sine wave. If this error is white, then its in-band power would be reduced by a factor of $OSR$. However, since it is likely that this error would be correlated with the input signal, the error is likely to manifest itself as a distortion product, and thus the ADC would probably be limited to the equivalent of 9-bit linearity. $SNR = 6.02 \times 9 \text{ bit} + 1.76 = 56 \, dB$

*not resolution?*

A solution to the above problem is to allocate a fixed amount of time for the quantizer to perform its job. Since delaying the feedback affects the sampled pulse response of the loop, the modulator's coefficients need to be adjusted to account for this change in feedback timing. Failure to account for the excess loop delay will yield an NTF which differs from the desired one, and the difference may be great enough to yield an unstable modulator. As will be seen in Section 9.4, the desire to maintain exact correspondence between a CT system with delayed feedback and a DT prototype may require the addition of extra feedback DACs as well as an adjustment of the loop filter coefficients.

Another important consideration in CT modulator design is clock jitter. As illustrated in Fig. 6.28a, sampling a sine wave of amplitude $A$ and frequency $f$ with an rms time jitter of $\sigma_t$ results in an rms error of $\sqrt{2}\pi A f \sigma_t$, which is proportional to

a) sine wave

$v(t) = A\sin\omega t$ :

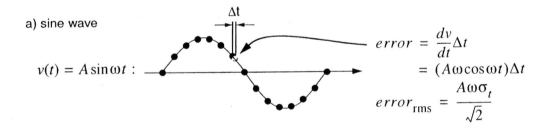

$$error = \frac{dv}{dt}\Delta t = (A\omega\cos\omega t)\Delta t$$
$$error_{rms} = \frac{A\omega\sigma_t}{\sqrt{2}}$$

b) 1-bit $\Delta\Sigma$ waveform

$v(n)$:

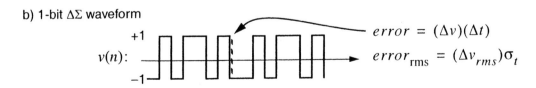

$$error = (\Delta v)(\Delta t)$$
$$error_{rms} = (\Delta v_{rms})\sigma_t$$

**Figure 6.28: Jitter-induced errors.**

both the signal's amplitude and its frequency. As illustrated in Fig. 6.28b, the rms error due to jitter in the feedback of a 1-bit CT modulator is $(\Delta v)_{rms} \sigma_t$, where $(\Delta v)_{rms}$ is the rms change in the 1-bit feedback. If we assume that $(\Delta v)_{rms} \approx 1$, the jitter-induced error in a 1-bit CT modulator is approximately equivalent to the error associated with sampling a full-scale sine wave whose frequency is $0.2 f_s$! Clearly, a 1-bit CT modulator is at a jitter disadvantage relative to a 1-bit DT modulator.

There are two ways to overcome this disadvantage. The most obvious is to increase the number of levels in the feedback DAC. With a sufficient number of levels[†] the feedback waveform approximates the input signal and thus the rms sample-to-sample change in the feedback waveform is approximately equal to the rms sample-to-sample change of the input signal. Under these circumstances, a CT modulator has a jitter sensitivity that is comparable to that of its DT counterpart.

A more elegant solution to the jitter problem is to change the feedback from a timing-dependent signal to a timing-insensitive signal. To see how this is possible, consider the simplified first-order CT modulator illustrated in Fig. 6.29a. Over a clock period $T$, the feedback signal induces a change $IT/C$ in the output voltage of the integrator. Jitter effectively causes variation in the value of $T$, and hence an error in the integrator's state. If a switched-capacitor branch replaces the current-source feedback as illustrated in Fig. 6.29b, then the change in the integrator's state over one clock period due to the comparator feedback is always $\pm V_{ref}$, regardless of the amount of jitter. This insensitivity to clock jitter is a simple consequence of the fact that the amount of charge fed into the integrator is now independent of the duration of a clock period.

## 6.7 Conclusions

In Chapter 6, several implementation issues important for $\Delta\Sigma$ ADCs were discussed. The topics discussed included the relative advantages of using multi-bit, rather than single-bit, internal quantization in $\Delta\Sigma$ ADCs. This was followed by a

---

[†] It is assumed that the aggressiveness of the NTF is not increased as the number of levels is increased.

Conclusions

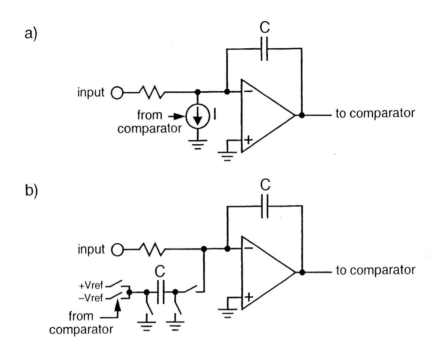

**Figure 6.29: a) A first-order CT modulator. b) A modulator with reduced sensitivity to clock jitter.**

description of several schemes for reducing the effects of the nonlinear distortion which the multi-bit internal DACs inherently introduce. The discussed strategies included the use of dual quantization, i.e., the inclusion of both single-bit and multi-bit quantizers in the converter. Also, various mismatch error shaping algorithms, which suppress the noise due to nonlinear distortion within the signal band and shift it out of band, were described. An alternative strategy, in which the DAC nonlinearity errors are acquired in digital form, either at power-up or during operation, and then corrected in the digital ADC output, was also discussed, and various implementations described.

Finally, an alternative realization of the loop filter in a $\Delta\Sigma$ ADC was discussed. This uses continuous-time circuitry, and thus has inherent anti-aliasing properties. This allows simplification of the overall system, and avoids noise folding which plagues switched-capacitor circuits. Also, since the internal amplifiers do not have to settle in a fast transient occurring in every clock period as in a switched-capacitor circuit, faster operation is possible, albeit with reduced linearity. The theory and key properties of ΔΣ ADCs with continuous-time loop filters were discussed in the context of the second-order continuous-time modulator MOD2-CT, and some design issues, potential problems and their solutions were considered.

## References

[1] B. P. Brandt and B. A. Wooley, "A 50-MHz multibit sigma-delta modulator for 12-b 2-MHz A/D conversion," *IEEE Journal of Solid-State Circuits*, vol. 26, no. 12, pp. 1746-1756, December 1991.

[2] A. Hairapetian and G. C. Temes, "A dual-quantization multi-bit sigma-delta A/D converter," *Proceedings of the 1994 IEEE International Symposium on Circuits and Systems*, vol. 5, pp. 437-440, May 1994.

[3] L. R. Carley, "A noise-shaping coder topology for 15+ bit converters," *IEEE Journal of Solid-State Circuits*, vol. 24, pp. 267-273, April 1989.

[4] S.R. Norsworthy, R. Schreier and G.C. Temes, *Delta-Sigma Data Converters*, Sec.8.3.3, IEEE Press, 1997.

[5] M. J. Story, "Digital to analogue converter adapted to select input sources based on a preselected algorithm once per cycle of a sampling signal," U.S. patent number 5138317, Aug. 11 1992 (filed Feb. 10 1989).

[6] R. T. Baird and T. S. Fiez, "Improved ΔΣ DAC linearity using data weighted averaging," *Proceedings of the 1995 IEEE International Symposium on Circuits and Systems*, vol. 1, pp. 13-16, May 1995.

[7] I. Fujimori, L. Longo, A. Hairapetian, K. Seiyama, S. Kosic, C. Jun and S.L. Chan, "A 90-dB SNR 2.5 MHz output-rate ADC using cascaded multibit delta-sigma modulation at 8x oversampling ratio," *IEEE Journal of Solid-State Circuits*, vol. 35, no. 12, pp. 1820-1828, December 2000.

[8] F. Chen and B. H. Leung, "A high resolution multibit sigma-delta modulator with individual level averaging," *IEEE Journal of Solid-State Circuits*, vol. 30, no. 4, pp. 453-460, April 1995.

[9] R. Schreier and B. Zhang, "Noise-shaped multibit D/A convertor employing unit elements," *Electronics Letters*, vol. 31, no. 20, pp. 1712-1713, Sept. 28 1995.

[10] I. I Galton, "Spectral shaping of circuit errors in digital-to-analog converters," *IEEE Transactions on Circuits and Systems II*, vol. 44, pp. 808-817, Oct 1997.

[11] M. Sarhang-Nejad and G. C. Temes, "A high-resolution multi-bit ΣΔ ADC with digital correction and relaxed amplifier requirements," *IEEE Journal of Solid-State Circuits*, vol. 28, pp. 648-660, June 1993.

[12] J. Silva, X. Wang, P. Kiss, U. Moon, G. C. Temes, "Digital techniques for improved ΔΣ data conversion," *Proceedings of the IEEE 2002 Custom Integrated Circuits Conference*, pp. 183-190, May 2002.

[13] R. Schreier and B. Zhang, "Delta-sigma modulators employing continuous-time circuitry," *IEEE Transactions on Circuits and Systems I*, vol. 43, no. 4, pp. 324-332, April 1996.

# CHAPTER 7  Delta-Sigma DACs

Up to this point, the concepts of oversampling and noise shaping were (with the exception of the brief discussions in Section 2.3) applied only to ADCs, and not to digital-to-analog converters (DACs). In fact, delta-sigma DACs are commercially as important as their ADC counterparts, if not more so, and their implementation is often just as difficult as the implementation of a $\Delta\Sigma$ ADC. This chapter will be devoted to the specific issues involved in the design of delta-sigma DACs.

The motivation to use noise shaping in D-to-A conversion is the same as in A-to-D conversion. For a DAC with a 3-V full-scale and an 18-bit resolution, the LSB voltage is only about 12 $\mu$V. Hence, the permissible deviation of the DAC output levels from their ideal values is of the order of 6 $\mu$V, which cannot be achieved in a conventional DAC without expensive trimming and/or extremely long conversion time. Therefore, the trade-off earlier discussed in connection with delta-sigma ADCs, wherein oversampling and additional digital hardware are applied to allow the use of robust and simple analog circuitry, is attractive for high-accuracy DACs as well. The actual structures implementing this trade-off will be discussed next.

## 7.1 System Architectures for ΔΣ DACs

Fig. 7.1 illustrates the basic system diagram of the ΔΣ DAC. As indicated, the front end (containing a digital interpolation filter IF and a noise-shaping loop NL) contains digital circuitry, while the output stages (the internal DAC and the reconstruction filter) are analog.

The spectra of the signals processed by the system are shown in Fig. 7.2. The input signal $u_0(n)$ is a multi-bit data stream with a word-length $N_0$ (typically, 15-24 bits) sampled near the Nyquist rate $f_N$. Its spectrum is illustrated in Fig. 7.2a.

The roles of the interpolation filter IF are (1) to raise the sampling frequency to $OSR \cdot f_N$ and thereby allow subsequent noise shaping, and (2) to suppress the spectral replicas centered at $f_N, 2f_N, \ldots, (OSR-1)f_N$. The purpose of this sideband suppression is to reduce digitally the out-of-band power of the input of NL without affecting the baseband signal spectrum. This improves the dynamic range of the noise-shaping loop, since larger signals can thus be accommodated. It also eases the task of the analog output filter, since it needs to suppress less out-of-band noise, and also its linearity requirements can be somewhat more relaxed since the amount of intermodulated out-of-band noise folding down into the signal band is reduced. The suppression need not be very accurate, since the truncation error generated in the noise-shaping loop will introduce unwanted noise in the same frequency range anyway. The ideal spectrum of the IF output signal is shown in

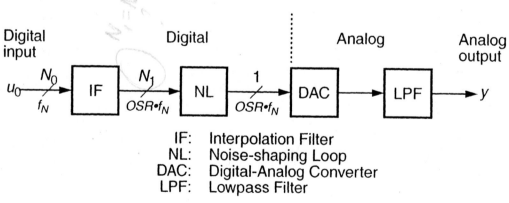

Figure 7.1: Block diagram of a ΔΣ DAC.

Fig. 7.2b. The word-length of this signal can remain about the same as that of the input data $u_0(n)$.

The noise-shaping loop NL reduces the word-length of its input signal to a few (1-6) bits. If a single-bit NL output is used, then (as discussed in Chapters 4 and 6 for the internal DACs in $\Delta\Sigma$ ADCs) the linearity requirements for the DAC following the NL can be relaxed. If the output data are multi-bit, then the techniques discussed in Sections 6.2 to 6.5 may be utilized to filter out or cancel the unavoidable DAC nonlinearity errors, and thus to achieve linear conversion. (The pros and cons of using multi-bit DAC loops will be discussed in Section 7.3). In any case, the NL output must contain a faithful reproduction of the input signal $u_0(n)$ in the baseband, but it will also include the filtered truncation noise caused by the reduction of the word-length in the loop. The spectrum of the NL output signal is schematically illustrated in Fig. 7.2c.

The next block in the system is the embedded DAC. As discussed above, it may have a single-bit input, in which case its output will be a two-level analog signal.

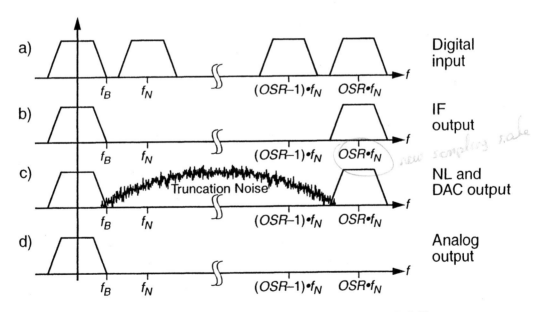

Figure 7.2: Signal and noise spectra in a $\Delta\Sigma$ DAC.

The structure of such a 1-bit DAC will be very simple, and its linearity will be theoretically perfect (although some practical precautions need to be observed to achieve good linearity). However, the high slew rate of the single-bit DAC output signal, and the large amount of out-of-band noise power it contains, make the design of the subsequent analog smoothing filter LPF a difficult task.

By contrast, for a multi-bit DAC, additional circuitry is required for the filtering or cancellation of the DAC nonlinearity error, which results in a more complex DAC. However, the reduced slew rate and out-of-band noise power of the DAC output signal allows reduced performance requirements, and hence simpler implementation, for the smoothing filter. Usually, the overall trade-off in complexity, chip area and power dissipation favors the multi-bit structure.

Ideally, the DAC will reproduce the digital signal at its input in an analog form without any distortion. Hence, the output spectrum of the DAC will be (except for a constant factor corresponding to the reference voltage or current of the DAC (and for a $(\sin x)/x$ frequency-dependent factor corresponding to the frequency response of a zero-order hold), the same as that shown in Fig. 7.2c for the output signal of the NL.

Finally, the task of the analog smoothing or reconstruction filter LPF is to suppress most of the out-of-band noise power contained in its input signal. Hence, the ideal spectrum of its output signal should be as shown in Fig. 7.2d. As already mentioned, it is relatively easy to achieve good noise suppression without introducing additional distortion for a multi-bit DAC output signal, but the task is usually quite difficult for a single-bit signal. The design of the analog post-filter will be discussed in Section 7.5.

## 7.2 Loop configurations for $\Delta\Sigma$ DACs

As was the case the $\Delta\Sigma$ ADCs, there is also a wide variety of loop architectures available for the designer of $\Delta\Sigma$ DACs. The function of the loop is similar to that of the noise-shaping loop of the $\Delta\Sigma$ ADC, namely to reduce the resolution of the input signal[†] to a few bits without significantly affecting its inband spectrum in the process. Since the reduced word-length means that quantization or truncation error is introduced, the loop must suppress the power spectrum of this added noise in the signal band. The only significant differences between ADC and DAC loops are that

1. In the DAC loop all signals are digital, and hence no internal data conversion is required.
2. For the same reason, the signal processing in the loop can be made highly accurate, and we need not take any analog imperfections into account when predicting the actual behavior of the loop. As we shall see, this permits the use of some efficient configurations which are impractical for ADC loops.

Some typical loop configurations will be discussed next.

### 7.2.1 Single-Stage Delta-Sigma Loops

All loop architectures discussed for ADCs in Section 4.4, and shown in Figs. 4.17 to 4.21, remain applicable for delta-sigma DACs. Thus, the structure containing cascaded integrators with distributed feedback and input coupling (CIFB), illustrated in Fig. 4.17; the circuit using cascaded resonators with distributed feedback and input coupling (CRFB), shown in Fig. 4.19; as well as the structures containing cascaded integrators or resonators with feedforward coupling (CIFF), shown in Figs. 4.20 and 4.21, respectively, can also be used in delta-sigma DAC loops. Of course, the component blocks are now accumulators rather than integrators, and they are implemented by digital adders and multipliers, rather than op amps, capacitors and switches as in the ADC loops.

The designer is still faced with some of the same problems (e.g., stability issues) as were encountered for analog loops, and also with some new ones. In finding the proper configuration and order for the loop, and calculating the coefficients needed, the noise-shaping and signal transfer specifications must be satisfied, and stability needs to be ascertained under all anticipated conditions. Also, the conditions for optimum dynamic range must be met, and the over- or underflow of any block avoided. Finally, the word-lengths of all coefficients and operations should

---

†. For an analog signal, the resolution can be regarded as infinite.

be carefully determined so that, on the one hand, the required accuracy in signal transfer and noise suppression is maintained, and, on the other hand, the complexity of the circuit is minimized subject to these accuracy conditions.

Qualitatively, the sensitivity considerations discussed for ADC loops remain valid for DAC loops, except that the errors generated here are due to coefficient truncations and to the round-off errors of the digital operations (additions and multiplications), rather than element-matching errors and finite op-amp-gain effects. Thus, the coefficient and round-off errors must be kept small in all signal paths connecting to the input node, but they may increase progressively as the signal propagates towards the output of the loop. Hence, the word-length needed may vary considerably with the location of the block in the loop.

Hardware can also be saved by choosing simple coefficients which contain only a few terms, each term an integer power of 2. This may alter the signal and noise transfer functions slightly, but the effect on the NTF and STF are usually small and, in the case of the STF, can often be corrected the blocks preceding or following the loop. The signal transfer function of the low-distortion architecture discussed in Chapter 4 tends to be less susceptible to coefficient truncation than that of competing architectures.

The detailed design process will be illustrated in the example discussed in Section 9.6, later in the book.

### 7.2.2 The Error Feedback Structure

A configuration which is not practical for ADC loops, but is highly efficient for DACs, is illustrated in Fig. 7.3 for a 1-bit loop. Here, rather than feeding back the MSB retained in the output signal as was done in the delta-sigma loop discussed in Section 7.2.1, the discarded LSBs (representing the truncation error $e(n)$) are filtered and fed back to the input. The loop filter $H_e$ used to filter $e(n)$ is now located in the feedback path.

In an ADC loop, this structure would be overly sensitive to the imperfections of the analog loop filter and of the analog subtraction needed to generate $e(n)$, since the errors generated in either enter the input terminal directly. Hence, this architecture is never used in an ADC. In a digital realization, however, if sufficient accuracy is used in the digital implementation of the $H_e$ filter, the circuit will perform well. Linear analysis shows that the output is given by

$$V(z) = U(z) + [1 - H_e(z)]E(z) \qquad (7.1)$$

Hence, the STF is 1, and the NTF equals $1 - H_e(z)$.

For low-order loops, the error feedback loop can usually be very simply realized. For a first-order loop, $NTF = 1 - z^{-1}$, and hence $H_e(z) = z^{-1}$, i.e., it is simply a delay. For a second-order loop with a double zero of the NTF at dc,

$$H_e(z) = 1 - (1 - z^{-1})^2 = z^{-1}(2 - z^{-1}) \qquad (7.2)$$

Thus, the loop can be realized from two delays, a shift in the binary point to implement the factor 2, and two adders (Fig. 7.4).

Higher-order error-feedback loops can, of course, also readily be designed, subject to stability considerations. As in the delta-sigma type loop, instability causes the input signal $y(n)$ of the quantizer (here, the truncator) to grow beyond the operating range of the digital logic. Depending on the arithmetic used, this may just cause the saturation of $y(n)$ at its largest possible value, or it may cause a wrap-around, where the output $v(n)$ suddenly decreases with increasing $y(n)$ at overflow. While saturation is usually acceptable, wrap-around causes large errors, and hence must be prevented, e.g. by including a digital limiter in the loop (Fig. 7.5) at

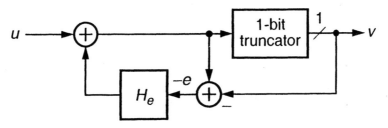

Figure 7.3: Error feedback structure.

the input of the truncator [1]. The limiter should saturate before overflow can occur.

### 7.2.3 Cascade (MASH) Structures

To achieve high-order noise shaping without the stability problems inherent in the design of higher-order loops, cascade structures may be used for $\Delta\Sigma$ DACs as well as for $\Delta\Sigma$ ADCs [2]. (Note that in fact cascaded DAC stages were proposed before ADC ones.) Fig. 7.6 illustrates the architecture of a two-stage cascade DAC. In a typical structure, both stages may contain second-order loop filters, resulting in a fourth-order noise shaping overall, while preserving the robust stability properties of second-order loops.

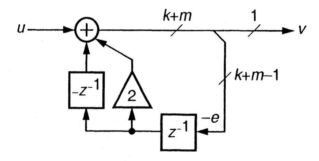

**Figure 7.4:** Second-order error-feedback noise-shaping loop.

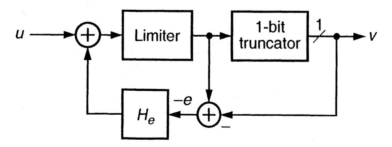

**Figure 7.5:** Error feedback with limiter.

A design issue, which did not occur for MASH ADCs but appears for cascade DACs, concerns the optimum location of the internal DACs in the structure. Assume at first that all signal processing in the structure of Fig. 7.6 is performed digitally. As explained in the discussions of Section 4.5, the post-filter $H_1$ usually replicates the signal transfer function $STF_2$ of the second stage. Often, $STF_2$ is simply a single or multiple delay, and hence $H_1$ can easily be implemented digitally without increasing the word-length $n_1$ of the first stage output $v_1$. By contrast, $H_2$ usually reproduces the noise transfer function $NTF_1$ of the first stage, and hence, if implemented digitally, it increases the word-length $n_2$ of $v_2$. Adding $H_1 \cdot V_1$ and $H_2 \cdot V_2$ digitally will increase the output word-length even further. Hence, such a structure will produce a multi-bit output $v(n)$, which then needs to be accurately converted in a multi-bit, and hence complex, internal DAC.

An alternative is to use a separate DAC in each stage, and combine their outputs using analog circuitry, as illustrated schematically in Fig. 7.7. This allows the use of less complicated DACs. A mismatch between the gains of the two paths caused by analog errors will introduce leakage of the first-stage truncation error, but will not affect the linearity of the signal conversion, which is limited only by the linearity of the first-stage DAC.

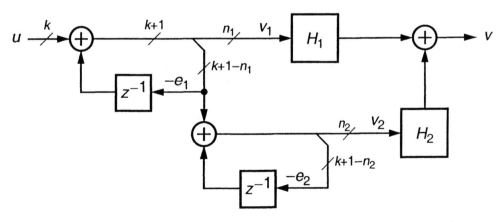

**Figure 7.6:** Cascade structure for a second-order noise-shaping loop.

It is also possible to place DAC2 ahead of the $H_2$ filter, which then must be realized by analog circuitry. This has two advantages: first, the resolution of DAC2 can be reduced, since it only needs to convert $V_2$, not $H_2 \cdot V_2$ which has longer words. In fact, for $n_1 = n_2 = 1$, both DACs can be single-bit ones. For multi-bit DACs, $H_2$ is now going to shape the noise introduced by the inherent nonlinearity of DAC2, along with the truncation noise of the second stage. A disadvantage of this modified scheme is that the analog implementation of $H_2$ cannot reproduce the digital $NTF_1$ as accurately as a digital filter could. (Note, however, that any zeros of $H_2$ at dc can be very accurately realized in an analog circuit, by having series capacitors in the signal path.)

Another option is to split the $H_2$ block into a digital stage preceding DAC2, and an analog one following it. This way, the larger truncation noise receives full shaping by $NTF_2$ and $H_2$, and the much smaller noise caused by DAC2 errors will be shaped only by the analog part of $H_2$. This scheme will realize the required replica of $NTF_1$ more accurately than a fully analog $H_2$ can.

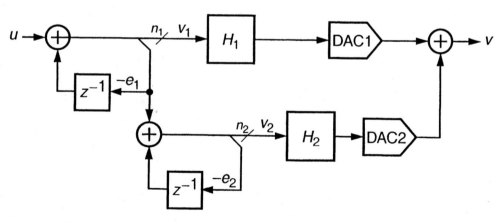

Figure 7.7: A cascade DAC using analog recombination.

## 7.3 ΔΣ DACs Using Multi-Bit Internal DACs

The parameters of the digital noise-shaping loops used in ΔΣ DACs are much more accurately controlled than those of the analog loops required in ΔΣ ADCs, and some of the basic arguments for using multi-bit quantization in ADCs (such as the ones based on op-amp slew rate, power dissipation, nonlinearity, clock jitter, etc.) are not valid for DAC loops. Nevertheless, the stability considerations presented in Section 6.1 remain valid, and an additional powerful reason for multi-bit operation emerges: the relaxed requirements on the analog smoothing filter LPF following the internal DAC. For a single-bit DAC, the input signal of this filter is a two-level fast-slewing analog signal (usually voltage), with most of its power contained in the large high-frequency quantization noise. This fast signal needs to be filtered by LPF so that almost all its high-frequency noise is removed. This must be accomplished without distorting the signal or even the out-of-band noise. (Distorting the noise will cause the folding down of the large noise spectrum from the region around $f_s/2$ into the signal band.)

In addition, due to the steep slopes of the two-level analog signal, any clock jitter is translated into substantial amplitude noise at the output of the filter.

In conclusion, the analog problems due to single-bit truncation which appeared in the noise-shaping loop in ΔΣ ADCs do not disappear in ΔΣ DACs; they are just shifted to the analog post-filter!

In earlier single-bit implementations [3], to overcome these difficulties, the smoothing filter was realized as a cascade combination of a high-order switched-capacitor (SC) filter, an SC buffer stage and a continuous-time post filter. It required a considerable chip area and dc power. The motivation for paying such a high price for the one-bit system was to avoid the inherent nonlinearities of a multi-bit internal DAC.

In recent years, various techniques (dual quantization, mismatch-error shaping and digital correction) became available for reducing the DAC nonlinearity effects, and hence multi-bit DAC structures became favored over single-bit ones. These DAC linearization methods are quite similar to their counterparts used in ADCs, which were discussed earlier in Sections 6.2 to 6.5. In the next Section, these schemes will be examined.

### 7.3.1 Dual-Truncation DAC Structures

The general principle of dual-truncation DACs is similar to that of dual-truncation ADCs: use single-bit truncation in the D/A conversion of the signal, and use multi-bit truncation where only the truncation errors are converted. A simple implementation, which is similar to the Leslie-Singh structure of Section 4.5.1, is shown in Fig. 7.8. [4]. As shown, the signal $u(n)$ is reduced to a single-bit data stream in a noise-shaping loop. This can be converted linearly in a 1-bit DAC. The large truncation error $-e_1$ is truncated to $M$ bits ($M > 1$), and converted in an $M$-bit internal DAC. It is then filtered, and added to the output of the 1-bit DAC to cancel $e_1(n)$. The spectrum of the nonlinearity error $d_M$ of the $M$-bit DAC is shaped by the analog filter $H_2(z)$, which duplicates the NTF of the 1-bit loop, and suppresses its inband power.

A more sophisticated and effective structure, shown in Fig. 7.9, uses a noise-shaping loop in both stages, with 1-bit truncation in the first stage and $M$-bit truncation

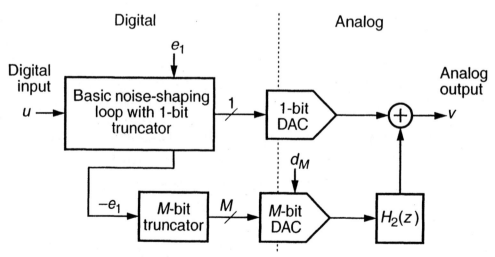

Figure 7.8: A dual-truncation DAC system.

## ΔΣ DACs Using Multi-Bit Internal DACs

in the second one. Fig. 7.10 shows the implementation of a third-order DAC [4] based on this structure. In the diagram, the switched-capacitor branch containing $C_1$ realizes the 1-bit DAC, while $C_2$, $C_3$, $C_4$ and their switches perform as the analog filter $H_2(z)$. Both loops use error feedback; the first loop has a second-order loop filter with a pole at $z = 0.5$ for improved stability, while the second loop uses a simple first-order filter.

It is also possible to realize a single-stage dual-truncation DAC (Fig. 7.11). This is similar to the Hairapetian ADC structure [5] discussed in Section 6.2.2. The single-bit output is fed to a 1-bit DAC, and is also fed back to all but the last stage in a cascade of integrators. An $M$-bit output is also generated; it is converted in a multi-bit DAC, and also entered into the last integrator. The input signal in this structure is converted by the 1-bit DAC in a potentially linear fashion, while the $M$-bit circuitry is used to cancel the large 1-bit truncation error in the output $v(n)$, and replace it with a much smaller $M$-bit error. An error cancellation logic, consisting of the analog filters $H_4$ and $H_5$, carries out this operation. Mismatch between the

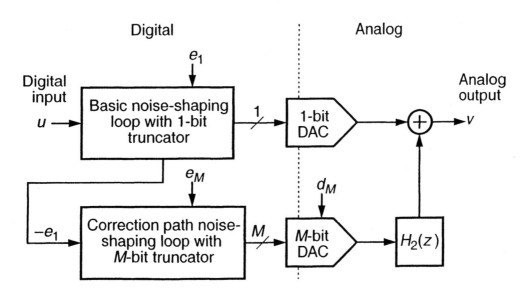

**Figure 7.9: A dual-truncation MASH structure.**

gains of these filters will degrade the cancellation of the 1-bit truncation error and hence the SNR, but will not introduce nonlinear signal distortion.

### 7.3.2 Multi-bit Delta-Sigma DACs with Mismatch Error Shaping

As mentioned earlier, the mismatch-error-shaping techniques discussed in Section 6.4 (data-weighted averaging, individual level averaging, vector-based mismatch shaping, tree-structure element selection) remain applicable to the internal DACs in multi-bit D/A converters. Again, however, there are new possibilities and trade-offs to consider for $\Delta\Sigma$ DACs.

In a multi-bit $\Delta\Sigma$ ADC, the number of bits $N$ used in the output is generally limited to about 4, since for $N = 5$ the internal ADC already needs 32 comparators with associated circuitry, and hence requires substantial supply power and chip area.

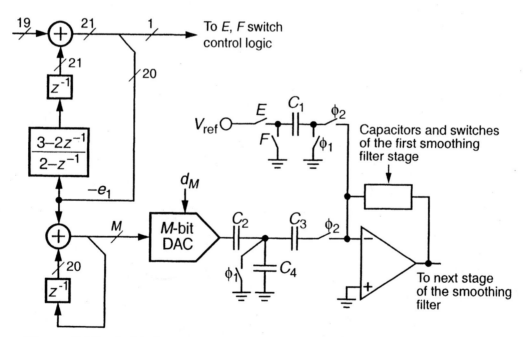

Figure 7.10: A third-order dual-truncation MASH noise-shaping stage.

For $N = 2\text{-}4$, the complexity of the DAC itself, and that of digital circuitry implementing the necessary mismatch shaping, are both relatively low, and no special schemes are needed to simplify them.

By contrast, in a multi-bit $\Delta\Sigma$ DAC, no internal ADC is required, and hence values of $N$ higher than 4 may be chosen. However, since the complexity of the DAC and its error-correction circuitry grow exponentially with $N$, they may then require too much chip area and bias power. We may use $2^N$ as a complexity index; generally, for $N > 4$, its value is impractically high. This problem, and its solution, will next be discussed in terms of a second-order 6-bit delta-sigma DAC.

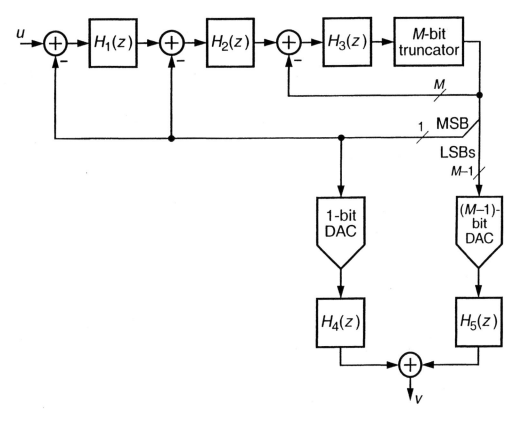

**Figure 7.11:** A single-stage dual-truncation D/A loop.

An obvious solution to having too many bits (here, 6 bits) in the DAC input signal is to use segmentation, i.e., to split the 6-bit input data stream into two 3-bit segments: an MSB signal and an LSB one. The two segments can then be separately encoded into thermometer-coded words, scrambled and converted into analog signals (Fig. 7.12). The weighted sum of the two analog outputs provides the overall output signal. The effective complexity index of this system is $2 \cdot 2^3 = 16$, lower (by a factor of 4) than that of a direct realization of the 6-bit DAC, which is $2^6 = 64$.

The problem with this approach is that both the MSB and LSB segments contain large distortion components, which ideally cancel if the two are recombined exactly, with the weight factor 8 needed to scale the MSB analog output. However, if this factor is inaccurate, then the unfiltered quantization noise and distortion components contained in the MSB and LSB outputs will not cancel perfectly, and this will significantly degrade the linearity and SNR performance. This degradation will occur even if the scramblers in both paths use mismatch shaping, since the noise is already contained in the scrambler inputs B and C, and not generated by the internal DACs.

A way to overcome this accuracy problem is illustrated in Fig. 7.13 [6]. An additional first-order $\Delta\Sigma$ loop is cascaded with the main modulator, and it compresses

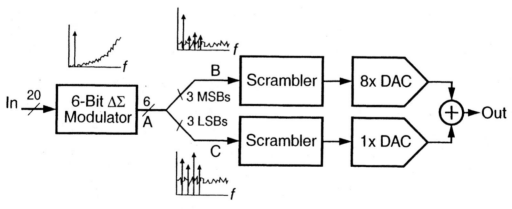

**Figure 7.12: Segmentation.**

the word-length of its 6-bit input $A$ into 4 bits. Denoting the NTF of MOD1 by $H_1$, and its quantization error by $E_1$, the two segmented signals are then the 4-bit output $B = A + H_1 E_1$ of the first-order loop, and $C = -H_1 E_1$, which is the negative of its shaped 3-bit quantization error. $C$ is generated by subtracting the input $A$ of MOD1 from its output $B$. Both signals $B$ and $C$ are next thermometer-coded, scrambled and D/A converted, and then added, using a scale factor 4 for $B$ to make up for the shift of the binary point when $C$ is thermometer coded. Ideally, the analog output is therefore $B + C = A$, as required. The complexity index is $2^4 + 2^3 = 24$, still much lower than the value $2^6 = 64$ associated with the unsegmented system.

In the system of Fig. 7.13, both $B$ and $C$ are noise-shaped signals, and hence if there is an error in the analog scale factor 4, so that $C$ is not completely cancelled, the resulting output error will be only some additional shaped noise. For sufficiently high oversampling ratio (say, 128), a 1% DAC element matching error will still allow a 110 dB SNR [6].

Another segmentation scheme for $\Delta\Sigma$ DACs is illustrated in Fig. 7.14. [7]. Here, the $L$ LSBs of the input data stream are compressed to shorter ($B$-bit, $B < L$) words by an error-feedback noise-shaping loop, and fed to a digital adder. The $M$ MSBs, by contrast, are directly entered into the adder. Since the addition is digital, it can

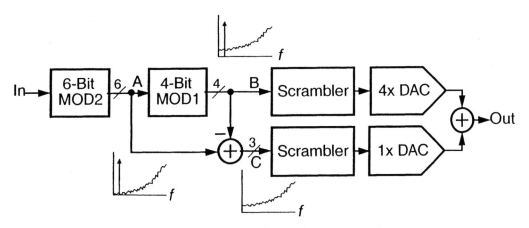

**Figure 7.13:** Noise-shaped segmentation.

be highly accurate. For a 6-bit input, the 4 LSBs may be compressed into 2 bits, and combined with the 2 MSBs to result in a 4-bit DAC circuit. For sufficiently large OSR, the accuracy can be satisfactory. (Note that this system is just a first-order modulator; it does not split the data into an MSB stream plus a noise-shaped LSB stream like the system of Fig. 7.13 does.)

*7.3.3 Digital Correction of Multi-Bit Delta-Sigma DACs*

As already mentioned in Section 6.5.1, a power-up calibration is readily available for multi-bit $\Delta\Sigma$ DACs. The block diagram is shown in Fig. 7.15. The RAM stores the digital equivalents of the actual analog outputs of the DAC for all possible input codes. The feedback loop forces the inband spectral components of the RAM output to follow the digital input $u(n)$, and since the inputs of the RAM and the DAC are the same, the output of the DAC follows that of the RAM. In conclusion, the inband part of the DAC output signal will be the analog version of the input signal $u(n)$.

As discussed in Section 6.5.1, the calibration (i.e., the storing of the appropriate numbers in the RAM) can be performed at power-up, using an auxiliary 1-bit delta-sigma ADC (Fig. 7.16). In the calibration process, a digital counter produces

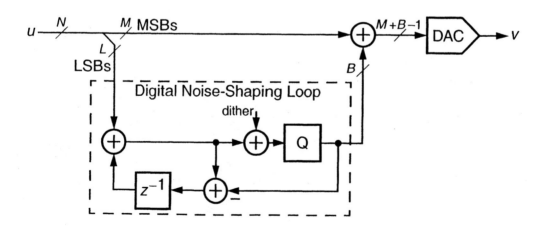

**Figure 7.14: A hardware-reduced first-order modulator with dither.**

sequentially all input codes for the DAC. For an *M*-bit DAC, the counter will thus count up to $2^M$. Each code from the counter is held at the DAC input for at least $2^N$ clock periods, where *N* is the required DAC linearity (in bits). The DAC output is converted by the ADC into a 1-bit data stream, whose dc average is linearly related to the DAC output. A digital lowpass filter recovers this dc value, which is then stored in the RAM at the address given by the counter output.

Background calibration is also possible. In a classical scheme [9] implemented for a current-switching DAC, the DAC contains two more unit current sources than necessary for conversion. One is used as a reference. In each clock period, a new

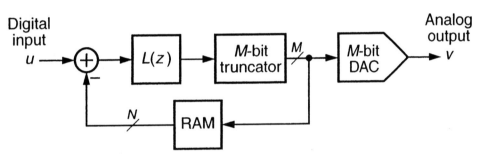

Figure 7.15: A digitally-corrected *M*-bit DAC.

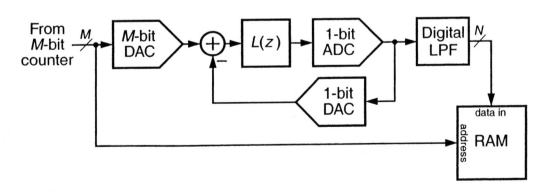

Figure 7.16: Calibration scheme for digital correction.

unit source is selected for calibration, and the reference current is copied into it, while the remaining sources perform the data conversion. Thus, by selecting the calibrated source in a rotating pattern, every source can be recalibrated in each $2^M$ clock periods.

Another background calibration scheme for current-mode DACs was described in [10]. Here, the current sources are measured and adjusted against a reference source, using an auxiliary DAC and a 1-bit $\Delta\Sigma$ ADC.

A charge-based calibration scheme similar in principle to that of [9], but suitable for switched-capacitor DACs, was described in [11]. Here, the charges delivered by the unit-element capacitors are adjusted sequentially, using a variable reference voltage for each element, until all charges match a fixed reference charge. This scheme also requires extra unit elements.

### 7.3.4 Comparison of Single-Bit and Multi-Bit $\Delta\Sigma$ DACs

Comparison of single-stage $\Delta\Sigma$ DACs with single-bit or multi-bit internal truncation shows the following relative advantages of the two schemes:

**Single-bit truncation:** Much simpler internal DAC structure can be used, without the need for thermometer coding, unit elements and digital mismatch-shaping logic.

**Multi-bit truncation:** Several advantages can be obtained, including

1. Simpler digital noise-shaping loop, since more aggressive NTF may be used, and since the truncation noise is reduced by at least $N - 1$ bits.
2. Less (or no) dithering, since tones are less likely to be generated, and since typically the amplitude of dithering is about 1/2 LSB, which is smaller in a multi-bit quantizer.
3. Much simpler analog smoothing filter, since the slewing and out-of-band noise in the DAC output are both reduced. Also, the sensitivity to clock jitter is reduced, due to the reduced step size in the DAC output signal.[†]

Generally, the advantages of multi-bit truncation outweigh those of single-bit truncation, and hence it is preferable to design $\Delta\Sigma$ DACs with multi-bit internal DACs.

As an illustration, ref. [3] describes a $\Delta\Sigma$ audio DAC using single-bit internal truncation, while ref. [8] discusses a 5-bit DAC with comparable performance. The 1-bit DAC needed a $5^{th}$-order noise-shaping loop; the 5-bit DAC required only a $3^{rd}$-order loop. The 1-bit DAC used an analog smoothing filter containing a $4^{th}$-order switched-capacitor (SC) filter, followed by an SC buffer stage and a $2^{nd}$-order continuous-time active filter. By contrast, in the 5-bit system, the SC analog filter was effectively merged with the DAC itself, and required no extra operational amplifiers. On the other hand, the mismatch-shaping applied in the 5-bit DAC needed a fairly elaborate digital circuitry.

## 7.4 Interpolation Filtering for $\Delta\Sigma$ DACs

Efficient implementation of the digital interpolation filter IF preceding the noise-shaping loop (Fig. 7.1) usually requires a multi-stage structure. A typical architecture and the role of the individual filter stages will next be discussed. We use as an example the IF of a classical 18-bit audio DAC [3]. The block diagram of the DAC is illustrated in Fig. 7.17; the structure of the IF in Fig. 7.18. The IF contains three cascaded finite-impulse-response (FIR) filter stages, followed by a digital sample-and-hold register. (A similar DAC IF was implemented in the more recent audio

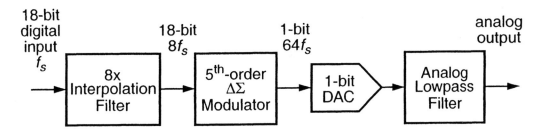

Figure 7.17: An 18-bit D/A converter architecture.

---

†. The sensitivity to clock jitter can also be reduced if the analog smoothing filter has switched-capacitor input stages [3].

DAC discussed in [8].) Fig. 7.19 shows schematically the spectra of the signals appearing at the inputs and outputs of the individual IF stages, as well as the final output of the DAC.

As explained in Section 7.1, the purpose of the IF is to take advantage of the increased clock frequency, and to suppress all unnecessary replicas of the signal spectrum occurring between the baseband and $OSR \cdot f_s$. This will improve the dynamic range of the noise shaping loop, and ease the selectivity and linearity requirements of the analog output filter. As was also mentioned in Section 7.1, the unwanted sidebands need not be totally erased, since truncation noise will be introduced in their place anyway in the noise-shaping loop NL.

In principle, it is possible to raise the sampling frequency immediately to $OSR \cdot f_s$, and then to carry out all filtering at this elevated clock rate. However, this would require all digital circuitry to function at high speed, and it would hence dissipate an unnecessarily large amount of power. It would also generate more digital activity, and thus more digital noise, than necessary. Therefore, it is preferable to perform the increasing of the clock frequency and the filtering in parallel steps, with most of the signal processing performed at a low clock rate.

The first stage of the filter is operated at $2f_s$ (Fig. 7.18), and is used to suppress the odd-order images. Thus, it removes the first replica of the baseband, assumed to extend approximately from $f_B$ to $3f_B$, as well as the third replica between $5f_B$ and $7f_B$, etc. The operation is illustrated in Figs. 7.19a and b, where the first curve shows the spectrum of the Nyquist-sampled input signal, and the second the desired spectrum of the first-stage output. Notice that the requirements on this

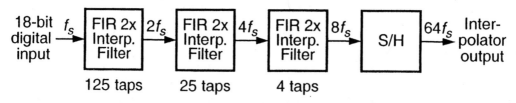

Figure 7.18: Interpolation filter architecture.

# Interpolation Filtering for ΔΣ DACs

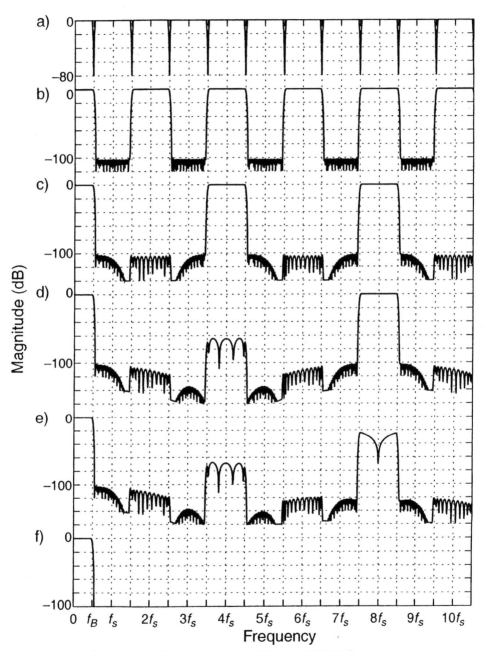

**Figure 7.19:** Spectra within a ΔΣ DAC system.

stage are very demanding: it needs to have a flat passband with extremely small (here, about 0.001 dB) gain variation in the 0 to $f_B$ frequency range, and a very sharp cutoff in order to suppress the adjacent image which is quite close. In the filter discussed in [3], this stage was realized by a 125-tap half-band FIR filter. (Half-band filters are FIR structures, which allow every second tap weight (except the center one) to be zero, and hence are very economical. A half-band filter can only realize, however, a frequency response which has a skew symmetry around its midpoint at $f_s/4$, as shown in Fig. 7.20. Thus, the pass- and stop-band limit frequencies must be symmetrically located, and the ripples must be the same in the two bands. These restrictions are usually acceptable in the interpolation-by-2 filtering task performed here.)

The second stage of the IF has a clock frequency of $4f_s$. Its task is to remove the images between $3f_B$ and $5f_B$, $7f_B$ and $9f_B$, etc., as shown in Fig. 7.19c. Its cutoff needs to be much less abrupt than that of the first stage. In the system described in [3], this task required a 24-tap half-band FIR filter. The third stage, operated at $8f_s$, was a 4-tap half-band FIR filter. It reduces the first, third, etc. of the remaining images (Fig. 7.19d).

Finally, a digital sample and hold operation was implemented by simply raising the sampling rate to $64f_s$, and repeating each output sample of the third IF stage 8

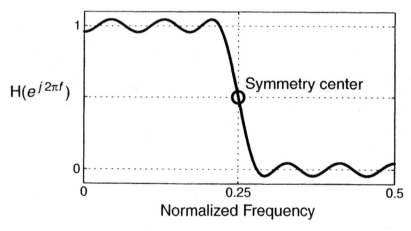

Figure 7.20: Frequency response of a half-band filter.

times. This S/H operation introduced a sinc function which had zeros at $8f_s$, $16f_s$, $24f_s$, etc., thus contributing slightly to the filtering at no added cost, as illustrated in Fig. 7.19e. The final OSR was thus 64.

Note that the IF is designed to provide most of its noise suppression just above the signal band, where the analog filter following the DAC has the most difficulty in removing noise.

In the noise-shaping loop NL following the IF, some truncation noise is added to the residual out-of-band noise (not shown in Fig. 7.19). Ideally, the resulting spectrum is then accurately reproduced at the DAC output, and finally all noise is removed by the analog LPF, giving the output spectrum illustrated in Fig. 7.19f.

The word-length used for data in the IF was 18 bits. 19-bit accuracy was used for the constant coefficients. The overall truncation noise was 107 dB below the full-scale sine-wave signal power, consistent with an approximately 18-bit performance.

Note that FIR filters are commonly used in $\Delta\Sigma$ systems since these filters can have perfectly flat group delay, and also since the required hardware can be clocked at the lower of the input and output data rates. IIR filters are less common, but have the advantage of being able to provide greater stopband attenuation with a given hardware complexity.

## 7.5 Analog Post-Filters for $\Delta\Sigma$ DACs

As discussed earlier, difficult analog circuit problems may arise in the design of the post-filter of the $\Delta\Sigma$ DAC. This filter, as illustrated in Fig. 7.19, needs to remove all out-of-band portions of the output signal of the internal DAC, and must not introduce appreciable nonlinear distortion into the signal while doing so. This task is particularly difficult for single-bit DACs, where a large two-level analog signal enters the post-filter.

Depending on the application, it may also be necessary for the post-filter to provide exactly or approximately linear phase characteristics. Alternatively, it may be designed with mildly nonlinear phase, and the phase error compensated in the digital interpolation filter.

In this Section, the post-filter design issues arising for single-bit and for multi-bit DACs will be discussed separately, and illustrated with examples from commercial chips.

### 7.5.1 Analog Post-Filtering in Single-Bit ΔΣ DACs

The block diagram of a typical post-filter for a single-bit DAC is shown in Fig. 7.21. The functions of the various blocks will be explained next.

As mentioned above, the input signal $x(t)$ of the post-filter in a single-bit DAC is a large two-level signal. The minimum swing of $x(t)$ is restricted by the condition that the inband component (i.e., the useful signal) needs to be much larger than the thermal and other noises introduced by the filter itself. This necessitates a large amplitude, since most of the power in the DAC output signal is out of band. Hence, if this signal were to be entered into a conventional active filter, the active component (op-amp or transconductance) would need an impractically high slew rate to avoid slew-rate-limited operation which generates harmonic distortion.

A more subtle linearity problem is due to the slew-rate limited slopes and imperfect symmetry of the waveform of $x(t)$ itself, since it is generated by an imperfect internal DAC. Also, the exact shape of the waveform may depend on its previous values. Thus, while the periodic samples $x(nT)$ of $x(t)$ may correctly and linearly reproduce the useful signal, the Fourier transform of the continuous-time $x(t)$ will usually contain harmonics.

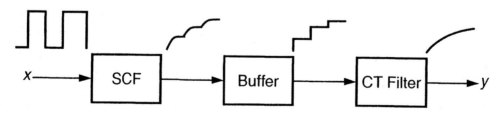

Figure 7.21: Post-filter for a 1-bit ΔΣ DAC and associated signals.

To alleviate both problems, it is customary to use switched-capacitor filter (SCF) stages as the input stages of the post-filter. A SCF with sampled-data input and output needs only the samples $x(nT)$ of $x(t)$ as its input signal, and it can remove most of the high-frequency power from the signal (and thus reduce its step size) without requiring high op-amp slew rate. Once the step size of the waveform is small enough, such that the slew rate required for its linear continuous-time (CT) processing is acceptably low, it can then be filtered by a CT active filter.

To understand the basic difference between the slew-rate requirements of CT and SC filters, consider the SC integrator shown in Fig. 7.22a. During phase $\phi_1$, the input voltage $x(t)$ charges $C_1$; at the end of the $n^{th}$ clock phase $\phi_1$, for properly designed switches, the charge will very accurately equal $C_1 \cdot x(nT)$. At the same time, $C_3$ samples the output voltage $y(t)$ of the op-amp. This voltage has abruptly changed when $\phi_2$ went high; as illustrated (in an exaggerated fashion) in

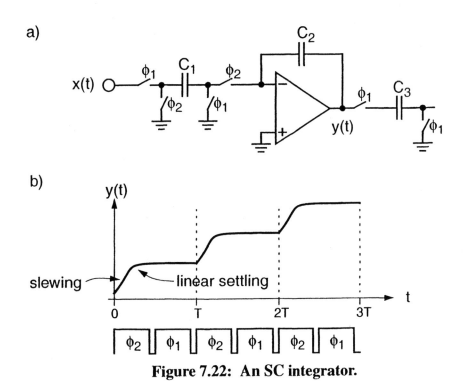

Figure 7.22: An SC integrator.

Fig. 7.22b, it underwent a slewing and settling process. The settling continues during $\phi_1$. For properly designed op-amp and switches, the final value $y(nT)$ will be very close to the theoretical value $y(nT-T) + (C_1/C_2)x(nT-T)$, and the charge in $C_3$ will be very close to $C_3 \cdot y(nT)$. Thus, the sampled signal processing is essentially unaffected by the nonlinear effects introduced by the slewing and (possibly nonlinear) settling of the op amp. Hence, the slew rate of the op amp does need to be very high, just high enough to allow accurate settling of $y(t)$ in the allocated time period. This makes the SCF particularly well suited for the task at hand.

Since the signal/noise ratio of the SC filter in a $\Delta\Sigma$ DAC must often be extremely high, its susceptibility to internal noise sources is an important design factor. This may result in the choice of unconventional architectures for the SCF. The most commonly used configuration, a cascade of biquads, has inferior noise-gain properties, as will be demonstrated next. Consider the block diagram of such a filter, shown along with its noise sources $n_{ij}$ in Fig. 7.23a. Clearly, $n_{11}$ has the same gain to the output as the input. When $n_{12}$ is referred back to the input, its power is divided by $|I_{11}|^2$, where $I_{11}$ is the transfer function of the first integrator. This division is equivalent to differentiating (high-pass filtering) the noise, and it thus reduces the inband noise introduced by $n_{12}$.

Consider now the first noise source $n_{i1}$ of the $i^{th}$ biquad. When it is referred back to the input, its power is divided by the factor $|H_1 H_2 ... H_{i-1}|^2$, where $H_k$ is the transfer function of the $k^{th}$ biquad. Dynamic range scaling causes $|H_1 H_2 ... H_{i-1}| \leq 1$ and hence when $n_{i1}$ is referred back to the input, its inband power, is not reduced. The power gain of $n_{i2}$ is $1/|I_{i1}|^2$ times that of $n_{i1}$, and hence it is first-order noise shaped.

In conclusion, for high oversampling ratios, the input-referred noise power is the weighted sum (with weight factors larger than or equal to 1) of the unshaped input noise powers of all biquads in the SCF, a decidedly unhappy situation.

Consider, by contrast, the structures shown in Figs. 7.23b and c. Simple analysis of the first one (sometimes called "inverse follow-the-leader" structure) shows that referring noise source $n_j$ to the input is equivalent to multiplying its noise power by $1/|I_1 I_2 ... I_{j-1}|^2$. Thus, all noise sources (except $n_1$) are shaped, and for high OSR the noise is dominated by $n_1$, with all other sources contributing negligible in-band noise power.

For the structure of Fig. 7.23c, analysis shows that $n_1$ remains unshaped, $n_2$ first-order shaped, and all other noise sources are suppressed by a second- or third-order shaping. Hence, again $n_1$ dominates the overall noise, for a very good noise performance. (Notice that— like the structure of Fig. 7.23a, but unlike that of Fig. 7.23b— the architecture of Fig. 7.23c allows the realization of finite transmission zeros, which improves its selectivity capabilities.)

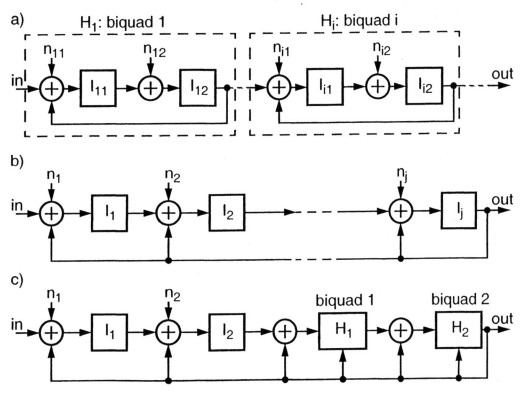

**Figure 7.23:** Reconstruction filter architectures.

In conclusion, in high-accuracy DACs, the architectures of Figs. 7.23b and c may be preferable to that of Fig. 7.23a, or other commonly used SCF structures. (Note that our discussions ignored the different sensitivities of the various configurations to element value variations. This is usually less important in the present context, since the variations tend to be small, and the exact response not too critical.)

Figs. 7.24-7.27 compare the realizations and noise shaping properties of two Bessel SCFs realizing the same transfer function [12]. Fig. 7.24 shows a biquad realization, and Fig. 7.25 its noise transfer functions from source to output. Fig. 7.26 and Fig. 7.27 show the same for the inverse follow-the-leader structure. The curves demonstrate the superior noise shaping properties of the latter architecture.

After sufficient filtering, the step size of the SCF waveforms can be greatly reduced. However, the waveforms will still exhibit op-amp-induced transients representing nonlinear distortion. Hence, it is necessary to use a buffer stage which is

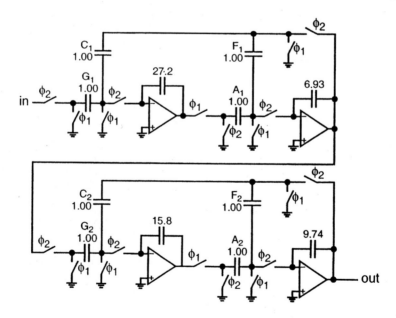

Figure 7.24: A $4^{th}$-order Bessel filter implemented with a cascade of biquads.

Analog Post-Filters for ΔΣ DACs

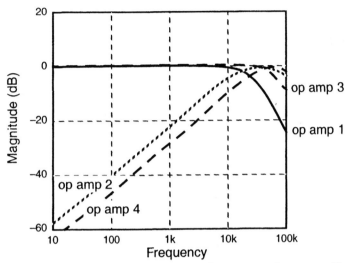

Figure 7.25: Noise gains from each op-amp input to the output for the circuit of Fig. 7.24.

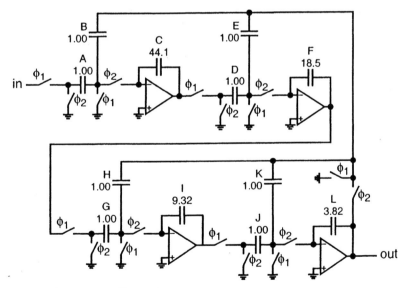

Figure 7.26: A 4$^{th}$-order Bessel filter implemented with the inverse follow-the-leader topology.

249

driven by the samples $y(nT)$ of the SCF output, and which provides a waveform free of such transients. This can be achieved by using a direct-charge-transfer (DCT) stage [13]. A low-pass DCT stage is shown in Fig. 7.28a. It samples the input signal $x(t)$ at the end of phase $\phi_1$, storing it in $C_1$. As $\phi_2$ goes high, $C_1$ is switched across $C_2$, and the two capacitors share charges (Fig. 7.28b). Since the left terminal of the parallel combination is floating at this time, no external charge enters the branch during the charge transfer; in particular, the op-amp does not need to contribute to the high impulsive current flowing. Thus, this transient is governed by a simple first-order differential equation, with only the switch on-resistances and the capacitance $C_1 + C_2$ determining the time constant. This way, a fast and clean transient may be obtained, which does not exhibit the slewing and nonlinear settling behavior which the op-amp would normally exhibit.

The output of the buffer stage can now be fed to the CT filter. This filter needs to eliminate the remaining noise above $f_B$. Typically, it is a second- or third-order active-RC circuit, often using the Sallen-Key configuration [14].

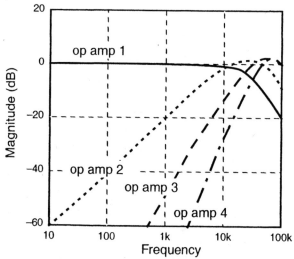

**Figure 7.27:** Noise gains from each op-amp input to the output for the circuit of Fig. 7.26.

## 7.5.2 Analog Post-Filtering in Multi-Bit ΔΣ DACs

For multi-bit ΔΣ DACs, the tasks, and hence the design, of the post-filter can be much easier. The out-of-band noise power is reduced due to the smaller step size; the remaining power decreases exponentially with the number of bits $N$ retained after truncation. The corresponding simplification in the SCF is hence also greatly dependent on $N$.

Two examples will be shown to illustrate the design of post-filters for multi-bit ΔΣ DACs. For the first one [8], $N \sim 5$ (31 levels) was used in the truncation. A single SC stage (Fig. 7.29.) was used to perform the functions of the internal DAC and the SCF. Since it was a direct-charge-transfer circuit, it also made the extra SC-to-CT buffer stage unnecessary. Thus the DAC, SCF and buffer functions, which would have required 5~6 op-amps for a single-bit DAC, were all performed by a single op-amp for the 5-bit system. In the circuit of Fig. 7.29, the DAC operation is

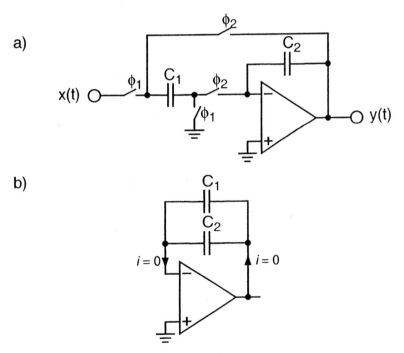

**Figure 7.28: A direct-charge-transfer (DCT) stage.**

performed during $\phi_1$, by pre-charging some of the 4x31 small capacitors in the four capacitor arrays to $V_{DD}$ and discharging others. By duplicating the capacitor arrays, and driving the secondary arrays with a delayed digital signal, a first-order (2-tap) FIR filter function is realized. During $\phi_2$, all capacitors are connected in parallel with the feedback capacitors $C_{FB}$ in a DCT operation. The overall transfer function between the digital input and the samples of the output signal, normalized to the full-scale output, is hence

$$H(z) = \frac{1}{2}\left(\frac{1+z^{-1}}{1+r-rz^{-1}}\right) \tag{7.3}$$

where

$$r = \frac{C_{FB}}{2(C_1 + C_2 + \ldots + C_{31})}. \tag{7.4}$$

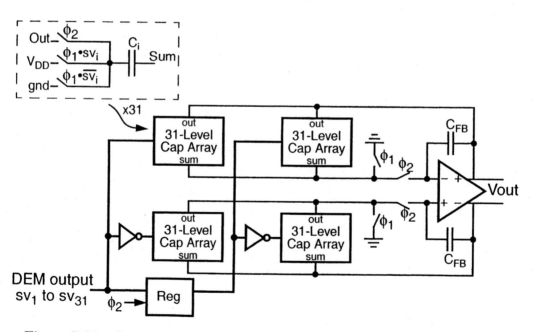

Figure 7.29: Combined DAC, DCT and filter for a multi-bit $\Delta\Sigma$ DAC [8].

## Analog Post-Filters for ΔΣ DACs

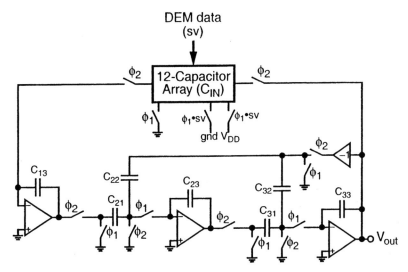

**Figure 7.30:** Another ΔΣ DAC with merged DAC, DCT and SCF filter functions [15].

This simple first-order IIR filter was adequate to prepare the signal for CT filtering, performed off-chip by a Sallen-Key filter.

Another ΔΣ DAC, described in [15], used a 13-level ($N \sim 3.7$ bits) truncation. Its SCF was a third-order Chebyshev filter. It is shown in Fig. 7.30 as a single-ended circuit (the actual implementation was fully differential). The DAC action is performed during $\phi_1$, by charging to $V_{DD}$ or discharging the 12 capacitors in the input array $C_{IN}$. During $\phi_2$, the circuit is configured as a third-order inverse follow-the leader SC filter, with the charge acquired by $C_{IN}$ during $\phi_1$ acting as its input signal. A simple first-order active-RC stage was used to perform both the CT filtering and the differential-to-single-ended conversion.

253

## 7.6 Conclusions

Chapter 7 was devoted to the design of delta-sigma DACs. First, the general principles and basic DAC architectures were discussed, and then various structures available for realizing their noise-shaping loops were described. Due to the high accuracy made possible by the all-digital loop, some novel loop architectures (not practical for ADC loops) exist in delta-sigma DACs. These were introduced, along with some variants of the conventional MASH configurations specific to DACs.

As was the case for $\Delta\Sigma$ ADCs, either single-bit or multibit internal quantization may be used in $\Delta\Sigma$ DACs. The relative merits of these two options were compared, and various methods were discussed for the filtering or compensation of the error signals introduced by the unavoidable nonlinearity of a multibit internal DAC. Again, some of these schemes are similar to those applicable to multibit ADCs, which have already been discussed in Chapter 6; others are specifically aimed at $\Delta\Sigma$ DACs, and were described in this chapter.

Next, the design issues of the digital interpolation filter were discussed, and illustrated by an example. The example chosen was an efficient multi-stage filter used in a commercial 18-bit audio delta-sigma DAC.

Finally, the design of the analog postfilter used in $\Delta\Sigma$ DACs was discussed. The two different situations arising for single-bit and for multi-bit truncations were contrasted, and filter design techniques were described for both systems, along with some typical examples.

## References

[1] P. J. Naus, E. C. Dijkmans, E. F. Stikvoort, A. J. McKnight, D. J. Holland, and W. Brandinal., "A CMOS stereo 16-bit D/A converter for digital audio," *IEEE Journal of Solid-State Circuits*, vol. 22, pp. 390-395, June 1987

[2] J. C. Candy and A. Huynh, "Double integration for digital-to-analog conversion," *IEEE Transactions on Communications*, vol. 34, no. 1, pp. 77-81, January 1986.

[3] N. S. Sooch, J. W. Scott, T. Tanaka, T. Sugimoto and C. Kubomura, "18-bit stereo D/A converter with integrated digital and analog filters," *presented at the 91st convention of the Audio Engineering Society*, New York, October 1991, preprint 3113.

[4]  X. F. Xu and G. C. Temes, "The implementation of dual-truncation ΣΔ D/A converters," *Proceedings of the IEEE International Symposium on Circuits and Systems*, pp. 597-600, May 1992.

[5]  A. Hairapetian, G. C. Temes and Z. X. Zhang, "A multibit sigma-delta modulator with reduced sensitivity to DAC nonlinearity," *Electronics Letters*, vol. 27, no. 11, pp. 990-991, May 23 1991.

[6]  R. Adams, K. Nguyen and K. Sweetland, "A 113dB SNR oversampling DAC with segmented noise-shaped scrambling," *IEEE Journal of Solid-State Circuits*, vol. 33, no. 12, pp. 1871-1878, December 1998.

[7]  S. R. Norsworthy, D. A. Rich and T. R. Viswanathan, "A minimal multibit digital noise shaping architecture," *Proceedings of the IEEE International Symposium on Circuits and Systems*, pp. I-5 to I-8, May 1996.

[8]  I. Fujimori, A. Nogi and T. Sugimoto, "A multibit Δ-Σ audio DAC with 120-dB dynamic range," *IEEE Journal of Solid-State Circuits*, vol. 35, pp. 1066-1073, August 2000.

[9]  D. Groeneveld et al., "A self-calibration technique for monolithic high-resolution D/A converters," IEEE J. Solid-State Circuits, vol. 24, pp. 1517-1522, Dec. 1989.

[10] A. R. Bugeja and B.-S. Song, "A self-trimming 14-b 100 MS/s CMOS DAC," *IEEE Journal of Solid-State Circuits*, vol. 35, pp. 1841-1852, Dec. 2000.

[11] U. K. Moon et al., "Switched-capacitor DAC with analogue mismatch correction," *Electronics Letters*, vol. 35, pp. 1903-1904, Oct. 1999.

[12] M. Rebeschini and P. F. Ferguson, Jr., "Analog Circuit Design for ΔΣ DACs," in S. Norsworthy, R. Schreier and G.C. Temes, *Delta-Sigma Data Converters*, Sec. 12.2.3, IEEE Press, 1997.

[13] J.A.C. Bingham, "Applications of a direct-transfer SC integrator," *IEEE Transactions on Circuits and Systems*, vol. 31, pp. 419-420, Apr. 1984.

[14] See, e.g., R. Schaumann and M.E. Van Valkenburg, *Design of Analog Filters*, pp. 161-163, Oxford University Press, 2001.

[15] M. Annovazzi et al., "A low-power 98-dB multibit audio DAC in a standard 3.3-V 0.35-μm CMOS technology," *IEEE Journal of Solid-State Circuits*, vol. 37, pp. 825-834, July 2002.

# 7 Delta-Sigma DACs

# CHAPTER 8 — High-Level Design and Simulation

This chapter describes the use of several key functions found in the Delta-Sigma Toolbox for MATLAB as well as some of the supporting theory. Since it is not necessary to understand algorithmic details before making use of the toolbox, the reader may choose to skip over sections that contain such material. A complete reference manual for this freeware toolbox, including instructions for obtaining it, may be found in Appendix B.

## 8.1 NTF Synthesis

The first step in the design of a $\Delta\Sigma$ modulator is the selection of the NTF. The modulator order, the number of quantization levels, and the choice of lowpass, bandpass or quadrature modulation are all design parameters. Figures 4.14 to 4.16 of Chapter 4 depict the SQNR limits for lowpass modulators of orders 1 to 8, with 1-bit to 3-bit quantization, as functions of the oversampling ratio. These plots can be used to estimate the number of quantization levels and the modulator order needed to achieve a particular OSR-SNR target. Once approximate values for these parameters have been selected, the designer can perform a detailed examination of the NTFs surrounding this point in the design space.

As an example, consider the goal of achieving an SNR of 100 dB with an OSR of 64. Aiming for an SQNR of 120 dB will ensure that quantization noise is negligible and will also provide plenty of design margin. According to the plots in Chapter 4, either a $5^{th}$-order single-bit modulator, a $4^{th}$-order 2-bit modulator, or a $3^{rd}$-order 3-bit modulator could be used to achieve this SQNR. In practice, each of these architectures should be tried and evaluated in terms of power and/or area efficiency, but for the sake of simplicity we will consider only the first configuration.

Fig. 8.1 shows the MATLAB code which synthesizes a $5^{th}$-order NTF, and which then creates plots of the pole-zero distribution and the frequency response. (Code for formatting the figures is not shown.) The function which synthesizes the NTF is synthesizeNTF. The first two arguments to synthesizeNTF specify the order of the NTF and the oversampling ratio, while the third argument (opt) is a flag which specifies whether or not the NTF zeros are to be optimized for maxi-

```
order = 5;
OSR = 64;
opt = 1;
H = synthesizeNTF(order,OSR,opt);
plotPZ(H);
f = linspace(0,0.5,1000);
z = exp(2i*pi*f);
plot( f, dbv(evalTF(H,z)) );
sigma_H = dbv( rmsGain(H,0,0.5/OSR) );
```

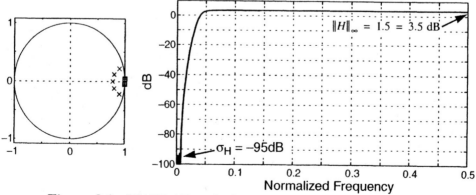

**Figure 8.1: MATLAB code for creating a lowpass NTF.**

mum attenuation of quantization noise in the band of interest. Optional fourth and fifth arguments specify the NTF's out-of-band gain (Hinf) and, for bandpass modulators, the center frequency (f0). Since the out-of-band gain was not specified in this example, $\|H\|_\infty$ defaults to 1.5. Likewise, since the center frequency was not specified, $f_0$ defaults to zero. In the documentation for synthesizeNTF (on page B-393), these default values are indicated in the synopsis of the function as values following an equals sign (=) in the parameter list:

```
ntf = synthesizeNTF(order=3,OSR=64,opt=0,H_inf=1.5,f0=0)
```

The same notational convention applies to all other toolbox functions.

The rms gain of the NTF in the band of interest is calculated using the function rmsGain, and the result is converted to units of dB using the function dbv. (dbv($v$) = $20\log_{10}(v)$ is the dB-equivalent of a voltage $v$.) From this calculation, we find that the chosen NTF provides 95 dB of suppression for the quantization noise which falls in the passband. If we assume that the power of the quantization noise is $\sigma_e^2 = 1/3 = -4.8 \text{ dB}$, then the power of the shaped in-band quantization noise is expected to be $\sigma_q^2 = \sigma_H^2 \sigma_e^2 / OSR$, which gives $-95 - 4.8 - 10\log_{10}(64) = -118$ dB.

Use of synthesizeNTF's optional arguments is illustrated in Fig. 8.2, which contains both the MATLAB code for designing the NTF of a bandpass modulator and for producing the plots shown. As this example demonstrates with the fourth argument (Hinf), an optional argument retains its default value if the empty matrix ([]) is supplied as a place-holder; not-a-number (NaN) can also be used for this purpose. This example uses opt=2, which specifies that the NTF zeros are to be optimized, under the constraint that the NTF must have at least one zero at band-center. Ensuring that at least one NTF zero is at band-center (at DC for lowpass modulators) can be helpful when the signal energy is known to be concentrated in the middle of the band. Since this NTF achieves an rms attenuation of 80 dB for quantization noise which falls in the passband, the expected SQNR for this design with a –6 dBFS input is $80 + \text{dbp}(3 \times 64) - 6 = 97$ dB (dbp($p$) = $10\log_{10}(p)$ is the toolbox function for the dB-equivalent of a power $p$).

## 8.1.1 How synthesizeNTF works

This section describes some of the internal details of synthesizeNTF, and thus may be skipped by those readers who wish to focus on putting the toolbox to work, rather than understanding how the toolbox itself functions.

The purpose of synthesizeNTF is to find an NTF of the specified order and out-of-band gain, whose zeros follow the formulae of Table 4.1 and whose poles are those of a maximally-flat all-pole transfer function. The reason for the latter requirement is primarily one of convenience: when the CRFB (or CIFB) topology is used with a single feed-in, the signal transfer function has the same frequency response as that of the all-pole transfer function created from the NTF's poles. (See Section 4.4.1.) In the authors' experience, the performance of an NTF is pri-

```
order = 8;
OSR = 64;
opt = 2;
f0 = 1/8;
H = synthesizeNTF(order,OSR,opt,[],f0);
plotPZ(H);
f = linspace(0,0.5,1000);
z = exp(2i*pi*f);
plot( f, dbv(evalTF(H,z)) );
sigma_H = dbv(rmsGain(H,f0-0.25/OSR,f0+0.25/OSR));
```

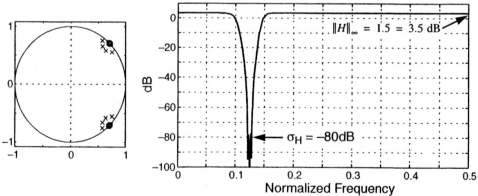

**Figure 8.2: MATLAB code for designing a bandpass NTF.**

marily a function of its out-of-band gain and its zero locations, whereas the form of the pole distribution is secondary.

The NTF poles which yield a maximally-flat all-pole transfer function can be derived as follows. Let $P(z)$ be an $n^{\text{th}}$-order polynomial in $z$ whose roots are such that $|P(e^{j\omega})|$ is maximally flat around $\omega = 0$. ($P(z)$ will be the denominator polynomial for the NTF.)

Since the coefficients of $P(z)$ are real,

$$|P(e^{j\omega})|^2 = P(e^{j\omega})[P(e^{j\omega})]^* = P(z)P(1/z)\big|_{z = e^{j\omega}} \qquad (8.1)$$

Thus the requirement that $|P(e^{j\omega})|$ be maximally flat around $\omega = 0$ is equivalent to the condition that $P(z)P(1/z)$ be maximally flat around $z = 1$, i.e.

$$P(z)P\left(\frac{1}{z}\right) = P(1) + a(z-1)^n\left(\frac{1}{z}-1\right)^n, \qquad (8.2)$$

where $a$ is some constant which controls the sharpness of the filter. For example, $a = 0$ results in a transfer function that is constant (and so is completely flat); larger values of $a$ result in lowpass transfer functions with increasing sharpness. Negative values of $a$ result in polynomials whose magnitudes decrease as one moves away from $z = 1$. Since this latter behavior is the opposite of what we want, $a$ must be positive.

Taking $P(1) = 1$ without loss of generality and simplifying (8.2) results in

$$P(z)P\left(\frac{1}{z}\right) = 1 + a\frac{(z-1)^{2n}}{(-z)^n} = \frac{a(z-1)^{2n} + (-z)^n}{(-z)^n}. \qquad (8.3)$$

Thus, the roots of

$$a(z-1)^{2n} + (-z)^n = 0 \qquad (8.4)$$

are the poles (and their inverses) of the sought-after NTF. These roots are given by the roots of the $n$ complex quadratic equations

$$z^2 + b_k z + 1 = 0, \quad k = 0 \ldots n-1, \tag{8.5}$$

where

$$b_k = \frac{e^{j\frac{(2k+1)\pi}{n}}}{a^{1/n}} - 2. \tag{8.6}$$

Since the constant term in the quadratic is 1, the product of the two roots of the quadratic is 1. Thus one root lies inside the circle while the other lies outside. Collecting the $n$ roots that are inside the unit circle yields the poles of the desired NTF.

### 8.1.2 Limitations of synthesizeNTF

Since synthesizeNTF chooses the zeros of the NTF according to Table 4.1 (i.e. the zeros are chosen independently of the poles), the zero locations are optimal only if the denominator polynomial is constant in the band of interest. For most practical situations, the fact that the denominator polynomial is that of a maximally-flat all-pole transfer function ensures that this requirement is met. However, when $OSR$ is low, the $-3$-dB cut-off frequency of the denominator polynomial may approach the edge of the passband, causing the zeros to no longer be optimal. For example, the cut-off frequency of the denominator polynomial associated with a $5^{\text{th}}$-order NTF designed using the default value of $\|H\|_\infty = 1.5$ falls close to the edge of the passband when $OSR = 12$.

Another pair of limitations on synthesizeNTF relate to the range of $\|H\|_\infty$-values supported by this function. Note that $\|H\|_\infty$ increases as the bandwidth of the denominator polynomial increases, reaching a maximum when the denominator polynomial has all its roots at $z = 0$. Since all the zeros of the NTF are clustered around $z = 1$, $\|H\|_\infty$ is equal to the magnitude of $H$ at $z = -1$, which can be no greater than $2^n$. Although a general $n^{\text{th}}$-order NTF can be constructed such that $\|H\|_\infty > 2^n$, such an NTF does not have poles corresponding to a maximally-flat all-pole transfer function and so cannot be found with synthesizeNTF.

Similarly, values of $\|H\|_\infty$ near 1 can also be troublesome if optimized zeros are used. When a low value of $\|H\|_\infty$ is requested, the poles of $H$ converge on $z = 1$. If the zeros of $H$ are all at $z = 1$, then $\|H\|_\infty$ approaches 1 and there is no problem. However, if the zeros have been assigned their "optimal" locations, $\|H\|_\infty$ does not converge to 1 as the poles converge to $z = 1$. This limitation is a result of the current strategy which separates the pole and zero optimizations, and may be corrected in a future release of the toolbox.

## 8.2 NTF Simulation, SQNR Calculation and Spectral Estimation

Once an NTF is in hand, the designer can immediately begin to evaluate the modulator via simulation. The toolbox function `simulateDSM` computes the output of a modulator which realizes the desired NTF, assuming the STF is unity. Since the input-output behavior of an ideal $\Delta\Sigma$ modulator is determined solely by the NTF, the STF, the number of quantizer levels and the initial state, the behavior predicted by `simulateDSM` in terms of stable input range, tone performance and SNR is the same as it would be for any realization of the desired NTF. Internal details of the loop filter are only relevant when one is concerned with coefficient errors, saturation of internal nodes and other non-ideal effects.

The primary arguments to `simulateDSM` are the input sequence and the NTF. Secondary arguments include the number of quantizer levels (`nlev`), which defaults to 2, and the modulator's initial state (`x0`), which defaults to zero. A code fragment for simulating a 3-level lowpass modulator[†] and for plotting portions of its input and output waveforms is shown in Fig. 8.3, along with the associated graphical output. The input signal was chosen as a half-scale sine wave in order to avoid overloading the modulator, and the frequency of the sine wave was chosen such that it would fall into one of the in-band frequency bins of an 8192-point FFT. The waveforms of Fig. 8.3 provide a dramatic illustration of the (apparently) poor correspondence between the input of a $\Delta\Sigma$ modulator and its output. As this figure shows, the only similarity between the input and output waveforms in the time

---

†. Although the design point considered in the preceding section was a 5th-order binary modulator, this section examines a 5th-order ternary modulator in order to illustrate the use of the toolbox with a multi-bit modulator.

domain is that when the input is positive, the output is usually either +2 or 0, and that when the input is negative, the output is usually either −2 or 0.

However, when viewed from the frequency domain, the near-equivalence of the input and output (in the band of interest) becomes more clear. Fig. 8.4 plots the FFT of a length-8192 output stream and superimposes the expected PSD. Noise-shaping is clearly evident, but the notch is much shallower than the expected PSD and consequently the observed SQNR is about 40 dB lower than our expectations.[†]

```
OSR = 64;
H = synthesizeNTF(5,OSR,1);
nLev = 3;
Nfft = 2^13;
tone_bin = 57;
t = [0:Nfft-1];
u = 0.5*(nLev-1)*sin(2*pi*tone_bin/Nfft*t);
v = simulateDSM(u,H,nLev);
n=1:150;
stairs(t(n),u(n),'g');
hold on;
stairs(t(n),v(n),'b');
```

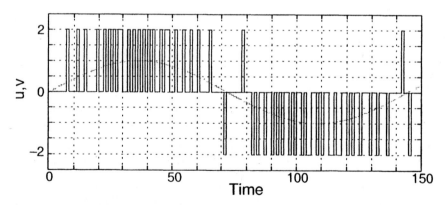

Figure 8.3: MATLAB code for simulating a ΔΣ modulator and for plotting a portion of the output waveform.

---

†. Since the power of the unit-amplitude sine wave input is −3 dB and the expected in-band noise power is -118 dB (see page 259), the expected SQNR is 115 dB.

The discrepancy between the simulation result and our expectations is resolved by *windowing*. When the input data to an FFT consists solely of tones located in FFT bins, there is no need to window the data since no spectral smearing due to the finite length of the time sequence can result. However, the aperiodic nature of a $\Delta\Sigma$ modulator's output stream will often suffer spectral smearing unless it multiplied by an appropriate window. Fig. 8.5 demonstrates the use of a Hann window. Note the changes to the scale factors of both the FFT and the expected PSD (via the *noise bandwidth*, NBW) when a Hann window is used. The spectrum contained in this figure exhibits a notch depth, and hence an SQNR, that are in good agreement with our expectations. Appendix A describes other aspects of spectral estimation and gives a more detailed discussion of the importance of windowing and the significance of NBW.

```
spec = fft(v)/(Nfft*(nLev-1)/2);
snr = calculateSNR(spec(1:ceil(Nfft/(2*OSR))+1),tone_bin);
NBW = 1/Nfft;
Sqq = 4 * (evalTF(H,exp(2i*pi*f))/(nLev-1)).^2 / 3;
f = linspace(0,0.5,Nfft/2+1);
plot( f, dbv(spec(1:Nfft/2+1)), 'b')
hold on;
plot( f, dbp(Sqq*NBW), 'm', 'Linewidth', 1 );
```

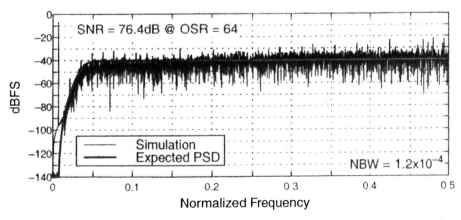

**Figure 8.4:** Crude MATLAB code for calculating a spectrum and comparing it against the desired NTF.

A single simulation is often insufficient to evaluate the performance of a ΔΣ modulator. Numerous simulations using excitations with varying amplitudes and frequencies are usually required. The `simulateSNR` function provides support for this task, allowing an amplitude sweep to be conducted with a single function call. Code illustrating the use of this function is given in Fig. 8.6.

## 8.3 NTF Realization and Dynamic Range Scaling

Once the designer is satisfied with the performance of the NTF, the next step is to realize that NTF with a particular modulator structure. Although the toolbox can represent just about any modulator structure, code for converting an NTF into coefficients is currently only available for four of the most popular structures. These four structures are the feedback (FB) and feedforward (FF) versions of loop filters containing either only delaying integrators or alternately delaying and non-delaying integrators. As discussed in Section 4.4, loop filters of the former type are

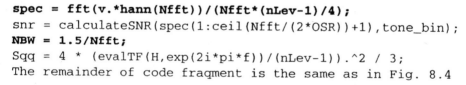

```
spec = fft(v.*hann(Nfft))/(Nfft*(nLev-1)/4);
snr = calculateSNR(spec(1:ceil(Nfft/(2*OSR))+1),tone_bin);
NBW = 1.5/Nfft;
Sqq = 4 * (evalTF(H,exp(2i*pi*f))/(nLev-1)).^2 / 3;
The remainder of code fragment is the same as in Fig. 8.4
```

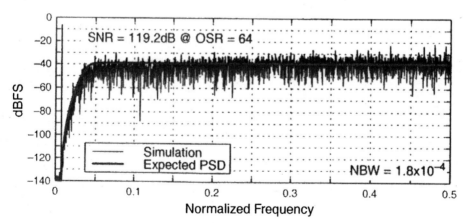

Figure 8.5: Code for calculating a windowed spectrum and comparing it against the expected PSD.

termed (somewhat inappropriately) cascade-of-integrator (CI) filters, whereas the latter are termed cascade-of-resonator (CR) filters[†] and thus the four supported topologies which are fully supported by the toolbox bear the abbreviations CIFB, CIFF, CRFB and CRFF.

Fig. 8.7 shows code for realizing a $5^{\text{th}}$-order NTF with the CRFB structure, which is also depicted in Fig. 8.7 for ease of reference. The coefficients of the structure are specified by the a, g, b, and c coefficient vectors. The coefficients computed by synthesizeNTF are such that the NTF of the resulting modulator is the desired

```
amp = [-130:5:-20 -17:2:-1];
snr = simulateSNR(H,OSR,amp,[],nLev);
plot(amp,snr,'-b',amp,snr,'db');
[pk_snr pk_amp] = peakSNR(snr,amp);
```

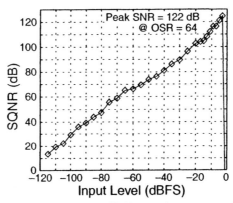

**Figure 8.6:** Generating an SNR vs. input amplitude curve.

---

[†]. In early versions of the toolbox, the CI filters were only allowed to implement NTF zeros at $z = 1$, whereas the CR filters were used to implement arbitrary NTF zeros. Thus the CI filters truly were cascaded (delaying) integrators, while the CR filters truly were (most often) cascaded resonators. However, since feedback around a pair of delaying integrators can realize zeros that are close to the unit circle if the zeros are low-frequency zeros, and since this trick has been used in real lowpass modulators, the CI structure was generalized to support such internal feedback terms and thus the term "cascade-of-integrators" became a misnomer. Since the resulting pseudo-resonators can only realize NTF zeros along the line $\text{Re}(z) = 1$, a warning message is issued if one tries to realize an NTF possessing general zeros with a CI filter.

NTF and the STF is unity. Setting all feed-in (b) terms except the first to zero results in the aforementioned maximally-flat all-pole STF. The frequency response of this STF is depicted in Fig. 8.8. The STF is calculated by the toolbox function `calculateTF` from a generic loop-filter description (the *ABCD matrix*), which is computed by the `stuffABCD` function from the topology-specific *a*, *g*, *b* and *c* coefficients. The structure of the ABCD matrix is described in Section 8.3.1.

The coefficients returned by `realizeNTF` are those of an *unscaled* modulator, i.e. a modulator whose internal states occupy an unspecified range. In order to restrict the state range to known (and practical) values, *dynamic-range scaling* must be performed. Dynamic-range scaling can be applied to any linear system on a state-by-state basis as depicted in Fig. 8.9, which shows how an internal state is scaled down by a factor $k$ simply by reducing all incoming branches to that state by a factor of $k$, and by making up for the attenuation so introduced by multiplying all outgoing branches by the same factor $k$.

The toolbox function `scaleABCD` uses simulations to determine the required scaling factors for each state of a delta-sigma modulator. The modulator is simulated with inputs of various amplitudes in order to determine the maximum stable input amplitude (`umax`) as well as the maximum value that each state achieves for input

```
H = synthesizeNTF(5,64,1);
form = 'CRFB';
[a,g,b,c] = realizeNTF(H,form);
b(2:end) = 0;   % for a maximally flat STF
```

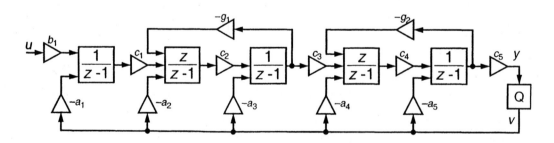

**Figure 8.7:** Realizing an NTF with the CRFB structure.

amplitudes up to umax. The ABCD matrix of the modulator is then subjected to dynamic range scaling so that the maximum value of each state equals the specified limit (xlim, which defaults to 1). The toolbox function mapABCD then trans-

```
ABCD = stuffABCD(a,g,b,c,form);
[Ha Ga] = calculateTF(ABCD);
f = linspace(0,0.5,100);
z = exp(2i*pi*f);
magHa = dbv(evalTF(Ha,z));
magGa = dbv(evalTF(Ga,z));
plot(f,magHa,'b', f,magGa,'m', 'Linewidth',1);
```

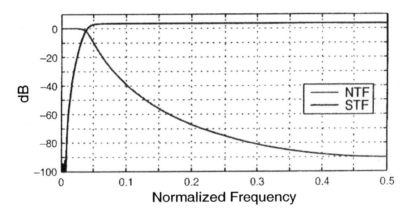

Figure 8.8: NTF and STF calculation for the modulator realization of Fig. 8.7.

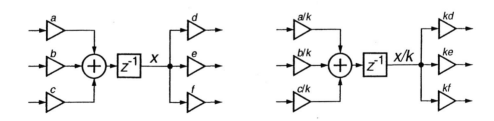

Figure 8.9: Scaling an individual filter state, *x*, to *x/k*.

lates the scaled ABCD matrix back into the coefficients for the chosen topology. Fig. 8.10 shows the associated code fragment, and also plots the observed state maxima, after scaling as a function of the input level for our $5^{th}$-order 3-level modulator. As the plot demonstrates, each state stays below the specified limit of 0.9 for inputs from 0 to 78% of full-scale.†

```
nLev = 3; xLim = 0.9; f0 = 0;
[ABCDs umax] = scaleABCD(ABCD,nLev,f0,xLim);
[a g b c] = mapABCD(ABCDs,form);
```

Scaled Coefficients:

| $i$ | $a_i$ | $g_i$ | $b_i$ | $c_i$ |
|---|---|---|---|---|
| 1 | 0.0383 | 0.0034 | 0.0383 | 0.1473 |
| 2 | 0.0748 | 0.0045 | 0 | 0.2045 |
| 3 | 0.0969 | – | 0 | 0.4007 |
| 4 | 0.1754 | – | 0 | 0.4385 |
| 5 | 0.1709 | – | 0 | 3.2533 |

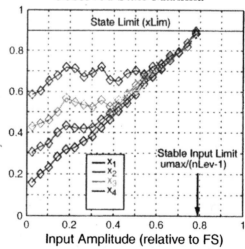

**Figure 8.10:** Code for performing dynamic range scaling, plus a plot of the state magnitude as a function of the input level for the scaled modulator.

---

†. The observed value of umax is 1.56, which is 78% of the nLev−1 = 2 full-scale input level.

## 8.3.1 The ABCD Matrix

This section describes the structure and interpretation of the ABCD matrix, and so may be skipped by readers who do not wish to delve into the internal details of the toolbox.

In order to make the toolbox as general as possible, the ABCD matrix is used for generic operations such as modulator simulation and dynamic-range scaling. Topology-specific functions are only needed to translate an NTF into coefficient vectors and to convert between coefficient vectors and the ABCD matrix.

The ABCD matrix is a conglomeration of four matrices, $A$, $B$, $C$ and $D$, which can be used to describe the dynamics of any discrete-time linear system. Simply assemble the state equations into the following form

$$x(k+1) = Ax(k) + B\begin{bmatrix}u(k)\\v(k)\end{bmatrix}$$

$$y(k) = Cx(k) + D\begin{bmatrix}u(k)\\v(k)\end{bmatrix} \qquad (8.7)$$

where $x(k)$ is the $n \times 1$ state vector at time $k$ ($n$ is the order of the system), $A$ is an $n \times n$ matrix describing the interconnections within the loop filter, $B$ is an $n \times 2$ matrix describing how the input $u(k)$ and the fed-back output $v(k)$ are applied to the loop filter, while $C$ and $D$ are $1 \times n$ and $1 \times 2$ matrices describing how $y(k)$, the output of the loop filter, is computed from $x(k)$, $u(k)$ and $v(k)$.[†]

For example, consider the $5^{th}$-order CRFB system shown in Fig. 8.7. By examining the expanded forms of the delaying and non-delaying integrators shown in Fig. 8.11, the following state equations can be written by inspection:

---

†. The dimensions given assume a single-quantizer modulator with a single input. For an $n_t$-input, $n_q$-quantizer modulator, the dimensions of the $A$, $B$, $C$ and $D$ matrices are $n \times n$, $n \times (n_t + n_q)$, $n_q \times n$ and $n_q \times (n_t + n_q)$, respectively.

## 8 High-Level Design and Simulation

$$
\begin{aligned}
x_1(k+1) &= x_1(k) + b_1 u(k) - a_1 v(k) \\
x_2(k+1) &= x_2(k) + c_1 x_1(k) - g_1 x_3(k) - a_2 v(k) \\
x_3(k+1) &= x_3(k) + c_2 x_2(k+1) - a_3 v(k) \\
&= (1 - c_2 g_1) x_3(k) + c_2 x_2(k) + c_2 c_1 x_1(k) - (a_3 + c_2 a_2) v(k) \\
x_4(k+1) &= x_4(k) + c_3 x_3(k) - g_2 x_5(k) - a_4 v(k) \\
x_5(k+1) &= x_5(k) + c_4 x_2(k+1) - a_5 v(k) \\
&= (1 - c_4 g_2) x_5(k) + c_4 x_4(k) + c_4 c_3 x_3(k) - (a_5 + c_4 a_4) v(k) \\
y(k) &= c_5 x_5(k)
\end{aligned} \tag{8.8}
$$

and thus the ABCD matrix for this structure is

$$
\left[\begin{array}{c|c} A & B \\ \hline C & D \end{array}\right] = \left[\begin{array}{ccccc|cc}
1 & 0 & 0 & 0 & 0 & b_1 & -a_1 \\
c_1 & 1 & -g_1 & 0 & 0 & 0 & -a_2 \\
c_1 c_2 & c_2 & 1 - c_2 g_1 & 0 & 0 & 0 & -a_3 - a_2 c_2 \\
0 & 0 & c_3 & 1 & -g_2 & 0 & -a_4 \\
0 & 0 & c_3 c_4 & c_4 & 1 - c_4 g_2 & 0 & -a_5 - a_4 c_4 \\
\hline
0 & 0 & 0 & 0 & c_5 & 0 & 0
\end{array}\right]. \tag{8.9}
$$

In the general form of this structure, the input signal $u(k)$ would be fed to the input of every integrator and the quantizer through coefficients $b_2...b_6$, and these coefficients would populate the column whose first element is $b_1$. The ABCD matrices of other structures can be derived by a similar procedure, and new topologies may be added to the toolbox simply by adding the appropriate conversion routines to mapABCD, stuffABCD and realizeNTF.

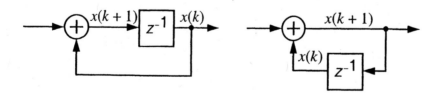

Figure 8.11: Delaying and non-delaying integrators expanded to show the associated state variables.

## 8.4 Creating a SPICE-Simulatable Schematic

Once the designer has obtained a scaled block diagram, the next step towards a transistor-level realization usually involves constructing a schematic that can be simulated in behavioral form. This schematic can be used to study such effects as switch resistance, comparator hysteresis, op-amp nonlinearity, finite op-amp gain, bandwidth, slew-rate and so on, in order to obtain specifications for the various sub-blocks before embarking on transistor-level design. When transistor-level implementations are available, these may be substituted for the behavioral elements, either individually or all at once, so that time-domain simulations can verify the modulator's functionality. This section describes how the designer translates a MATLAB description of the modulator into a behavioral description that can be simulated with a SPICE-like simulator.

In principle, translating a block diagram into a behavioral schematic is straight forward: simply connect the appropriate behavioral elements such as switches and amplifiers as indicated by the block diagram. In practice, details such as voltage scaling, polarity and timing complicate the process, and the numerous connections in the schematic make the process error-prone. As an example of the translation process, we shall next show how to construct a schematic that implements the block diagram of Fig. 8.7 with the coefficients listed in Fig. 8.10.

### 8.4.1 Voltage Scaling

Assume that the reference voltage is 1 V and that both the full-scale input and the swing available at the outputs of the op-amps are 2 $V_{pp,differential}$. By this assumption, the full-scale input of $\pm 1$ V matches the 1-V reference voltage[†], so the input capacitors may be used for both sampling the input and feeding back the reference, as shown in Fig. 8.12. The DAC function in this stage is implemented by connecting each of the two halves of the sampling capacitor to either the positive reference voltage (v+) or the negative reference voltage (v−), according to the state of the two feedback bits (v0 and v1).

---

† In practice, the modulator will be driven to instability if the input voltage is close to the reference voltage, and thus the allowable input voltage will be a few dB lower than the reference voltage.

Also, since the op-amp swing is ±1 V, the scaling of the modulator states for a swing of ±0.9 in MATLAB provides 10% of design margin, provided we equate the value of a MATLAB signal with the differential voltage of the corresponding signal in the behavioral version of the modulator. Stated more succinctly, 1 (in MATLAB) equals 1 V.

Since the full-scale input of a 3-level modulator in MATLAB is ±2 (recall the toolbox convention that the outermost quantizer levels are at $\pm(nLev-1)$), the $b_1$ and also the $a_1$ coefficients given in Fig. 8.10 must be doubled in order for a ±1 V input signal to occupy full scale. Conveniently, the input coefficient for the modulator of Fig. 8.12 is $2C_1/C_2$, and thus these factors of 2 cancel, leaving $a_1 = b_1 = C_1/C_2$. This fortuitous cancellation will occur with any value of $M$, as long as the voltage scaling factor is unity and the DAC levels are implemented with individual capacitors as shown in Fig. 8.12.

### 8.4.2 Timing

According to Fig. 8.7, the first integrator in this $5^{th}$-order modulator is a delaying integrator. If we adopt the convention that one time period consists of phase 1 (p1) followed by phase 2 (p2), then a delaying integrator can be constructed by sam-

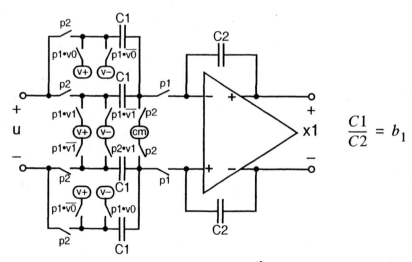

**Figure 8.12:** The first integrator of the $5^{th}$-order modulator.

pling on phase 2 and integrating on phase 1, as shown in Fig. 8.12. As a final step in the justification for the topology depicted in Fig. 8.12, the reader should verify that the polarity of the various connections is correct.

Referring back to Fig. 8.7 once more, we next observe that the first resonator in this modulator consists of a non-delaying integrator followed by a delaying integrator. Following the same reasoning as used in the preceding paragraph leads to the switch phasing shown in Fig. 8.13 for these two integrators. Since the feedback coefficients ($a_2$, $a_3$) match neither the interstage ($c_1$, $c_2$) nor the resonator ($g_1$) coefficients, separate capacitors have been used to implement these terms. The equations governing the capacitor selection follow directly from the difference equations and so are simply listed in Fig. 8.13 without further justification.

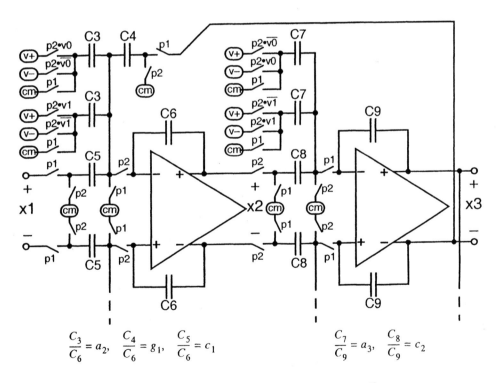

Figure 8.13: **A realization of the first resonator in a $5^{th}$-order modulator.**

However, the phasing of the feedback elements (the capacitors labelled $C_3$ and $C_7$) merits further explanation. Fig. 8.14 shows a timing diagram for the loop filter and quantizer. On phase 1, the values of $x_1(k)$, $x_3(k)$ and $x_5(k)$ are calculated. The input branches for these integrators follow the expected form for an integrator which samples on phase 2 an updates on phase 1. However, on phase 2, $x_2(k+1)$ and $x_4(k+1)$ are calculated, and these calculations make use of $v(k)$, which has itself only just been computed. (See Eq. (8.8) on page 272.) Hence, the feedback elements for $x_2$ and $x_4$ must acquire their data-dependent values during phase 2. Since the resulting switched-capacitor branch is inverting, the polarity of the data used for feedback is the opposite of what is used for the non-inverting feedback branches of the $x_1$, $x_3$ and $x_5$ states. The topology of the second resonator would be essentially the same as that of the first.

In this modulator, the quantized version of $x_5$ must be available shortly after $x_5$ has been computed. In order to allow sufficient time for regeneration in the comparator (and for the delay of the element selection logic, if present), the comparator may be strobed part-way through phase 1, before $x_5$ has settled fully. Strobing the comparator early usually results in only a slight increase in the quantization noise power and is an effective way to accommodate quantizer delay.

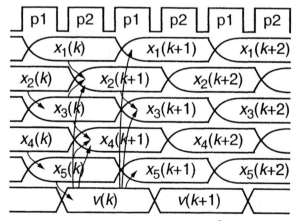

Figure 8.14: Timing diagram for the $5^{th}$-order modulator.

Alternatively, the modulator could be redesigned with an explicit delay assigned to the quantizer. The loop filter would be similar to that of Fig. 8.7, except that the delaying integrators would become non-delaying integrators and vice versa, so that there would still be precisely one unit of delay through the quantizer plus loop filter and back to the quantizer input. The feedback coefficients of this architecture would have to be recalculated, since they would not be the same as those listed in Fig. 8.10, but the quantizer (and element selection logic) would have a half-cycle in which to determine $v$.

Now that the loop filter has been converted into schematic form, it is possible to perform the first verification step: an impulse response check. In this step, the quantizer is replaced with a pulse source (so that the loop is effectively open), and the resulting response is compared against a MATLAB prediction. Fig. 8.15 shows the plot of the simulated waveform as well as the MATLAB-predicted samples. As this figure shows, the correspondence between the two is essentially perfect. Note

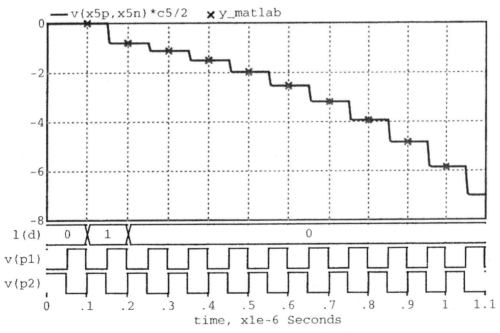

**Figure 8.15: Loop filter verification.**

that since the behavioral model uses a quantizer step (which corresponds to an impulse of height $\Delta=2$) to excite the loop filter, and since the relation between $y$ (the input of the MATLAB quantizer) and $x_5$ is $y = c_5 x_5$, the observed response at $x_5$ must be multiplied by $c_5/2$ in order to match the MATLAB prediction.

If there is a discrepancy between the simulated and predicted responses, the designer can identify the problem by using MATLAB to predict the evolution of each state within the modulator in response to an impulse input. Comparing these predictions against the observed responses usually pinpoints the source of the error.

In a binary modulator, it is often not possible to excite the loop filter with an impulse, since zero is not a valid feedback value. The impulse response can still be simulated by taking the difference between the responses due to a baseline input (such as $-1, -1, \ldots$) and an input with one sample incremented ($+1, -1, -1, \ldots$). In a multi-bit modulator, the DACs can be verified by exciting each level individually.

Once the feedback path through the loop filter has been tested, the quantizer can be verified. In MATLAB, the quantizer quantizes $y = c_5 x_5$, with an LSB size of 2. In the behavioral model, the quantizer quantizes $x_5$ with an LSB size of $2/c_5$. Since we are employing 3-level quantization, the two thresholds are therefore at $\pm(0.5 \text{ LSB}) = \pm 1/c_5 = \pm 0.31 \text{ V}$. Verification of the quantizer can be performed by replacing the loop filter with a ramp source and checking that the quantizer thresholds are in the right place, and that the timing is also correct.

If the loop filter and the quantizer pass the above tests, the loop should behave properly when it is closed. The most common error which causes the loop to malfunction at this stage is incorrect feedback or sensing polarity, a catastrophic condition that causes the loop to become unstable almost immediately, even with zero input. To verify that the loop is functioning properly, and that the input signal has the correct full scale, a longer time-domain simulation may be used. Fig. 8.16 and Fig. 8.17 depict the resulting spectrum when the modulator is supplied with a $-3$ dBFS input and clocked at 10 MHz for 4096 cycles. The noise spectrum follows the NTF closely, the signal is indeed at $-3$ dBFS, and the calculated SQNR is 116 dB. (Since increasing the number of simulated cycles to 8192 increases the

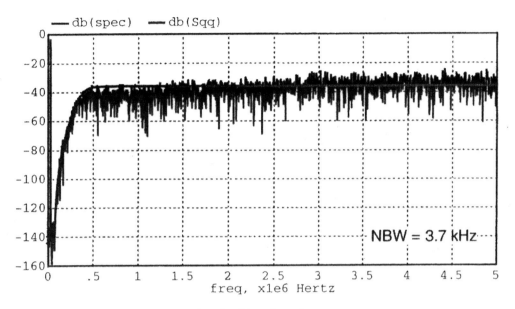

**Figure 8.16: Simulated spectrum.**

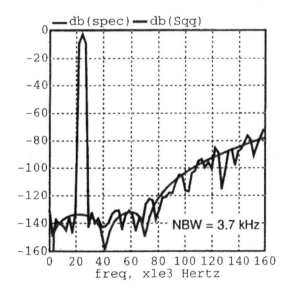

**Figure 8.17: Simulated spectrum in the passband.**

observed SQNR to 120 dB, whereas decreasing the number of simulated cycles to 1024 decreases the observed SQNR to 96 dB, this slight SQNR discrepancy may be assumed to be the result of insufficient simulation time.)

### 8.4.3 kT/C Noise

The final step in the realization of this modulator which we will perform is the sizing of the capacitors in the first stage. The sampled thermal noise ($kT/C$ noise) is a major limitation of switched-capacitor circuits, and must be taken into account in the design. The theory of $kT/C$ noise is briefly discussed in Appendix C. Let us choose capacitances such that $kT/C$ noise from the first integrator is 3 dB lower than what would be required for SNR = 100 dB with a −3 dBFS input. If we assume that the op-amp itself contributes no excess noise, then the noise associated with the input capacitors is

$$\frac{2kT}{2C_1} \times 2 = \frac{2kT}{C_1}, \tag{8.10}$$

where the factor of 2 in the numerator comes from the fact that each capacitor is subjected to 2 charging/discharging events per period, while the factor of 2 in the denominator is a result of the fact that the sampling capacitor is effectively $2C_1$. The final factor of 2 comes from the fact that the circuit is differential.

Since a −3 dBFS signal has a power of 0.25 V$^2$, 103 dB SNR requires the noise voltage to be

$$v_n = \sqrt{\frac{0.25}{10^{103/10}}} = 3.5\ \mu V_{rms}, \tag{8.11}$$

which in turn requires

$$C_1 = \frac{2kT}{v_n^2}/(OSR) = 10\ \text{pF}. \tag{8.12}$$

and thus

$$C_2 = \frac{C_1}{b_1} = 270 \text{ pF}. \tag{8.13}$$

Since these capacitances are rather large, the designer should consider some of the alternative designs before committing to the realization of this design. In particular, Section 9.2 will show that using a feed-forward topology reduces the required capacitance substantially.

Now that the architecture has been translated into behavioral form, the designer can easily investigate practical impairments (op-amp non-linearity, slew rate and finite gain, etc.) and thereby derive specifications for the various sub-blocks.

## 8.5 Conclusions

This chapter introduced the use of the Delta Sigma Toolbox for NTF synthesis, modulator simulation, modulator realization and dynamic-range scaling. The toolbox is best suited to single-quantizer, discrete-time, lowpass and bandpass modulators, but it can also be applied to the design of continuous-time and multi-quantizer modulators. Several of these systems will be studied in the next chapter. The reader is referred to Appendix B for a more complete guide to the Delta Sigma Toolbox.

# CHAPTER 9
# Example Modulator Systems

This chapter illustrates the design principles presented in the preceding chapters through a study of five $\Delta\Sigma$ ADC systems and one DAC system. The first example demonstrates the basics of $\Delta\Sigma$ ADC design and also presents example circuits for several building blocks commonly found in $\Delta\Sigma$ ADCs. Subsequent ADC systems illustrate increasingly complex concepts and techniques, culminating in a hybrid continuous-time/discrete-time bandpass ADC.

## 9.1 SCMOD2: General-Purpose Second-Order Switched-Capacitor ADC

This ADC system is intended to digitize low-frequency signals and so could be used as a voltage monitor or as part of a low-speed on-chip calibration engine. For such applications, a signal bandwidth of 1 kHz is more than adequate. With such a small signal bandwidth, a clock rate of only 1 MHz yields an oversampling ratio of 500 and thus a signal-to-quantization-noise ratio well in excess of 100 dB for a second-order modulator. This high SQNR gives the ADC enough resolution to allow its use in very demanding calibration/measurement applications. To accommodate the applications envisioned, the ADC is required to accept a single-ended input ranging from 0 to $V_{DD} = 3$ V. These specifications are summarized in Table 9.1.

In keeping with the emphasis on simplicity in this design, we will assume that the positive supply rail is used as the reference voltage. In practice, using the supply voltage as the reference requires careful filtering in order to achieve high conversion accuracy in the face of supply noise. A suitable circuit utilizing a two-step coarse/fine charging scheme and an external capacitor to effectively filter the supply voltage with a 2-Hz lowpass filter is described in [1].

### 9.1.1 System Design

Fig. 9.1 shows a block diagram of the system under consideration. The reader may recognize the modulator topology as a standard CIFB (cascade of integrators, feedback) structure. As discussed in Chapter 8, coefficients for such standard structures are readily found using the $\Delta\Sigma$ Toolbox. The code fragment shown in Fig. 9.2 yields the parameters $a_1 = b_1 = 0.2665$, $a_2 = 0.2385$, $c_1 = 0.3418$ and $c_2 = 5.18$. Rounding these to simple ratios then gives $a_1 = b_1 = a_2 = 1/4$ and $c_1 = 1/3$. (The $c_2$ coefficient is unimportant since the quantizer is single-bit.)

| Parameter | Symbol | Value | Units |
|---|---|---|---|
| Bandwidth | $f_B$ | ~1 | kHz |
| Sampling Frequency | $f_s$ | 1 | MHz |
| Signal-to-Noise Ratio | SNR | 100 | dB |
| Supply Voltage | VDD | 3 | V |

Table 9.1. Specifications for the example 2$^{nd}$-order modulator.

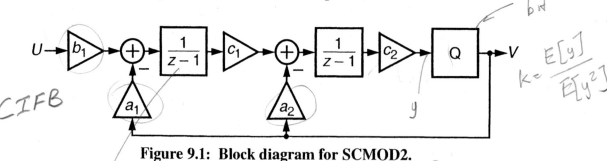

Figure 9.1: Block diagram for SCMOD2.

The impact of this coefficient quantization is minimal. Calculations using the toolbox show that the maximum out-of-band gain of the NTF increases from 2 to 2.25, and simulations indicate a peak SQNR of approximately 115 dB. Furthermore, the effective quantizer gain (including the $c_2$ coefficient) is found from simulations to be $k \approx 16/3$. Using this value of $k$ as an added argument to the $\Delta\Sigma$ toolbox function calculateTF (see page 403 in Appendix B) yields the simple NTF

$$NTF(z) = \left(\frac{z-1}{z-1/3}\right)^2. \quad (9.1)$$

The $u_{max}$ variable, which indicates the stable input range (normalized to the 3-V reference voltage), is 0.9, so we should expect the modulator to function perfectly for inputs in the range 0.15 V to 2.85 V. Since the modulator is only second order, inputs which are outside this range will most likely cause saturation in the integrators (rather than drive the modulator into a self-sustaining oscillation) but the modulator output will nonetheless be an unsatisfactory representation of the input. For the sake of simplicity, we will consider the naturally-available input range to be acceptable.[†] With a scaled block diagram now in hand, we can turn to the practical problem of implementing it.

```
H = synthesizeNTF(2,500,0,2);
form = 'CIFB';
[a,g,b,c] = realizeNTF(H,form);
b(2:end) = 0;
ABCD = stuffABCD(a,g,b,c,form);
[ABCDs umax] = scaleABCD(ABCD);
[a,g,b,c] = mapABCD(ABCDs,form);
```

Figure 9.2: Toolbox code fragment for determining coefficients.

---

†. A simple transformation $u' = 0.9u + 0.15$ would allow input voltages $u$ in [0,3V] while ensuring that the modulator receives an input $u'$ which is within the stable input range. Such a transformation could be implemented by changing the input-sampling/reference-feedback network.

### 9.1.2 Timing

Perhaps the most common source of error in switched-capacitor modulator design is improper timing. It is imperative that the timing of the quantization and feedback operations be such that the loop filter follows the desired difference equations, otherwise the modulator will not function as desired, if it functions at all. Fig. 9.3 shows a simplified switched-capacitor schematic of our modulator, as well as the desired difference equations, the switch/clock timing and representative waveforms. The difference equations and the topology of the schematic follow directly from Fig. 9.1. To verify that the timing shown is correct, we need to verify that the loop filter implements the desired difference equations. Starting at the end of phase 2 (marked in the timing diagram with a dashed line), observe that $x_2(n)$ has just settled and thus strobing the comparator as phase 2 falls will implement the quantization operation implied by the third equation. The next rising edge of phase 2 causes $x_2(n+1)$ to be generated using $x_1(n)$ and $v(n)$ as dictated by the second difference equation. The subsequent phase 1 interval is used to generate $x_1(n+1)$ in accordance with the first difference equation. Since $v(n)$ is needed for both of these operations, a flip-flop clocked on falling phase 1 holds $v(n)$ over a phase-2/phase-1 interval and thus the timing shown is consistent with the desired difference equations. Many other timings and switch arrangements exist which

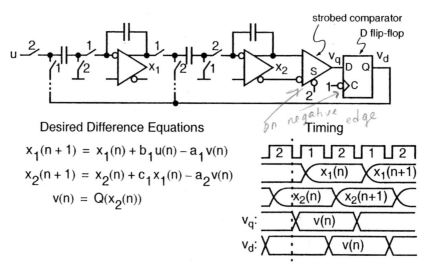

Figure 9.3: Timing check for SCMOD2.

## SCMOD2: General-Purpose Second-Order Switched-Capacitor ADC

also implement the desired difference equations. An advantage of the timing shown in Fig. 9.3 is that the comparator and both integrators have an entire clock phase, or nearly half of a clock period, in which to settle.

Fig. 9.4 shows the first integrator stage in more detail. A differential circuit with a single-ended input is assumed. Since $a_1 = b_1$, the same physical capacitor $C_1$ can be used to implement both coefficients. As shown in the schematic, the input signal is sampled onto the upper $C_1$ input capacitor on phase 2, and then the difference between the input and the reference is integrated onto the upper $C_2$ integrating capacitor on the next phase 1. The lower path feeds the inverted signal to the integrating capacitor also on phase 1, but samples the input on phase 1 and the reference on phase 2 in order to accomplish the inversion. The fact that the input is sampled twice per period adds a $(1 + z^{-1/2})/2$ factor to the STF, which creates a zero at $f_s$ but has little impact on the in-band portion of the STF. In this integrator, the reference is sampled on two clock phases, so we have to ensure that the feedback signal ($v$) has the same value in both phases. Fortunately, the flip-flop which holds $v$ constant from phase 2 to phase 1 realizes the required timing.

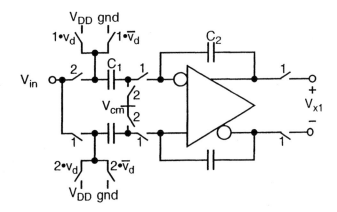

**Figure 9.4: First integrator for SCMOD2.**
(This circuit has signal-dependent loading on the reference.)

287

### 9.1.3 Scaling

Another common source of error in the translation of a block diagram into a practical circuit is the process of denormalization and scaling. The $\Delta\Sigma$ toolbox assumes that the input of a binary modulator ranges from $-1$ to $+1$; the default scaling is also such that the integrator states occupy the $[-1, 1]$ range. These ranges are given in normalized (unit-less) form, but in an analog circuit the ranges must have physical units and also need to match the capabilities of the circuits.

In this example, the full-scale input is 3 V, whereas the toolbox default value is 2. Let us assume that the amplifier supports a differential swing with the same numerical range as used in the Toolbox, i.e. the amplifier output swing is 2 $V_{pp}$. Lastly, assume that the digital $v_d$ signal is interpreted as 0 or 1. Thus the relationships between the circuit variables shown in Fig. 9.4 and the state variables used in the difference equations are

$$x_1 = \frac{v_{x1}}{1\text{ V}}, \tag{9.2}$$

$$u = \frac{v_{in} - 1.5\text{ V}}{1.5\text{ V}} \tag{9.3}$$

and

$$v = 2v_d - 1. \tag{9.4}$$

The difference equation that we need to implement is ← toolbox equation

$$x_1(n+1) = x_1(n) + b_1 u(n) - a_1 v(n), \tag{9.5}$$

where $a_1 = b_1 = 1/4$. In terms of the circuit variables, the desired difference equation becomes

$$v_{x1}(n+1) = v_{x1}(n) + \frac{b_1[v_{in}(n) - 1.5\text{ V}]}{1.5} - a_1[2v_d(n) - 1] \cdot 1\text{ V}$$

$$= v_{x1}(n) + \frac{v_{in}(n)}{6} - 0.5\text{ V} \cdot v_d(n) \tag{9.6}$$

If $v_{in}$ is assumed to be unchanged from phase 2 to phase 1, the equation implemented by the circuit in Fig. 9.4 is

$$v_{x1}(n+1) = v_{x1}(n) + \frac{2C_1}{C_2}v_{in}(n) - \frac{2C_1}{C_2}V_{DD}v_d(n) \tag{9.7}$$

Thus, in order for (9.6) and (9.7) to be equivalent, the ratio of the input and integrating capacitances needs to be

$$C_1/C_2 = 1/12. \tag{9.8}$$

With $V_{DD} = 3\text{ V}$, the above ratio also gives the correct coefficient for $v_d$.

Fig. 9.5 shows a schematic for the second integrator. In the first integrator, the input and reference feedback capacitors were shared since the associated coefficients ($b_1$ and $a_1$) were equal. However, the two input coefficients ($c_1$ and $a_2$) associated with the second integrator are not equal, and so are implemented with separate branches.[†] Following the same methodology as was used for the first integrator results in the capacitor ratios indicated in the figure.

### 9.1.4 Verification

At this point, we have a circuit which we expect will implement the desired difference equations. To verify that the topology, timing and capacitor ratios are indeed correct, several simulations need to be performed. The first simulations use an

---

[†]. If the $a_2$ and $c_1$ coefficients are different, it is still possible to share a portion of the associated capacitors. We use independent branches for the sake of schematic clarity.

open-loop configuration to demonstrate that the loop filter has indeed been implemented correctly. The second set of simulations operates the modulator in closed-loop form, and serves to confirm the input range, state scaling and NTF.

In the first set of simulations, the loop filter is initialized to a known state and the quantizer is replaced with a simple data source. By running two simulations, for example one in which the data source produces the sequence $(-1, +1, -1, +1, \ldots)$ and one in which the data source produces $(+1, +1, -1, +1, \ldots)$, the open-loop impulse response $l_1(n)$ may be determined. Since the difference between the two input sequences is $2\delta(n)$, where $\delta(n)$ is a discrete-time impulse, $l_1(n)$ is the difference in the loop filter's output between the two simulations, divided by 2.

Using the difference equations, we find that the impulse response of the loop filter is

$$l_1(n) = \left\{ 0, -\frac{3}{12}, -\frac{4}{12}, -\frac{5}{12}, \ldots \right\} V. \tag{9.9}$$

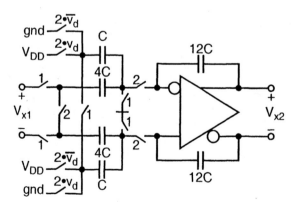

Figure 9.5: Second integrator in SCMOD2.
(In this circuit, the load on the reference is signal-independent.)

(The impulse response may also be calculated directly from the NTF given in (9.1) using the Toolbox function `impL1`, provided the result is divided by the assumed quantizer gain of 16/3.)

Fig. 9.6 shows the simulated output of the loop filter when ideal behavioral models are used for the analog components, as well as the desired impulse response given by (9.9). Since the observed response matches the desired response precisely, we can be confident that the circuit really does implement the desired difference equations. (More precisely, we can be confident that the circuit implements all terms in the desired difference equations, except possibly for the feed-in ($u$) terms.)

The second set of behavioral simulations verify that the modulator operates properly when the loop is closed. In order to get an accurate SQNR measurement, such simulations need to be run for a long time (on the order of $OSR \times 64$ clock cycles) and so tend to be prohibitive. Shorter simulations can be used to verify that the modulator follows the desired difference equations and that it starts up properly, but long simulations should be run at some point in order to verify the modulator's SQNR via simulation. In the early days of $\Delta\Sigma$, such simulations were impractical (especially since $OSR$ tended to be high), but nowadays such simulations are via-

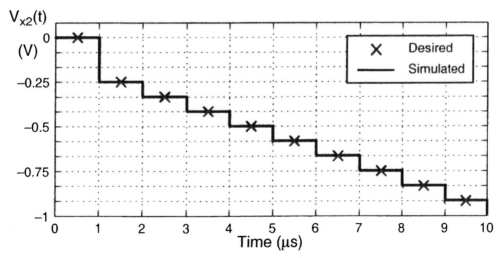

Figure 9.6: Impulse response of SCMOD2's loop filter.

ble. Fig. 9.7 shows the spectrum of the simulated output as well as the expected PSD (after taking into account the observed quantizer gain of $k = 16.6/3$), for a $2^{14}$-point simulation of the modulator with a $-3$ dBFS sine-wave input. The observed SQNR is 102 dB, which is dominated by the presence of a $3^{rd}$-harmonic of the input tone. If this harmonic is discounted, the SQNR is 114 dB and we can conclude that the modulator is indeed performing as desired.

## 9.1.5 Capacitor Sizing

At this point, we have a scaled block diagram which has been verified with ideal circuit elements. The next step towards a transistor-level implementation is capacitor sizing. Since we desire a fairly high SNR, the size of the capacitors in the first stage is likely to be dictated by thermal noise considerations.

If we assume that the amplifier noise is negligible, then the input-referred thermal noise of the circuit in Fig. 9.4 is

$$v_{n1}^2 = \frac{kT}{C_1}. \qquad (9.10)$$

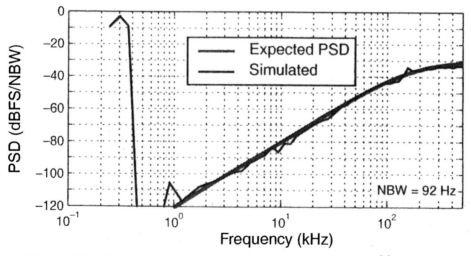

**Figure 9.7: Spectrum from a behavioral simulation; $2^{14}$-point FFT.**

## SCMOD2: General-Purpose Second-Order Switched-Capacitor ADC

(As discussed on page 430 of Appendix C, the switched capacitors in this circuit transfer a mean-square noise charge of $4kTC_1$ to the integrating capacitor in each clock period. However since this circuit double-samples its input, the charge transferred to the integrating capacitor due to an input voltage $v$ is $2C_1v$. Thus the input-referred noise is $(4kTC_1)/(2C_1v)^2 = kT/C_1$.)

This noise is evenly spread over the frequency range $[0, f_s/2]$ (i.e. the noise is white), so its in-band power is

$$v'^2_{n1} = \frac{v^2_{n1}}{OSR}. \tag{9.11}$$

In order to achieve $SNR = 100$ dB with a full-scale sine wave input, we require

$$v^2_n < 10^{-10} \cdot (1.5 \text{ V})^2 \cdot \frac{1}{2} \approx (10 \text{ μV})^2. \tag{9.12}$$

If the entire noise budget is allocated to the first integrator and if we take $T = 300\text{K}$, then we require

$$C_1 = \frac{kT}{OSR \cdot v^2_n} = 74 \text{ fF} \tag{9.13}$$

and thus $C_2 = 12C_1 = 0.88$ pF. Rounding $C_2$ to 1 pF, so that $C_1 = 83$ fF, gives a modest amount of margin to the design.

The noise at the input to the second integrator is (see Appendix C)

$$v^2_{n2} = \frac{kT}{4C} \cdot \left(1 + \frac{1}{4}\right) \cdot 4 = \frac{5kT}{4C}. \tag{9.14}$$

Since this noise is shaped by the inverse of the $\frac{2C_1/C_2}{1-z^{-1}}$ transfer function of the first integrator, the in-band power of this noise is

$$v'^2_{n2} = \frac{\pi^2}{3(OSR)^3}\left(\frac{C_2}{2C_1}\right)^2 v^2_{n2} \approx 10^{-6} v^2_{n2}. \tag{9.15}$$

The 60-dB suppression afforded by the first integrator makes the thermal noise of the second integrator wholly negligible in this application. For example, if we choose $C = 20$ fF (assuming that this is the lowest value which gives reasonable accuracy in the technology used), then the power of the in-band noise from the second integrator is below the $-125$ dBFS level, i.e. more than 20 dB lower than the first integrator's noise.

### 9.1.6 Circuit Design

Now that we have selected all absolute capacitor sizes, we can estimate some specifications for the op amp. Let's first consider the bias current. Assume that the folded-cascode op-amp topology depicted in Fig. 9.8 will be used. With a bias current of $2I$ for the differential pair, the slew current available at each output terminal is $I$. To estimate the slew requirements, note that the largest quantity of charge which may need to be transferred from the input capacitors to the integrating capacitors is $C_1 V_{DD}$. If we allocate 25% of a half-clock-period for slewing, then the required slew current is

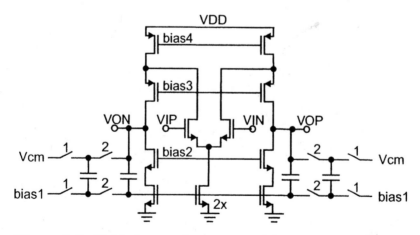

Figure 9.8: Folded-cascode op amp with switched-capacitor common-mode feedback.

$$I = 8f_{CK}C_1V_{DD} = 2 \ \mu A. \tag{9.16}$$

According to this calculation, the bias current of the op amp can be very low– a mere 8 µA!

Let us now consider the bandwidth requirement for the op amp. Suppose we require 10 time-constants of linear settling in the remaining 75% of the half-clock period. (Having 10 time-constants of linear settling makes the settling error nearly 87 dB lower than the error $e_0$ at the onset of linear settling. Since $e_0$ is only a fraction of the step, 100 dB SNR should be within reach.)

As shown in Appendix C, the settling time constant $\tau$ is given by $C_{load}/(\beta g_m)$, where $C_{load}$ is the total capacitance loading the output of the op amp, while $g_m$ is the transconductance of the input devices, and $\beta$ is the feedback factor. For the first integrator, $C_{load}/\beta \approx C_1$, and hence we require

$$g_m = 10 \cdot \frac{2}{0.75} \cdot f_{CK} \cdot C_1 = 2.3 \ \mu A/V. \tag{9.17}$$

This transconductance is readily achieved with a 2-µA drain current.

Lastly, let us consider the dc gain of the first op amp. Linear analysis indicates that the attenuation offered by the NTF begins to degrade if its zero moves inside the unit circle by about $\pi/OSR$. Since finite op-amp gain shifts the pole of the first integrator by $C_1/(AC_2)$, one might be tempted to conclude that $A = (OSR \cdot C_1)/(\pi C_2) \approx 13$ would be adequate. However, this linear analysis neglects the nonlinear errors which result from slewing and nonlinear DC gain. (Note that as long as the amplifier's gain is linear, finite DC gain does not cause distortion. Similarly, as long as the amplifier does not slew, settling errors caused by finite op-amp bandwidth do not cause distortion. However, these ideal conditions should not be assumed.)

Although it is not possible to obtain an accurate estimate of the required dc gain without knowledge of the nonlinear nature of the amplifier's transfer characteristic, it is possible to put an upper bound on the required gain. To do this, observe that the settled voltage at the input terminals of the op amp is $V_{out}/A$, and that (due to the double sampling used, as shown in Fig. 9.4) this voltage results in an error charge that is equivalent to an input voltage error of $V_{out}/(2A)$. This voltage consists of a signal component (the effect of which is benign), broadband noise (the impact of which would be reduced by an *OSR* factor), and distortion components. If we assume the error consists of distortion terms only (an assumption which is often pessimistic by 30 dB or more), then the requirement that distortion terms stay below –100 dBFS gives

$$\frac{V_{out}}{(2A)} < 10^{-5} V_{in}. \tag{9.18}$$

Using $V_{in} = 1.5$ V and $V_{out} = 1.0$ V leads to a worst-case gain requirement of $A > 90$ dB. In practice a gain of 60 dB is likely to be adequate, but this would need to be verified using simulations involving the actual nonlinear gain characteristic of the amplifier.

The above discussion dealt only with the op-amp requirements in the first integrator. We have already noted that noise at the input of the second integrator is greatly attenuated, owing to the high in-band gain of the first integrator. Other non-idealities in the second integrator also benefit from this attenuation mechanism, with the result that the distortion and settling requirements of the second op amp are greatly relaxed. In this example, the second op amp could simply be a copy of the first op amp, scaled down in size and current by, say, a factor of 4. More aggressive scaling is theoretically possible, but the additional power and area savings would be minimal.

To complete this modulator example, Fig. 9.9 shows the schematic of a common comparator, while Fig. 9.10 depicts a widely used clock generator. By using these blocks in conjunction with those shown earlier, a fully transistorized implementation of MOD2 can be constructed. References [2-3] describe systems similar to this example, and the reader may wish to consult these references for further design information.

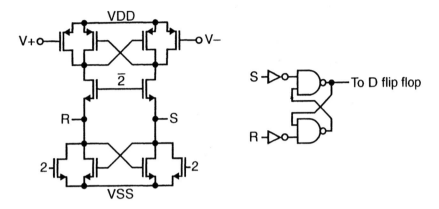

**Figure 9.9:** Latched comparator.

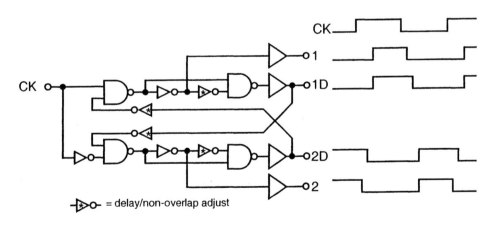

**Figure 9.10:** Non-overlapping clock generator.

## 9.2 SCMOD5: A Fifth-Order Single-Bit Noise-Shaping Loop

The specifications for our second ADC example are listed in Table 9.2. Since we are aiming to increase the bandwidth by a factor of 50 and the SNR by 10 dB relative to the previous example, while only increasing the clock rate by a factor of 8, we must increase the modulator order and/or the number of quantization levels. For this ADC example, we choose to implement a high-order single-bit modulator.

### 9.2.1 NTF and Architecture Selection

Fig. 4.14 on page 112 showed the achievable SQNR as a function of OSR for single-bit systems of various orders. According to Fig. 4.14, a $4^{th}$-order system can achieve $SQNR \approx 120$ dB with $OSR = 80$. Typically, we want the theoretical SQNR to be 10-20 dB more than the desired SNR in order to allow for the inevitable SQNR degradation due to circuit non-idealities and to leave as much of the noise budget for thermal noise as possible. Furthermore, since the SQNR limits shown in Fig. 4.14 correspond to modulators which are on the verge of instability, this figure should be viewed as defining the boundary of what is possible, rather than what is practical. For these reasons, we opt to use a $5^{th}$-order NTF.

Fig. 9.11 shows a code fragment for simulating the SQNR of a $5^{th}$-order system and Fig. 9.12 contains the resulting plot. As Fig. 9.12 shows, the peak SQNR of this system is 128 dB with a maximum stable input of –4 dBFS. In order to leave some margin for instability, we will scale the input such that a 2-$V_p$ differential

| Parameter | Symbol | Value | Units |
|---|---|---|---|
| Bandwidth | $f_B$ | 50 | kHz |
| Sampling Frequency | $f_s$ | 8 | MHz |
| Signal-to-Noise Ratio | SNR | 110 | dB |
| Supply Voltage | VDD | 3 | V |
| Input Voltage Range |  | [ –2, 2 ] | V |

**Table 9.2. Specifications for the example $5^{th}$-order modulator.**

signal corresponds to –6 dBFS, i.e. we will set the differential full-scale range to –4 V to +4 V[†]. With a theoretical SQNR of 128 dB, we could opt to decrease the peak out-of-band gain of the NTF ($\|NTF\|_\infty$) in a bid to gain increased input range, but for the sake of moving forward we will stick with the default value of $\|NTF\|_\infty$ and be content to use only half of the full-scale range.

```
OSR = 8e6/(2*50e3);% OSR = 80
NTF = synthesizeNTF(5,OSR,1);
amp = [-140:5:-15 -12 -10:0];
snr = simulateSNR(NTF,OSR,amp);
plot(amp,snr,'b-d');
```

**Figure 9.11:** Code for simulating the SQNR curve.

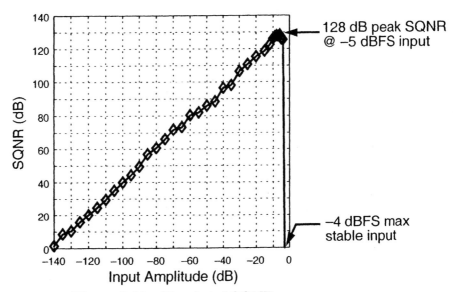

**Figure 9.12:** Simulated SQNR curve.

---

[†]. The reader should not be concerned that the ADC's full-scale is larger than can be supported by a 3-V supply voltage. As indicated in the text, the ADC is only intended to operate on signals that are smaller than half of full-scale, i.e. –2 V to +2 V.

# 9 Example Modulator Systems

The code for the next two steps in the design process, namely realization (with a cascade-of-resonators feedback, or CRFB, topology) and dynamic range scaling, is listed in Fig. 9.13 along with the resulting coefficients. As was observed in Section 8.4.3, we will find that the small value of $b_1$ leads to an unreasonably large integrating capacitance.

Consider the topology shown in Fig. 9.14 for use in the first integrator. This topology has the attractive feature that the amplifier does not experience a common-mode step if the input common-mode voltage and the reference common-mode

```
form = 'CRFB';
[a,g,b,c] = realizeNTF(NTF,form);
b(2:end) = 0;
ABCD = stuffABCD(a,g,b,c,form);
[ABCDs umax] = scaleABCD(ABCD);
[a,g,b,c] = mapABCD(ABCDs,form);
→ a = [0.09675   0.1628   0.2015   0.3386   0.2577]
  g = [0.00228 0.003687]
  b = [0.09675        0        0        0        0        0]
  c = [0.1261    0.1961   0.3715   0.3434   2.157]
```

**Figure 9.13:** Realization and dynamic range scaling– CRFB topology.

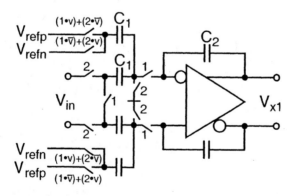

**Figure 9.14:** First integrator in SCMOD5.
(In this circuit, the load on the reference is signal-independent.)

voltage differ, and so can be used with both balanced and unbalanced inputs. However, this circuit's insensitivity to the input common-mode voltage comes at the cost of increased noise relative to a structure in which the input capacitors are also used to feed back the reference. According to the analysis of Appendix C, the rms input-referred noise voltage of this circuit is $\sqrt{8kT/C_1}$, assuming the amplifier noise is negligible.

Since the outer capacitors are switched between $V_{refp}$ and $V_{refn}$, or vice versa, a differential reference $V_{refp} - V_{refn} = 2$ V implements a full-scale range of $-4$ V to $+4$ V. However, since the input sampling capacitors are connected to either the input or to each other, a 2-V input corresponds to a half-scale signal, as called for in Table 9.2. The signal-to-thermal-noise ratio for this system with a half-scale input is therefore

$$SNR = \left(\frac{1}{OSR}\right)\left(\frac{(2\text{ V})^2/2}{8kT/C_1}\right), \qquad (9.19)$$

which leads to $C_1 = 21$ pF for $SNR = 110$ dB at $OSR = 80$. A 21-pF capacitor is rather large (on the order of 150 μm × 150 μm), but it pales in comparison to the size of the integrating capacitor $C_2$. With aggressive design, the output swing of the first integrator given a 3-V supply may be as large as 4 $V_{pp}$. However, even with this wide swing, the ratio $C_1/C_2$ must be $b_1/2$, or about $1/20$, when the voltage scaling of the modulator coefficients is taken into account. Thus, the integrating capacitance needs to be 420 pF, and since there are two such capacitors the area occupied by the first integrator is prohibitive. (A similar conclusion was reached in Section 8.4.3, where we found that a CRFB topology targeting $SNR = 103$ dB at $OSR = 64$ required a total integrating capacitance of more than 500 pF.)

A number of alternatives exist which make the desired specifications more readily achievable. For example, we could increase the sampling rate in order to take advantage of the $OSR$ factor in (9.19) and thereby reduce the size of all capacitors, while leaving the topology unchanged. Increasing $OSR$ also allows the specifications to be achieved with a lower-order modulator, which also helps save capacitor area since the $b_1$ coefficient tends to increase as the order is decreased. A factor of

10 improvement is possible, provided the designer has the freedom to set the sampling frequency arbitrarily.

A second alternative, which does not require increasing the sampling frequency, is to employ a feedforward topology. Fig. 9.15 lists the code for realizing the desired NTF with a CRFF topology and gives the resulting coefficients. Since $b_1$ is increased by a factor of about 3.5, the integrating capacitors are reduced by this factor. Although the resulting 120-pF integrating capacitors are still large, they are within the realm of practicality.

As discussed in Section 4.4.2, feeding the input signal forward to the input of the quantizer with a gain of $b_6 = 1$ ideally results in a unity STF. In order for this feedforward technique to work properly with a binary modulator, the coefficient of the feedforward term should be $b_6 = 1/k$, where $k$ is the effective quantizer gain. Using simulations to evaluate the effective quantizer gain results in the requirement $b_6 = 0.53$.

### 9.2.2 Implementation

Fig. 9.16 shows the topology for which we now have coefficients. The remaining steps for the implementation of SCMOD5 are similar to those followed in the implementation of SCMOD2. In brief, we need to perform voltage scaling, construct a behavioral schematic, verify the behavioral model, and then substitute transistor-level implementations for each behavioral block until the schematic has been fully realized.

```
form = 'CRFF';
[a,g,b,c] = realizeNTF(NTF,form);
ABCD = stuffABCD(a,g,b,c,form);
[ABCDs umax] = scaleABCD(ABCD);
[a,g,b,c] = mapABCD(ABCDs,form);
→ a = [1.561     1.75      1.355    0.9588   0.7931]
  g = [0.001586 0.01526]
  b = [0.3561    0         0        0        0       1]
  c = [0.3561    0.4025    0.2819   0.2215   0.08296]
```

**Figure 9.15:** Realization and dynamic range scaling– CRFF topology.

In preparation for these activities, it is wise to create a simplified behavioral schematic, such as that shown in Fig. 9.17, in order to determine the switch and quantizer timing which will be used to implement the desired difference equations. Fig. 9.17 lists the difference equations for SCMOD5 and illustrates a possible timing for the various analog computations. As shown in the figure, we have chosen to update even-numbered states on phase 2 and odd-numbered on the subsequent phase 1. This timing is convenient because the updated even-numbered states, namely $x_2(n+1)$ and $x_4(n+1)$, have terms which depend on the previous odd-numbered states, while two of the odd-numbered states ($x_3$ and $x_5$) have update equations which contain terms involving $x_2(n+1)$ and $x_4(n+1)$. Thus, with the timing shown, the inter-state coefficients will correspond directly to capacitor ratios.

Unfortunately, the situation is a little more complicated for the summation and quantization operations. To see this, note that although the update for $x_1(n+1)$ occurs on phase 1, the structure of the feedback branch for the first integrator

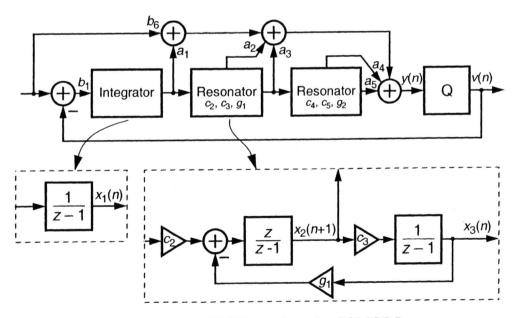

**Figure 9.16: CRFF topology for SCMOD5.**

shown in Fig. 9.14 requires the quantizer output $v(n)$ to be available during the preceding phase 2 and held over the subsequent phase 1.[†] However, according to the block diagram shown in Fig. 9.16, $y(n)$ (which is quantized to yield $v(n)$) is dependent on $x_2(n+1)$ and $x_4(n+1)$, which only become available at the *end* of phase 2. One solution to this interdependency problem is to expand the $x_2(n+1)$ and $x_4(n+1)$ terms as shown in Fig. 9.17. With this technique, $y(n)$ can be computed using the terms $x_1(n), x_2(n), ..., x_5(n)$, all of which are available in phase 1. Quantizing $y(n)$ at the end of phase 1, as indicated in Fig. 9.17, then yields $v(n)$ with the desired timing.

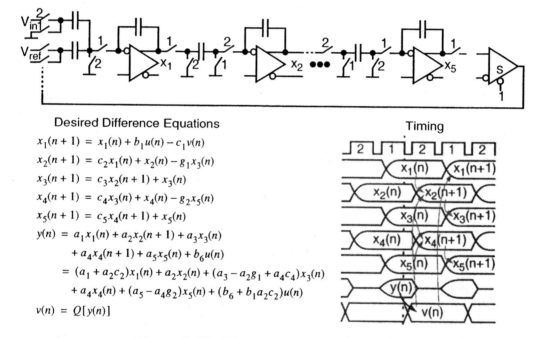

Desired Difference Equations

$x_1(n+1) = x_1(n) + b_1 u(n) - c_1 v(n)$
$x_2(n+1) = c_2 x_1(n) + x_2(n) - g_1 x_3(n)$
$x_3(n+1) = c_3 x_2(n+1) + x_3(n)$
$x_4(n+1) = c_4 x_3(n) + x_4(n) - g_2 x_5(n)$
$x_5(n+1) = c_5 x_4(n+1) + x_5(n)$
$y(n) = a_1 x_1(n) + a_2 x_2(n+1) + a_3 x_3(n)$
$\quad\quad + a_4 x_4(n+1) + a_5 x_5(n) + b_6 u(n)$
$\quad = (a_1 + a_2 c_2) x_1(n) + a_2 x_2(n) + (a_3 - a_2 g_1 + a_4 c_4) x_3(n)$
$\quad\quad + a_4 x_4(n) + (a_5 - a_4 g_2) x_5(n) + (b_6 + b_1 a_2 c_2) u(n)$
$v(n) = Q[y(n)]$

**Figure 9.17:** Timing check for SCMOD5.

---

† Alternatively, a switched-C branch involving cross-connections to the amplifier's inputs could have been used to eliminate this requirement. We use the integrator arrangement shown in Fig. 9.14 in order to demonstrate another way to resolve this timing conflict.

Table 9.3 lists the coefficients which result from scaling the original block diagram for an 8-$V_{pp}$ full-scale input range and a 4-$V_{pp}$ integrator swing. (An 8-$V_{pp}$ full-scale ensures that the half-scale input range is the desired [−2 V,+2 V] differential input range.) For the $a_1$, $a_3$, $a_5$ and $b_6$ coefficients, Table 9.3 lists the modified values resulting from the expansion given in Fig. 9.17.

Table 9.3 also lists rational approximations to these coefficients. Approximating the coefficients with rationals allows the use of unit capacitors in the implementation, which helps to ensure that the capacitor ratios will be insensitive to fringing effects as well as processing-induced length and width variations. Note that we have set the small $g_1$ coefficient to zero, and that although the $g_2$ term has been retained it has been altered by about 10%. The other coefficients have been approx-

| Coefficient | Value | Approximation | Capacitor Ratio |
|---|---|---|---|
| $a_1 \rightarrow (a_1 + a_2c_2)$ | 2.266 | 9/4 = 2.250 | $C_{13}/C_T$ |
| $a_2$ | 1.750 | 7/4 = 1.750 | $C_{14}/C_T$ |
| $a_3 \rightarrow (a_3 - a_2g_1 + a_4c_4)$ | 1.564 | 3/2 = 1.500 | $C_{15}/C_T$ |
| $a_4$ | 0.9588 | 1 | $C_{16}/C_T$ |
| $a_5 \rightarrow (a_5 - a_4g_2)$ | 0.7785 | 3/4 = 0.7500 | $C_{17}/C_T$ |
| $g_1$ | 0.0016 | 0 | $C_4/C_5$ |
| $g_2$ | 0.0153 | 1/60 = 0.0167 | $C_9/C_{10}$ |
| $b_1 = c_1$ | 0.3561 | 5/14 = 0.3571 | $C_1/C_2$ |
| $b_6 \rightarrow (b_6 + b_1a_2c_2)$ | 0.7840 | 3/4 = 0.75 | $C_{18}/C_T$ |
| $c_2$ | 0.4025 | 2/5 = 0.4000 | $C_3/C_5$ |
| $c_3$ | 0.2819 | 2/7 = 0.2857 | $C_6/C_7$ |
| $c_4$ | 0.2215 | 2/9 = 0.2222 | $C_8/C_{10}$ |
| $c_5$ | 0.08296 | 1/12 = 0.0833 | $C_{11}/C_{12}$ |

**Table 9.3.** Coefficients and capacitor ratios for SCMOD5. $C_T = \sum_{i=13}^{18} C_i$

imated with greater precision (typically with an error of a few percent), even though such high precision is typically not required in a single-loop $\Delta\Sigma$ modulator.

If the NTF zeros remain intact, slight changes to the coefficients usually have only a small effect on the modulator. The reason is that such changes only shift the NTF's poles, and the theoretical SQNR is a weak function of the NTF's poles. However, in this example we chose to set $g_1$ to zero, in order to avoid having to realize a large (600:1) capacitor ratio. This choice moves two NTF zeros down to dc, and so does degrade the theoretical SQNR.

To quantify the degradation, Fig. 9.18 compares the NTF associated with the original coefficients to the NTF associated with the quantized coefficients. Since the modulator is single-bit, both NTFs take into account the effective quantizer gain as determined by simulations. According to this plot, the new NTF has about 7 dB less quantization noise attenuation than the original NTF. Since the peak SQNR with the original NTF was about 128 dB, we would expect the new NTF to yield $SQNR_{peak} \approx 121$ dB. Simulations confirm this value. Since the desired SNR is 110 dB, quantization noise will still be a small portion (less than 10%) of the noise budget, and hence our choice to dispense with $g_1$ is a reasonable one.

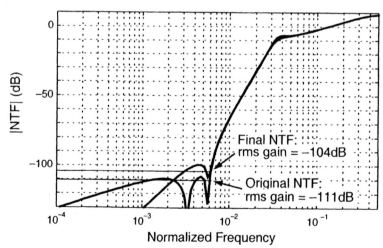

**Figure 9.18: Original and final NTFs for SCMOD5.**

If the SQNR loss associated with setting the $g_1$ coefficient to zero is considered excessive, a capacitive T network, such as that shown in Fig. 9.19 can be used to realize the required small coupling term. The charge provided to the op-amp input by this network equals that of a direct capacitive connection with a capacitance $C/(k+2)$. Thus, if $C$ is a unit capacitor and the integrating capacitance $C_5$ is made with 40 such units, then using $k = 13$ will yield $g_1 = 1/600$. Since parasitic capacitance on the T node will affect the accuracy of the implemented coefficient, the capacitors' top plates should be connected to node T and the switch grounding T should be made small.†

As a final confirmation that the quantized coefficients are acceptable, Fig. 9.20 plots the simulated integrator maxima as a function of the dc input level. Since the swing at the output of each integrator is below the 2-$V_p$ limit, the amplifiers within the modulator should not saturate for inputs in the $[-2V, +2V]$ input range.

Now that the modulator structure, timing and filter coefficients have been determined, a behavioral schematic needs to be constructed and verified. Fig. 9.21 shows a detailed schematic which follows from the simplified schematic of Fig. 9.17. For the sake of generality, this schematic shows the feedback capacitor $C_4$ associated with the first resonator even though we have already decided to omit this element. The summation operation which creates the output of the loop filter is performed by the passive capacitor network shown at the bottom of the figure. The phase 1 switches connected to the capacitor labelled CHOLD, and CHOLD itself, are

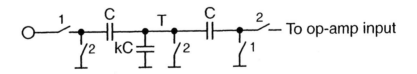

**Figure 9.19: A capacitive T.**

---

† A further caution is in order if the circuit of Fig. 9.19 is used in an ordinary switched-capacitor biquad. In this application, op-amp dc offset is more problematic with a T network than with an equivalent standard SC branch.

included for simulation convenience. In an actual implementation, these elements can be omitted since their only function is to improve the readability of the YP and YN waveforms by making them non-return-to-zero.

As with SCMOD2, the first step in the verification of this schematic is an impulse-response test. Fig. 9.22 compares the simulated impulse response with the MATLAB-calculated impulse response. Since these agree precisely, we can be fairly confident that the loop filter has been implemented correctly and can therefore proceed to closed-loop simulations. Fig. 9.23 shows the (smoothed) spectrum which results from an 8192-cycle simulation, and superimposes the desired NTF, scaled as prescribed in Appendix A to follow the expected noise density. The SQNR observed in this simulation is within a few dB of the MATLAB-simulated SQNR; doubling the number of simulated cycles closes the gap to less than 1 dB. Furthermore, since a check of the voltage excursions at the integrator outputs confirms that the swing at these nodes is consistent with the swings predicted in Fig. 9.20, the designer can be confident that the behavioral schematic is correct. Capacitor selection, switch sizing, and amplifier design remain to be performed, but will not be considered here. References [4-7] describe these and other considerations in the context of similar modulator systems.

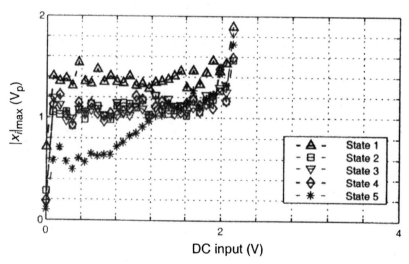

**Figure 9.20:** State maxima vs. input level for SCMOD5.

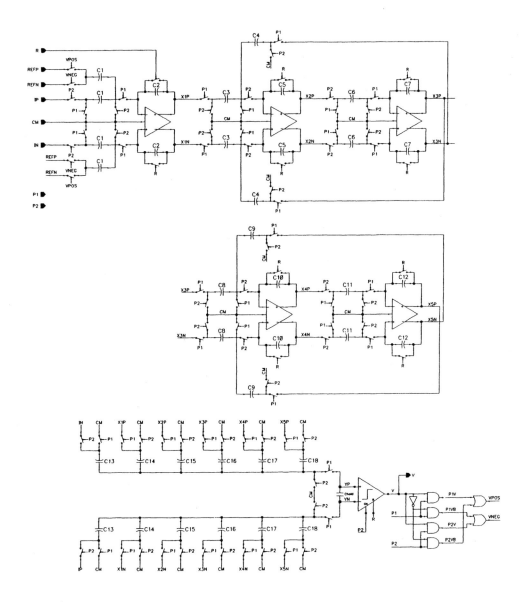

**Figure 9.21: SCMOD5 behavioral schematic.**

## 9 Example Modulator Systems

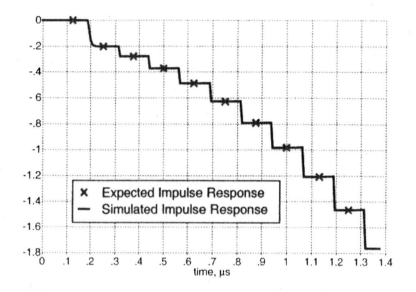

Figure 9.22: SCMOD5 impulse response check.

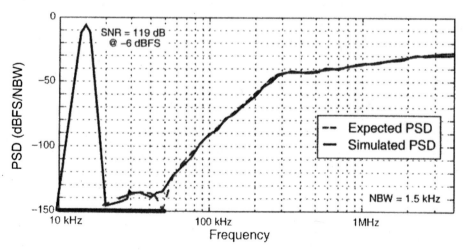

Figure 9.23: SCMOD5 spectrum from behavioral simulation.

### 9.2.3 Instability and Reset

In a single-bit system such as this one, it is important to consider the possibility of modulator instability and to provide means for initializing the modulator on power-up. The reset (R) switches shown in Fig. 9.21 set all integrator states to zero, and thereby both provide a safe starting state, and may return the modulator to a safe state in response to overload.

A reset event can be triggered by either analog or digital means. An effective analog technique for determining that the modulator needs to be reset is to monitor the input to the quantizer and reset the modulator whenever the magnitude of the quantizer's input exceeds a threshold. The threshold should be high enough that the modulator does not reset under normal operation, but low enough to activate a reset even if the amplifiers saturate.

Modulator instability can also be detected by digital means. Since the stable input range of this modulator is half of full-scale, any output value that is greater than half of full-scale indicates overload. By monitoring the output of the first stage of decimation for values which are, say, 75% of full-scale, modulator instability can be detected rapidly with a negligibly small false-alarm probability.

To minimize artifacts caused by reset, some modulator systems set the output of the decimation filter to full-scale until the modulator is once again operating properly, and the glitch caused by reset has propagated through the decimation filter.

## 9.3 A Wideband 2-0 Cascade System

The specifications of our third ADC example are listed in Table 9.4. The signal bandwidth is 1.25 MHz (25 times that of the preceding example), while the sampling frequency is 20 MHz (2.5 times that of the preceding example). Thus, the oversampling ratio for this modulator is one-tenth of the previous example, i.e. $OSR$ is only 8. In order to achieve $SNR = 90$ dB at such a low oversampling ratio, multi-bit quantization is now a necessity.

## 9.3.1 Architecture

Fig. 9.24 shows the architecture which we will examine. A second-order modulator is used as the first stage of the cascade in order to avoid the problems associated with a first-order first stage and to avoid the complexity of a third-order first stage. As noted in the previous example, keeping the modulator order low keeps the coefficient $b_1$ large and thereby saves capacitor area.

The number of bits allocated to the first stage involves a trade-off between the complexity of the quantizer and the sensitivity of the ADC system to analog non-idealities. If the first stage uses 12 bits, then there is no need for a second stage at all and so the deleterious effects of mismatch between the analog and digital transfer functions is not an issue. However, the internal ADC and DAC need to be very complex. On the other hand, if the first stage uses only a 1-bit quantizer, the raw

| Parameter | Symbol | Value | Units |
|---|---|---|---|
| Bandwidth | $f_B$ | 1.25 | MHz |
| Sampling Frequency | $f_s$ | 20 | MHz |
| Signal-to-Noise Ratio | SNR | 90 | dB |
| Supply Voltage | VDD | 5 | V |
| Input Voltage Range |  | [−2, 2] | V |

Table 9.4. Specifications for the example cascade modulator.

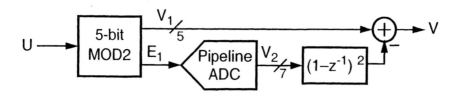

Figure 9.24: High-level system diagram for the 2-0 cascade.

performance of a second-order modulator at $OSR = 8$ is so low that we require 99.9% of its quantization noise to be cancelled digitally, which can only be achieved with stringent matching between the analog and digital domains. A 5-bit quantizer was selected as a compromise between these two extremes. The complexity and power consumption of a 5-bit quantizer were judged to be reasonable and, as we shall demonstrate shortly, 5-bit quantization yields a reasonably robust system.

Fig. 9.25 shows a code fragment for simulating the SQNR of the ideal system and Fig. 9.26 displays the resulting plot. As the latter figure shows, the $2^{nd}$-order modulator is able to realize a peak SQNR of 65 dB, whereas the composite system exhibits a peak SQNR which is 40 dB higher, namely 105 dB. As is commonly done in $\Delta\Sigma$ systems, the theoretical SQNR is set 15 dB higher than the SNR target (which is 90 dB) in order to make quantization noise an insignificant contributor to the system noise budget.

```
BW = 1.25e6; Fs = 20e6; OSR = Fs/(2*BW);
M = 32; nlev = M+1;
nb = 7; kpipe = 2^nb;
Ha = zpk([1 1],[0 0],1,1);
amp = [-120:5:-15 -12:2:-6 -5:0];
sqnr = zeros(2,length(amp));
N = 8192;
ftest = round(0.16/OSR*N);
u1 = M*sin(2*pi*ftest/N*[0:N-1]);
for i = 1:length(amp)
    [v1 junk1 junk2 y1] = simulateDSM(undbv(amp(i))*u1,Ha,nlev);
    v2 = ds_quantize(kpipe*(v1-y1),kpipe+1);
    v = v1 - filter([1 -2 1],1, v2/kpipe);
    spec1 = fft(v1.*hann(N))/(M*N/4);
    sqnr(1,i) = calculateSNR(spec1(1:ceil(N/2/OSR)),ftest);
    spec = fft(v.*hann(N))/(M*N/4);
    sqnr(2,i) = calculateSNR(spec(1:ceil(N/2/OSR)),ftest);
end
```

**Figure 9.25: Code for simulating the cascade system.**

## 9 Example Modulator Systems

Since the SQNR improvement is 40 dB, we are aiming to cancel 99% of the quantization noise, and thus would expect that a 1% gain error would result in a noticeable SQNR degradation. In order to verify this prediction, we can scale the input to the pipeline ADC by a factor of (1 + gain_mismatch):

```
v2 = ds_quantize(kpipe*(v1-y1)*(1+gain_mismatch),kpipe+1)
```

Fig. 9.27a plots the peak SQNR of the cascade system as a function of the gain mismatch. As this figure shows, a 1% gain error reduces SQNR by about 4 dB, which is in agreement with our intuition, while a gain mismatch of less than 0.5% results in negligible SQNR degradation.

Similarly, the effect of finite op-amp gain in the amplifiers of the modulator can be mimicked by shifting the zeros of the NTF to the left of $z = 1$:

```
Ha = zpk([1 1]*(1-1/gain),[0 0],1,1)
```

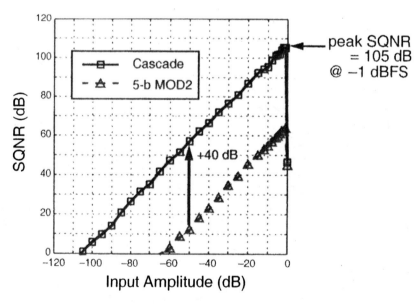

Figure 9.26: Simulated SQNR curves.

Fig. 9.27b plots the resulting curve. According to these simulation results, the amplifier gain needs to be at least 65 dB in order to have negligible SQNR loss.

*9.3.2 Implementation*

Fig. 9.28 shows the ADC system as implemented in [8]. As the diagram indicates, the 5-bit $2^{nd}$-order modulator (which makes use of dynamic element matching to reduce the effects of DAC mismatch errors) is followed by a 7-bit pipeline possessing 1 bit of redundancy at each residue stage. The data from the modulator and the pipeline stages are combined digitally in accordance with Fig. 9.24.

Fig. 9.29 shows the structure of the first integrator. Note that the sampling capacitors are divided into 32 units which accomplish the DAC function. As discussed earlier, this structure minimizes noise, but the amplifier must tolerate whatever common-mode step results from the difference between the input and reference common-mode voltages. References [9-11] describe implementations of similar multi-bit cascade modulator systems in more detail.

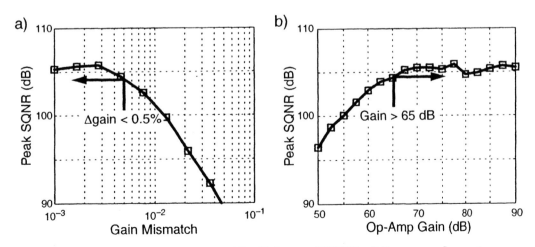

**Figure 9.27: Effect of two nonidealities on SQNR of the cascade system.**

9 Example Modulator Systems

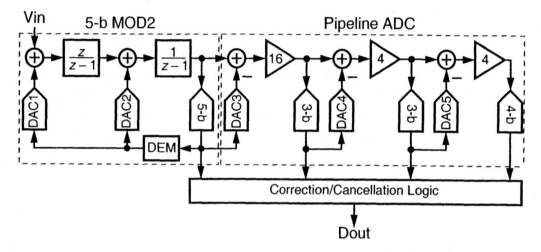

Figure 9.28: Expanded cascade system diagram.

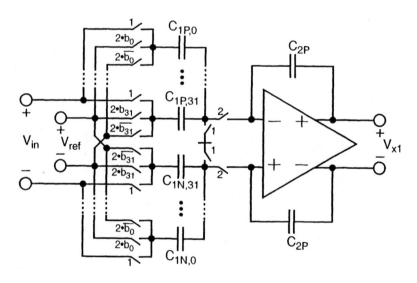

Figure 9.29: First integrator in the cascade system.
(In this circuit, the current drawn from the reference is signal-dependent.)

## 9.4 A Micropower Continuous-Time ADC

The preceding examples dealt with systems characterized by high dynamic range. In this example, the ADC is intended to interface directly to a microphone and our primary concern is micropower operation. To eliminate the need for an anti-alias filter and a preamplifier, the ADC will be continuous-time with a input range of ±20 mV. The signal bandwidth for this audio application is 100 Hz-10 kHz.

Our goal is to maximize dynamic range, while consuming less than 50 µA from a 1.8-V supply. Table 9.5 summarizes the design specifications for this ADC system.

Since the bandwidth is low, a high oversampling ratio is readily achieved and thus a low-order modulator can be used. For example, if the clock frequency is $13/2$ MHz = 6.5 MHz,[†] then $OSR \approx 300$. Since an ideal single-bit second-order modulator achieves $SQNR > 100$ dB with this OSR, this configuration provides

| Parameter | Symbol | Value | Units |
|---|---|---|---|
| Signal band | $B$ | 100-10k | Hz |
| Supply Voltage | VDD | 1.8 | V |
| Current Consumption | IDD | 50 | µA |
| Input Voltage Range | | −20 to 20 | mV |
| Sampling Frequency | $f_s$ | 6.5 | MHz |
| Signal-to-Noise Ratio | SNR | 80 | dB |

**Table 9.5. Specifications for the micropower modulator.**

---

† 13 MHz is a frequency that is readily available in many cellular telephones. A lower clock frequency could easily have been chosen, but as we shall see, employing a high sampling rate reduces the area occupied by the circuit. Dividing the clock by two ensures that the duty cycle of the clock is 50%, a condition which we implicitly assume when we set the DAC feedback delay to half of a clock period.

more than enough quantization noise suppression. Lastly, we choose to adopt a feedback topology in order to realize a lowpass STF having no out-of-band peaks.[†]

### 9.4.1 High-Level Design

Fig. 9.30 shows the modulator topology which we wish to implement. In contrast to the second-order system discussed in Section 6.6.1, the DAC timing for the current system is assumed to be [0.5, 1.5], i.e. the DACs are updated half a clock cycle after the comparator samples its input. This half-cycle delay gives the comparator sufficient time to resolve its input and also provides set-up time for the DACs. Although the clock frequency in this example is low enough that a modern CMOS process would not require a feedback delay of this magnitude, it is nonetheless instructive to see how this delay can be incorporated into the design process.

As with previous examples, the first task in the design process is to select the target NTF. For simplicity, we will use the same NTF as was used in the first design example, namely

$$NTF(z) = \left(\frac{z-1}{z-1/3}\right)^2. \tag{9.20}$$

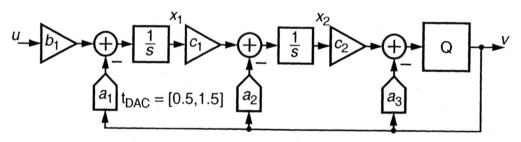

Figure 9.30: Continuous-time modulator system diagram.

---

[†]. As discussed on page 95, a discrete-time feedforward topology can have a a perfectly flat STF if a direct connection from the input to the quantizer is included. Unfortunately, realizing an STF which has no out-of-band peaks with a continuous-time feedforward topology is not so straightforward.

The second design task is to determine coefficients which realize this NTF. For this operation, the $c$ coefficients, which only affect internal scaling, can initially be set to unity. As described in Section 6.6, we must select the coefficients $(a_1, a_2, a_3)$ to match the sampled pulse response of the loop filter to the impulse response of our target discrete-time system's loop filter.

Since the loop filter is second-order with two poles at dc, its long-term pulse response is a ramp. The $a_1$ and $a_2$ coefficients control the slope and intercept of this ramp, so it is straightforward to find values for these terms which yield the desired long-term behavior. However, while the pulse is active, the system is responding as if the input were a step, and since this step occurs midway through the first feedback interval, the first sampled value of the system's output will not fall on the same ramp as the remaining samples. Fortunately, the $a_3$ coefficient affects only the first sample of the pulse response, and thus provides the means to compensate for the half-cycle feedback delay. In principle, any amount of delay in the main loop filter can be accommodated by adding a sufficient number of properly-timed feedback paths to the quantizer, but the sensitivity of the modulator to errors in the coefficients increases as the feedback delay is increased.

Fig. 9.31 lists MATLAB code which leads to the following coefficients for our half-cycle delay system:

$$(a_1, a_2, a_3) = (0.4444, 1.3333, 0.6111),^\dagger \qquad (9.21)$$

while Fig. 9.32 plots the STF associated with this modulator. The latter figure reinforces the discussion of Section 6.6.2, namely that the STF of a continuous-time modulator naturally attenuates frequencies which alias to the passband. For an ideal implementation, the attenuation experienced by frequencies which alias into the passband is over 130 dB. In practice, finite dc gain in the loop filter will limit

---

†. There are many ways to compute these coefficients. The method used in Fig. 9.31 is based on the fact that the pulse response of the loop filter is a weighted linear combination of the pulse responses from each DAC. By computing these pulse responses (yy in Fig. 9.31), the weightings (i.e. the $a_i$ coefficients) needed to achieve a desired response can be found by solving a set of linear equations, as is done in the last line of Fig. 9.31.

## 9 Example Modulator Systems

```
% Desired NTF and its impulse response
NTF = zpk([1 1],[1 1]/3,1,1);
n_imp = 10;
y_desired = impL1(NTF,n_imp)';
% State-space description of CT loop filter
% as a 3-input, 1-output system
Ac = [ 0 0; 1 0 ];
Bc = [ -1 0 0; 0 -1 0 ];
Cc = [ 0 1 ];
Dc = [ 0 0 -1];
td = 0.5;
sys_c = ss(Ac,Bc,Cc,Dc);
set(sys_c,'InputDelay',td*[1 1 1]);
% Discrete-time equivalent and associated impulse response
sys_d = c2d(sys_c,1);
yy = squeeze( impulse(sys_d,n_imp) )';
% Solve for coefficients s.t. a*yy = y_desired
a = y_desired/yy;
```

Figure 9.31: Code for computing CTMOD2's coefficients.

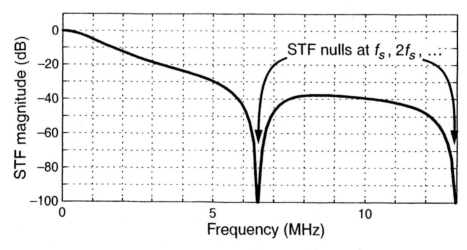

Figure 9.32: Signal transfer function for CTMOD2.

the depth of the NTF's and hence the STF's notches, as dictated by (6.35), but an alias suppression of over 80 dB is nonetheless readily achieved.

The third design step, namely dynamic range scaling, can be performed by using the discrete-time model of the modulator, but since the discrete-time model only computes the modulator's internal states at the sampling instants, it is advisable to use a behavioral version of the continuous-time modulator instead. Fig. 9.33 shows a behavioral schematic which can be simulated to determine the internal signal swings. The circuit parameters, along with the formulae used to calculate them, are tabulated in Table 9.6. Behavioral simulations indicate that the peak swings of the two integrators are 1.6 V and 3.5 V, respectively. Scaling the states for peak swings of 0.5 V, setting the full-scale level 25% higher than the desired 20-mV input range and choosing $I_1 = 10I_2 = 10I_3 = 10\mu A$ yields the scaled values shown in the fourth column of Table 9.6. These scaled values can be used to construct a transistor-level implementation.

Referring to Table 9.6, note that the scaled value of $C_1$ is 11 pF. Since in a typical process the capacitance per unit area is 1 fF/$\mu m^2$, each $C_1$ capacitor occupies a sizable (100 $\mu m$ × 100 $\mu m$) area. Since the value of $C_1$ is inversely proportional to the clock rate, it should now be apparent that excessive silicon real estate would have been consumed had the modulator's clock rate been much lower than the value chosen.

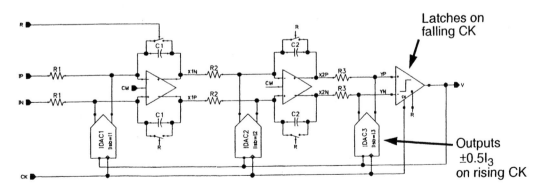

**Figure 9.33: CTMOD2 behavioral schematic.**

## 9.4.2 Circuit Design

Fig. 9.34 shows the schematic of DAC1, the first feedback DAC, along with the latch which precedes it. In order to minimize thermal noise, the DAC uses a complementary structure[†], and the current source devices MN1 and MP1 are sized to have as large a $V_{GS}$ as possible, while leaving only as much headroom for swing at their outputs as is necessary.[‡] Furthermore, the devices are physically large in

| Parameter | Formula | Starting Value | Scaled Value | Units |
|---|---|---|---|---|
| $V_{FS}$ | — | 1.0 | 0.025 | V |
| $C_1$ | — | 1.0 | 11 | pF |
| $I_1$ | $a_1 C_1$ | 2.9 | 10 | µA |
| $R_1$ | $V_{FS}/(b_1 f_s C_1)$ | 350 | 2.5 | kΩ |
| $C_2$ | — | 0.1 | 0.8 | pF |
| $I_2$ | $a_2 f_s C_2$ | 0.87 | 1 | µA |
| $R_2$ | $1/(c_1 f_s C_2)$ | 1540 | 420 | kΩ |
| $I_3$ | — | 1 | 1 | µA |
| $R_3$ | $a_3/(c_2 I_3)$ | 611 | 87 | kΩ |

**Table 9.6. CTMOD2 circuit parameters.**

---

[†]. For a fixed LSB size, a complementary current-mode DAC has a 3-dB noise advantage over a single-ended DAC with associated counterbalancing current sources since all currents in the complementary DAC are half the corresponding currents in the single-ended DAC and its associated fixed current sources.

[‡]. Since the noise density of the drain current in an MOS device is $\frac{8}{3} kT g_m$, minimizing noise requires minimizing $g_m$. Since $g_m = 2I_D/(\Delta V)$, where $\Delta V = V_{GS} - V_T$ and the drain current $I_D$ is fixed, we therefore desire to maximize $V_{GS}$.

order to minimize their $1/f$ noise. With the sizes shown, the (non-$1/f$) current noise density at the output of DAC1 is 0.4 pA/$\sqrt{\text{Hz}}$, which translates to an input-referred noise of 2 nV/$\sqrt{\text{Hz}}$. This noise density is approximately 12 dB lower than the $\sqrt{(4kT)(2R_1)} = 9$ nV/$\sqrt{\text{Hz}}$ density associated with the input resistors, and hence the DAC will be a relatively minor contributor to the noise of the modulator. If the resistors and the DAC were the only noise sources in the circuit, the expected SNR would be

$$SNR = 10\log\left[\frac{(0.5 \cdot (20 \text{ mV})^2)}{(2^2 + 9^2) \times (10^{-9})^2 \times (10^4 \text{ Hz})}\right] = 87 \text{dB}. \quad (9.22)$$

As anticipated, the above noise-limited SNR is much lower than the >100-dB theoretical SQNR of an ideal second-order modulator operated at an OSR of 300. Since doubling $I_1$ by a factor of 2 would halve $R_1$, and thereby increase the SNR given in (9.22) by 3 dB, it makes sense to spend as much of our current budget on DAC1 as we can afford. However, with the values chosen, and assuming the current in the bias network is equal to that of the main DAC branch, we have only allocated 20% of our power budget to DAC1.

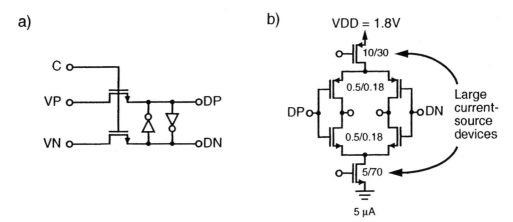

**Figure 9.34:** (a) Latch and (b) DAC schematics.

The reason that we cannot afford to allocate more current to DAC1 is that the first amplifier must possess low noise and must also be able to source or sink the sum of the input current and DAC1's output current. Both requirements imply that the bulk of our current budget must be allocated to the first amplifier.

Assuming that a class-A amplifier is used, the bias current in the output legs must be at least 5 µA in order to meet the output current requirements. To reduce the ratio of the maximum current to the minimum current in the output devices, a more practical bias current would be on the order of 10 µA per leg. Next, assume that we desire the amplifier's input-referred noise to be equal to that of the input resistors, namely $9 \text{ nV}/\sqrt{\text{Hz}}$. Then, assuming the $1/f$ noise is negligible, the transconductance of the devices in the input differential pair needs to be

$$g_m = \frac{\frac{8}{3}kT}{(9 \text{ nV}/\sqrt{\text{Hz}})^2/2} = 270 \text{ µA/V}. \tag{9.23}$$

Assuming a square-law model, $g_m = 2I_D/\Delta v$, and choosing $\Delta v = 100$ mV leads to $I_D = 14$ µA. Thus, we need to spend significant fractions of our 50-µA current budget on both the differential pair and the output legs. In a single-stage amplifier, the current in the differential pair flows in the output legs, and so it might appear that a single-stage amplifier would provide a power-efficient way to achieve both low noise and adequate drive current. Unfortunately, the transconductance of a single-stage CMOS amplifier biased with 25 µA is no more than 125 µA/V or so,[†] and thus the differential voltage at the input of the amplifier corresponding to an output current of 5 µA will be at least 40 mV, which is double our desired full-scale input voltage! Clearly the transconductance of a single-stage amplifier is much too low for this application, and thus we need to adopt a multi-stage architecture, such as the two-stage architecture depicted in Fig. 9.35. The bias currents shown reflect the above calculations, while the output devices are cascoded in

---

[†]. With a bias current of $2I_C = 25$ µA, a BJT differential pair achieves a transconductance of $g_m = 0.5I_C/V_T = 0.25$ mA/V. A CMOS differential pair can only have a $g_m$ which is a fraction of this amount. In the above, we generously assumed that this fraction is 50%.

order to achieve an output resistance which is comparable to $R_2 = 400$ kΩ. Compensation and common-mode feedback networks are not shown.

Since the second amplifier in the modulator is required to supply currents which are approximately 10% of those of the first amplifier, simply scaling the first amplifier down by a factor of 10 should yield a circuit which has negligible contributions to the noise and distortion of the system.

Lastly, Fig. 9.36 shows the schematic of a low-power comparator which is suitable for the modulator under consideration. This comparator may be operated with a low (1 µA) bias current, since the speed requirements are quite lax and since the modulator is tolerant of comparator offset. Simulations indicate that the power consumption of this circuit is less than 5 µW at a clock frequency of 6.5 MHz, so the power consumed by the quantizer will be a very small portion of our 90-µW budget.

The above discussion indicates that a modulator of the form considered here should be able to achieve a dynamic range of more than 80 dB over a 10–kHz bandwidth while consuming less than 100 µW, an expectation which is supported by the excellent work of [13] (which also used chopping to combat $1/f$ noise).

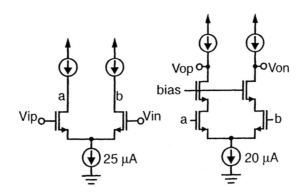

Figure 9.35: Two-stage amplifier.

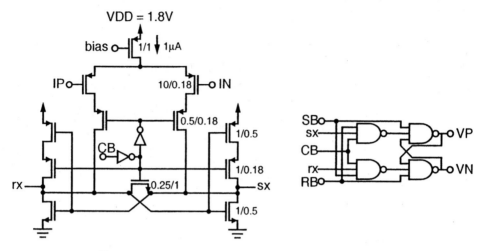

Figure 9.36: A low-power comparator.

## 9.5 A Continuous-Time Bandpass ADC

Section 5.4 described the structure of a bandpass $\Delta\Sigma$ ADC which is well-suited to the *dual-conversion superheterodyne receiver* illustrated in Fig. 9.37. In this receiver architecture, the incoming radio-frequency (RF) signal is filtered, ampli-

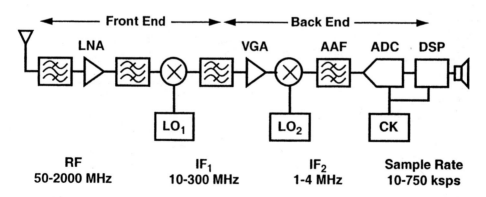

Figure 9.37: A dual-conversion superheterodyne receiver.

fied and mixed to a first *intermediate frequency* IF$_1$, which is in turn filtered, amplified and mixed to a second intermediate frequency IF$_2$ before being converted to digital form. Although a discussion of the advantages of this receiver architecture over competing architectures is beyond the scope of this text, it is worth noting that the venerable superheterodyne architecture is preferred when the system must achieve both high sensitivity and high selectivity. By converting an IF signal to digital form, low-frequency analog impairments such as $1/f$ noise and dc offset do not affect the signal, and thus this system is able to achieve a higher dynamic range than a system in which the signal is translated to dc before A/D conversion.

A straightforward implementation of the receiver back end as depicted in Fig. 9.37 is feasible, but suffers from the drawback of requiring several high-dynamic-range (and thus high-power) blocks, namely the variable-gain amplifier (VGA), the second mixer, the anti-alias filter (AAF) and the ADC. Since the VGA can be eliminated if the ADC has sufficient dynamic range, and since the AAF can be eliminated if the ADC is of the continuous-time variety, the system can be simplified greatly by using a power-efficient and high-dynamic-range continuous-time bandpass $\Delta\Sigma$ ADC.

Section 5.4 argued that such an ADC can be constructed if an LC resonator plus a current-mode feedback DAC (IDAC) are used in the first stage of the ADC. This architecture, first illustrated in Fig. 5.24 but rendered in slightly greater detail in Fig. 9.38, obtains high efficiency by exploiting the high output resistance of the active mixer in the construction of the ADC's first and most critical resonator. Since subsequent resonators require less dynamic range than the first, the only elements in this system which must process the full dynamic range of the signal are the LNA/Mixer and IDAC, thereby giving this arrangement a significant power advantage.

Section 5.4 gave an overview of a particular embodiment of this ADC architecture and provided justification for several of the architectural choices, in particular the choice to use active-RC and switched-capacitor resonators in the second and third resonator stages, respectively. In this section, we take this topology as a given and delve deeper into the design process and the underlying circuits.

### 9.5.1 Architecture/Analysis

Fig. 9.39 depicts a simplified schematic of the ADC, including timing details, which we will use as a starting point. Our first step is to transform this mixed continuous/discrete system into a discrete-time equivalent so that we can bring the power of the ΔΣ Toolbox to bear.

A state-space description of the continuous-time portion of the loop filter is

$$\begin{bmatrix} C_1 \dot{x}_{1c} \\ L_1 \dot{x}_{2c} \\ C_{t3} \dot{x}_{3c} \\ C_{t4} \dot{x}_{4c} \\ y_{1c} \end{bmatrix} = \begin{bmatrix} 0 & -1 & 0 & 0 & g_{1u} & g_{1v} \\ 1 & 0 & 0 & 0 & 0 & 0 \\ -g_{31} & 0 & 0 & -g_{34} & 0 & -g_{3v} \\ 0 & 0 & -g_{43} & 0 & 0 & -g_{4v} \\ 0 & 0 & \alpha & \beta & 0 & 0 \end{bmatrix} \begin{bmatrix} x_1 \\ x_2 \\ x_3 \\ x_4 \\ u_c \\ v_c \end{bmatrix}, \quad (9.24)$$

where the state variables $(x_{1c}, x_{2c}, x_{3c}, x_{4c})$ are as indicated in Fig. 9.39. (For example, $x_{1c}$ is the voltage across capacitor $C_1$ and $x_{2c}$ is current in inductor $L_1$.)

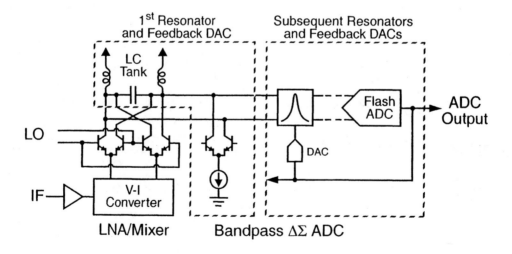

Figure 9.38: **Active mixer connected to a bandpass ΔΣ ADC.**

# A Continuous-Time Bandpass ADC

Also as indicated in Fig. 9.39, the output of the loop filter is sampled when the current-mode feedback DACs ($g_{1v}, g_{2v}, g_{3v}$) change state. Since these DACs feed current back for a full clock period, the appropriate specification for the DAC timing is $t_{dac} = [0, 1]$.

a) Loop Filter

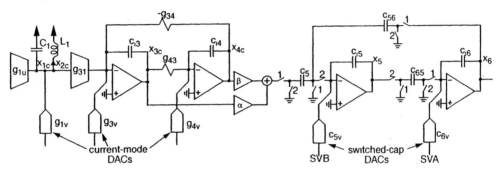

b) Quantization and Feedback

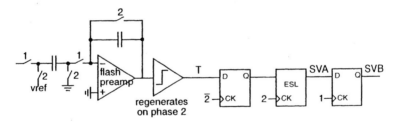

c) Timing

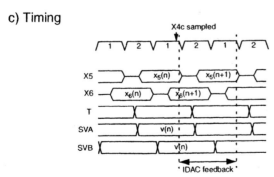

**Figure 9.39:** A single-ended representation of the bandpass system.

The next piece of information we need in order to construct a toolbox-compatible description of this modulator is the state-space description of the discrete-time portion of the loop filter. Following a procedure like that used earlier, while paying attention to the timing of Fig. 9.39, yields

$$\begin{bmatrix} x_5(n+1) \\ x_6(n+1) \\ y(n) \end{bmatrix} = \begin{bmatrix} \left(1 + \dfrac{c_{56}c_{65}}{c_{15}c_{16}}\right) & \left(\dfrac{c_{56}}{c_{15}}\right) & \left(\dfrac{c_5}{c_{15}}\right) & \left(\dfrac{c_{5v}}{c_{15}} + \dfrac{c_{56}c_{6v}}{c_{15}c_{16}}\right) \\ \dfrac{c_{65}}{c_{16}} & 1 & 0 & \dfrac{c_{6v}}{c_{16}} \\ 0 & 1 & 0 & 0 \end{bmatrix} \begin{bmatrix} x_5(n) \\ x_6(n) \\ y_1(n) \\ v(n) \end{bmatrix}. \quad (9.25)$$

The matrix descriptions of (9.24) and (9.25) specify a mixed continuous-time/discrete-time system of the form illustrated in Fig. 9.40a. To transform such a system into an equivalent system with a discrete-time loop filter as shown in Fig. 9.40b, we use a pulse-invariant transformation (toolbox function mapCtoD) to transform $ABCD_c$ into $ABCD_1$ for the feedback path, while keeping the signal path through $ABCD_c$ intact. The ABCD matrix which describes the discrete-time portion of the loop filter is then

$$ABCD = \left[ \begin{array}{cc|cc} A_1 & 0 & B_{11} & B_{12} \\ B_{21}C_1 & A_2 & B_{21}D_{11} & B_{21}D_{12} + B_{22} \\ \hline D_{21}C_1 & C_2 & D_{21}D_{11} & D_{21}D_{12} + D_{22} \end{array} \right]. \quad (9.26)$$

This ABCD matrix can be used to calculate the NTF using the toolbox function calculateTF.

To compute the STF, simply observe from Fig. 9.40b that
$$STF = L_{0c}H_1,$$

where $L_{0c}$ is the open-loop transfer function of the continuous-time system and $H_1$ is the closed-loop transfer function from $y_1$ to $v$. $L_{0c}$ may be evaluated from ABCDc using MATLAB's LTI (linear time-invariant) system capabilities. To compute $H_1$, the following state-space formulation may be used

$$\begin{bmatrix} A_1 + B_{12}D_{21}C_1 & B_{12}C_2 & B_{12}D_{21} \\ B_{21}C_1 + B_{22}D_{21}C_1 & A_2 + B_{21}D_{12}C_2 + B_{22}C_2 & B_{21} + B_{22}D_{21} \\ \hline D_{21}C_1 & C_2 & D_{21} \end{bmatrix}. \quad (9.27)$$

With the above formulae, the NTF and STF can be computed for a given set of parameter values. Thus, system *analysis* is readily accomplished. System *synthesis* (the inverse procedure of converting a desired NTF into a set of parameter values) is less straightforward, but may be accomplished iteratively. Once a set of parameter values which realize a desirable NTF and STF have been found, dynamic range scaling (using the results from behavioral simulations) and system denormalizaton need to be performed. The results of these operations for the system under consid-

a) Original System

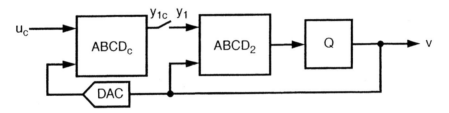

b) Discrete-Time Model, plus continuous-time prefilter

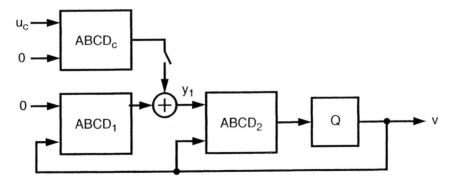

Figure 9.40: Transformation to a discrete-time model.

# 9 Example Modulator Systems

eration, denormalized for an 18-MHz clock rate, are tabulated in Table 9.7, while Fig. 9.41 and Fig. 9.42 plot the associated NTF and STF, respectively. The NTF has a peak out-of-band gain of 3, and the STF exhibits the inherent anti-aliasing

| Parameter | Value | Parameter | Value |
|---|---|---|---|
| $L_1$ | $2 \times 10$ μH | $c_5$ | 200 fF |
| $C_1$ | 250 pF | $c_{56}$ | −200 fF |
| $g_{31}$ | 21 μA/V | $c_{5v}$ | −16.1 fF |
| $g_{34}, g_{43}$ | 26.8 μA/V | $c_{i5}$ | 400 fF |
| $g_{1v}$ | 256 μA | $c_{65}$ | 280 fF |
| $g_{3v}$ | 5.2 μA | $c_{6v}$ | −36.6 fF |
| $g_{4v}$ | 0 μA | $c_{i6}$ | 240 fF |
| $c_{i3}, c_{i4}$ | 1.9 pF | Vref | 0.5 V |
|  |  | flash LSB | 106 mV |

Table 9.7. Parameter values for the bandpass modulator of Fig. 9.39.

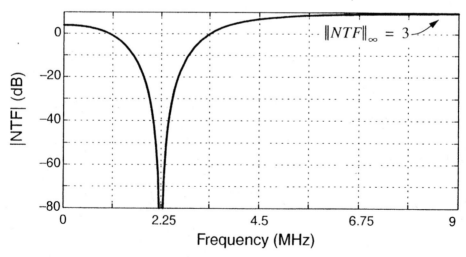

Figure 9.41: NTF for the bandpass example.

property which is characteristic of a continuous-time modulator. Although the STF nulls are not as deep in this system as they would be in a fully continuous-time system, signals which alias to the passband are nonetheless attenuated by at least 75 dB.

Fig. 9.43 shows the simulated SQNR vs. input amplitude curve for a 10-kHz signal bandwidth. The peak SQNR is 107 dB and the modulator is stable for inputs that are within a fraction of a dB of full-scale.

*9.5.2 Subcircuits*

Fig. 9.44 depicts the equivalent input-referred noise for a variety of noise sources in the system as a function of the full-scale setting. As described in Section 5.4, this ADC implements a variable full-scale by changing the full-scale current of the first feedback DAC in tandem with the $g_{31}$ transconductance. For this figure, the signal bandwidth is assumed to be 150 kHz; at lower signal bandwidths, the contributions of shaped noise sources such as DAC mismatch noise, quantization noise and SC noise are much reduced. Note that the noise due to the front end RF com-

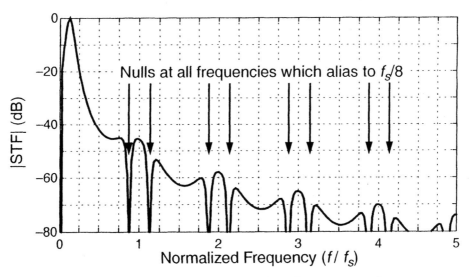

**Figure 9.42: STF for the bandpass example.**

9 Example Modulator Systems

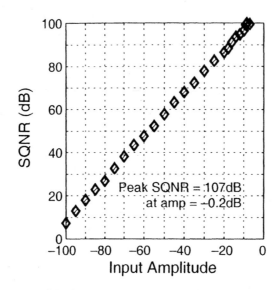

Figure 9.43: Simulated SQNR for the bandpass example.

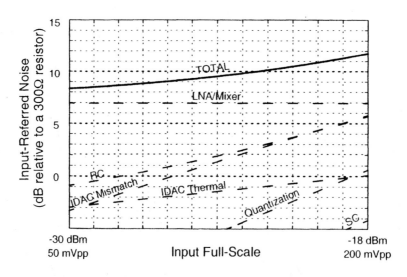

Figure 9.44: Noise contributions in the bandpass ADC system.

ponents, namely the LNA and the mixer, is independent of the ADC's full-scale setting and furthermore dominates the system noise. As the ADC's full-scale setting is reduced, i.e. as the effective gain of the ADC is increased, the input-referred equivalents of noise sources located after the $g_{31}$ transconductance are reduced at the rate of 1 dB per dB, simply by virtue of the fact that $g_{31}$ is proportional to the ADC's gain. (The slope of the RC noise is somewhat shallower than 1 dB/dB because the noise of the $g_{31}$ transconductor, which is included in the RC noise, increases with increasing $g_{31}$.) Note also that both the IDAC's thermal noise and its mismatch noise decrease as the full-scale limit of the IDAC is reduced. At first this may seem counter-intuitive, since one would expect that reducing the IDAC's full-scale should reduce the dynamic range. However, both statements are correct. As the full-scale of the IDAC is reduced, its dynamic range falls, but it falls at a rate of less than 1 dB/dB. In other words, as the full-scale of the IDAC is reduced, the *absolute* size of its input-referred noise (as opposed to its size relative to full-scale) is also reduced. The main advantage of a variable full-scale in the ADC is that the noise figure of the system at the minimum full-scale setting is within approximately 1 dB of the 7–dB noise figure of the LNA/Mixer.

Fig. 9.45 shows the structure of the active-RC resonator used for the second stage of the modulator. By making the op amps out of simple transconductors, the Q of this resonator can be very high, since the Q-degrading effect of finite gain (caused

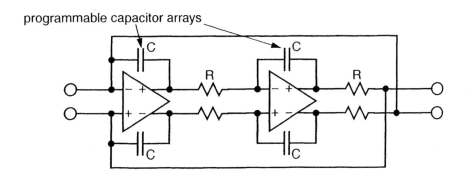

**Figure 9.45: Structure of the RC resonator.**

by resistive loading) turns out to be cancelled by the Q-enhancing effect of phase shift (caused by capacitive loading). In order to accommodate process tolerances and to allow some flexibility in the clock rate, this resonator needs to be tunable. Tuning is accomplished by transforming the resonator into an oscillator and adjusting the feedback capacitors until the oscillation frequency equals $f_s/8$.

Fig. 9.46 shows a single-ended representation of the switched-capacitor resonator. The structure is a standard LDI (lossless discrete integrator) topology. As indicated in the figure, the capacitors in each integrator are made with simple integer multiples of a unit capacitor. The realized center frequency is within a fraction of a percent of the desired center frequency, despite the use of the simple coefficients shown.

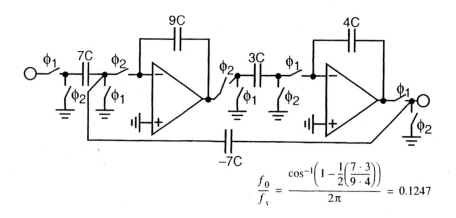

$$\frac{f_0}{f_s} = \frac{\cos^{-1}\left(1 - \frac{1}{2}\left(\frac{7 \cdot 3}{9 \cdot 4}\right)\right)}{2\pi} = 0.1247$$

**Figure 9.46: Structure of the SC resonator.**

## 9.6 Audio DAC

Table 9.8 lists the specifications of a representative audio $\Delta\Sigma$ DAC. In order to simplify the reconstruction filter, the modulator <u>assumes a passband</u> whose width is greater than the bandwidth of the input signal. In particular, the modulator's passband is set to twice the input sample rate (i.e. $f_B = 2f_{s,in} = 88.2 \text{ kHz}$, or more than 4 times the input bandwidth $f_{B0}$), and so the modulator's clock rate must be $256 f_{s,in} \approx 11.29 \text{ MHz}$ to provide an oversampling ratio of 64 for this extended passband. By extending the modulator's passband in this way, the quantization noise (which grows abruptly at the modulator passband limit frequency) need not be removed by an excessively sharp reconstruction filter. Note that the passband edge of the interpolation filter remains at $f_{B0} = 20 \text{ kHz}$. For this filter, we require the passband variation to be no more than 0.1 dB and the image attenuation, namely the attenuation beyond $f_{s,in}/2$, to be at least 90 dB. Lastly, we shall assume that the quantizer has $M = 8$ steps.

| Parameter | Symbol | Value | Units |
|---|---|---|---|
| Input Sample Rate | $f_{s,in}$ | 44.1 | kHz |
| Signal Bandwidth | $f_{B0}$ | 20 | kHz |
| Output Signal-to-Noise Ratio | SNR | 110 | dB |
| Passband Flatness |  | 0.1 | dB |
| Image Attenuation |  | >90 | dB |
| Modulator Bandwidth | $f_B = 2 f_{s,in}$ | 88 | kHz |
| Modulator Sample Rate | $f_s = 256 f_{s,in}$ | 11.29 | MHz |
| Modulator OSR | $OSR = f_s/(2f_B)$ | 64 |  |
| Number of Quantizer Steps | $M$ | 8 |  |

**Table 9.8.** Specifications for the DAC example.

The overall architecture of the DAC system is illustrated in Fig. 9.47. The following subsections will describe the modulator, the interpolation filter, and the reconstruction filter in detail.

## 9.6.1 Modulator Design

As discussed in Section 7.3.4, a multi-bit modulator is especially desirable in a DAC system since multi-bit quantization allows the modulator output to follow the desired signal more closely, and thereby ease the design of the reconstruction filter. For the modulator parameters given above, a third-order, 9-level modulator employing an NTF with $\|H\|_\infty = 2.5$ yields a peak SNR of nearly 115 dB over the extended passband. Since the NTF uses coincident zeros at dc, the SQNR in the true 0-to-20-kHz passband will be much greater than the desired 110 dB and we can therefore rest assured that the performance of the modulator will be more than sufficient for this application.

Fig. 9.48 shows a block diagram of the modulator. A standard CIFB (cascade-of-integrators, feedback) topology was selected so that the natural lowpass STF asso-

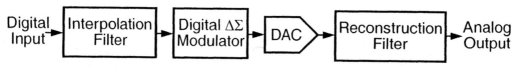

**Figure 9.47: DAC system.**

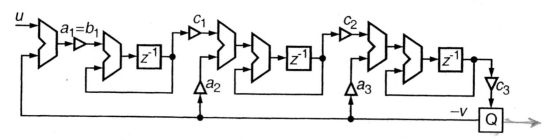

**Figure 9.48: Block diagram for the digital modulator.**

ciated with the modulator can simplify the interpolation filter. The coefficients for this structure were obtained by using the first code block given in Fig. 9.49 to provide first-cut coefficients (listed) which are scaled for dynamic range. These coefficients were then re-scaled or changed, as indicated in the code block on the right, to give power-of-two interstage coefficients or to simplify the feedback coefficients. Note that the interstage coefficients are rounded *downward* so that the maximum values of the internal states stay less than one. The final coefficients (listed in Table 9.9) are obtained from the values in Fig. 9.49 by quantizing the only non-power-of-two coefficients, namely $a_3$ and $c_3$, to two binary terms.

Fig. 9.50 compares the poles of the original NTF with the poles of the NTF after coefficient quantization. Since the poles are nearly identical, the performance of the modulator will not be significantly affected by coefficient quantization; toolbox simulations with the final NTF confirm this expectation.

The next step toward a hardware implementation involves determining the word lengths needed within the modulator loop. In performing this task, we will choose word lengths such that the in-band noise caused by finite precision arithmetic is below the –130 dBFS level.

```
form = 'CIFB';
[a,g,b,c]=realizeNTF(NTF,form);
b(2:end) = 0;
ABCD= stuffABCD(a,g,b,c,form);
ABCDs = scaleABCD(ABCD,M+1);
[a,g,b,c]= mapABCD(ABCDs,form);
% a = [0.0331 0.0807 0.1626]
% g = 0
% b = [0.0331 0 0 0]
% c = [0.6176 1.3672 10.4879]
```

```
% scale for a(1) = 1/32
k = a(1) * 32;
a(1) = a(1)/k;
b(1) = b(1)/k;
c(1) = c(1)*k;
% scale for c(1) = 0.5
k = c(1)/0.5;
c(1) = c(1)/k;
a(2) = a(2)/k;
c(2) = c(2)*k;
% scale for c(2) = 1
k = c(2)/1;
c(2) = c(2)/k;
a(3) = a(3)/k;
c(3) = c(3)*k;
```

**Figure 9.49: Toolbox code for determining coefficients.**

# 9 Example Modulator Systems

| Coefficient | Original Value | Transformed Value | Quantized Value | % Error |
|---|---|---|---|---|
| $a_1 = b_1$ | 0.0331 | 0.0312 | $\frac{1}{32}$ | 0 |
| $a_2$ | 0.0807 | 0.0617 | $\frac{1}{16}$ | 1 |
| $a_3$ | 0.1626 | 0.0909 | $\frac{1}{16} + \frac{1}{32}$ | 3 |
| $c_1$ | 0.6176 | 0.5000 | $\frac{1}{2}$ | 0 |
| $c_2$ | 1.3672 | 1.0000 | 1 | 0 |
| $c_3$ | 10.4879 | 18.7539 | 16+2 | -4 |

Table 9.9. Modulator coefficients.

```
% quantized coefficients
aq = [1/32 1/16 1/16+1/32];
gq = 0;
bq = [1/32 0 0 0];
cq = [1/2 1 16+2];
ABCDq = stuffABCD(aq,gq,bq,cq,form);
[NTFq STFq] = calculateTF(ABCDq);
figure(1); clf;
plotPZ(NTF,'m',10);
hold on;
plotPZ(NTFq,'b',10);
```

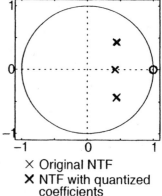

× Original NTF
× NTF with quantized coefficients

**Figure 9.50: NTF before and after coefficient quantization.**

For the first integrator, the input-referred in-band noise power associated with $N_1$-bit quantization is

$$\frac{1}{a_1^2} \cdot \frac{(2^{-N_1})^2}{3} \cdot \frac{1}{OSR}. \tag{9.28}$$

Since the power of a full-scale sine wave is $M^2/2$, the number of bits in the first stage must satisfy

$$\frac{(2^{-N_1})^2}{3a_1^2 OSR} < 10^{-13}(M^2/2) \tag{9.29}$$

or

$$N_1 > -\log_2(Ma_1\sqrt{1.5 \times 10^{-13} OSR}) = 20.3. \tag{9.30}$$

Hence, we require the use of 21-bit arithmetic in the first integrator.

Since noise added to the input of the second integrator is first-order shaped when referred to the modulator's input, the input-referred in-band noise power associated with $N_2$-bit quantization in the second integrator is

$$\frac{1}{(a_1 c_1)^2} \cdot \frac{(2^{-N_2})^2}{3} \cdot \frac{\pi^2}{3(OSR)^3}. \tag{9.31}$$

From (9.31), the number of bits required for a $-130$-dBFS noise contribution is $N_2 > -\log_2(Ma_1 c_1 \sqrt{0.45 \times 10^{-13} (OSR)^3}) = 16.2$, so we will use 17 bits in the second integrator.

Lastly, since the input-referred in-band noise power associated with $N_3$-bit quantization in the third integrator is

## 9 Example Modulator Systems

$$\frac{1}{(a_1 c_1 c_2)^2} \cdot \frac{(2^{-N_3})^2}{3} \cdot \frac{\pi^4}{5(OSR)^5}, \qquad (9.32)$$

we require $N_3 > -\log_2(M a_1 c_1 c_2 \sqrt{0.077 \times 10^{-13} (OSR)^3}) = 11.4$, i.e. $N_3 = 12$.

In order to verify that the above word-lengths are sufficient, a version of the modulator which uses integer arithmetic will be constructed (in C) and simulated.

Since the word-lengths of the integrators differ, the coefficients in the C model are not the same as the coefficients listed in Table 9.9. For example, the third integrator uses 12-bit arithmetic whereas the output of the second integrator has 17 bits, and thus a coupling coefficient of $c_2 = 1$ translates into a right-shift of 5 bits.[†] Similarly, since the toolbox convention is that both the input $u$ and the feedback $v$ occupy a range of $-M$ to $M$, these signals must also be scaled appropriately. Fig. 9.51 lists the code which simulates the operation of the modulator. Since each integrator is delaying, the updated states of the modulator can be computed without making use of temporary variables. Computing the output of the third integrator, then the second, and finally the first, implements the desired difference equations. For the sake of simplicity, the code does not include saturation for each summation operation, even though it would be wise to include such logic in the actual implementation. As the figure shows, the C model can be quite compact.

Lastly, Fig. 9.52 plots the spectrum of the output data when the C model is supplied with a −1 dBFS input. Since the "noise floor" in the 0 to 20 kHz passband is −153 dBFS/500 Hz, the integrated noise in this band is −137 dBFS. This value is small enough to make the effect of digital truncation noise negligible, and thereby validates our word-length choices.

---

†. Recall that the modulator coefficients are scaled for a dynamic range of unity. Thus the binary representations of these states should be viewed as fractional quantities, with a binary point located immediately after the sign bit. Aligning the binary point of a 17-bit word with a 12-bit word therefore requires a shift of 5 places to the right.
Alternatively, one can view the integer version of the modulator as having resulted from scaling the internal states of the original modulator by $2^{N_1-1}$, $2^{N_2-1}$ and $2^{N_3-1}$. As a result of this scaling, all coefficients of the modulator change by powers of 2.

```
#include <stdio.h>

main(){
    long u, v, x1=0, x2=0, x3=0;
    while( scanf( "%d", &u ) >0   ){
        v = (18*x3) >> 12;
        if( v > 4 )
            v = 4;
        if( v < -4 )
            v = -4;
        x3 += (x2>>5) - (v<<8) - (v<<7);
        x2 += (x1>>5) - (v<<13);
        x1 += u - (v<<16);
        printf( "%d\n", v );
        }
}
```

Figure 9.51: C model of the modulator.

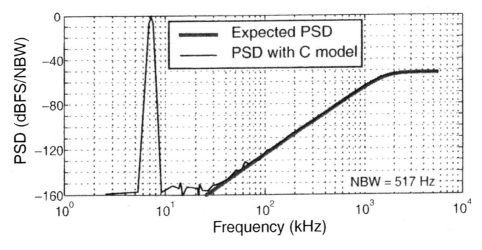

Figure 9.52: Simulated and expected PSDs.

### 9.6.2 Interpolation Filter Design

As discussed in Chapter 7, the task of the interpolation filter is to reduce the power of the unneeded spectral replicas centered at the Nyquist frequency $f_{s,in}$ and at its multiples up to $(OSR-1)f_{s,in}$. This reduces the noise power in the noise-shaping loop, improving its dynamic range. It also makes the task of the analog reconstruction filter easier. Since the quantization noise occupies the same frequency band, it is unnecessary to suppress the spectral images very rigorously; it is sufficient to reduce them to (or somewhat below) the quantization noise level.

As an aid to the design of the filter, Fig. 9.53 plots the STF for the modulator designed in the previous subsection along with the quantization noise's power spectral density (PSD). (The PSD plotted assumes a noise bandwidth $NBW = f_{s,in}/2$ in order to facilitate comparisons between the PSD and the STF.[†]) As this figure demonstrates, the STF attenuates any high-frequency images present in the modulator's input by about 20 dB; another 15 dB of attenuation must

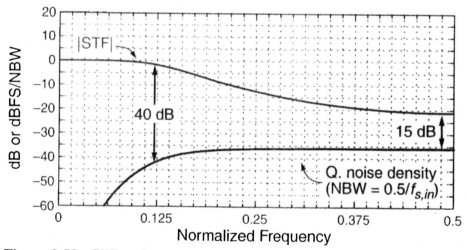

Figure 9.53: STF and quantization noise density for the DAC modulator.

---

† The expression plotted is $|NTF|$, multiplied by the scale factor $\left(\sqrt{\frac{f_{s,in}}{f_s}}\right)\left(\frac{\sqrt{2/3}}{M}\right)$. The reader is referred to Appendix A for a discussion of noise and its PSD.

be provided by the interpolation filter to bring the image power down to the level of the high-frequency quantization noise. Since the reconstruction filter must attenuate both the images and the quantization noise to acceptable levels, providing further image attenuation is generally neither necessary nor very helpful. Having less image attenuation is certainly possible, but may increase the burden on the reconstruction filter.

Since the interpolation factor is $OSR = 256 = 2^8$, it is convenient and efficient to realize the filter as the cascade of several stages, each interpolating by a power of 2. Such stages can be realized by the sinc filters earlier introduced as simple decimation filter stages in Section 2.9. As described there, the transfer function of the $\text{sinc}^k$ filter is

$$H(z) = \left[\frac{(1-z^{-N})}{N(1-z^{-1})}\right]^k. \tag{9.33}$$

Note that here $z$ is associated with the high (output) sampling frequency of the stage; $N$ is the interpolation factor, the ratio of the output and input sampling rates.

In the frequency domain, the transfer function becomes $[\text{sinc}(Nf)/\text{sinc}(f)]^k$. This function is close to 1 around dc, and zero at $f = f_s/N, 2f_s/N, \ldots$[†] Hence, a $\text{sinc}^k$ filter can be used to reduce the amplitudes of the spectral images centered around these frequencies.

A key issue in the design of the sinc filter is how well it can suppress the power at the lowest image of the signal band edge $f_{B0}$, located at $f_1 = f_s/N - f_{B0}$. This can be described by the *image gain*

$$G = \frac{H(e^{j2\pi f_1})}{H(e^{j2\pi f_{B0}})} \tag{9.34}$$

---

[†]. Here, $f_s$ is the sample rate at the output of the filter.

## 9 Example Modulator Systems

$G$ is a function of the interpolation factor $N$ (here equal to 2), and of the input oversampling ratio

$$OSR_{in} = \frac{f_s/N}{2f_{B0}} \qquad (9.35)$$

A plot of $G$ vs. $OSR_{in}$, shown in Fig. 9.54, indicates that $G$ is a strong function of $OSR_{in}$, but only a weak function of $N$.

Next, it will be shown that the last stage of the interpolation filter may be chosen as a simple $\text{sinc}^k$ filter with $k = 1$ and $N = 4$.[†] For this stage, $OSR_{in}$ is 64, and hence by the curve of Fig. 9.54 the replica attenuation is around 43 dB. Also, here $f_1 \approx f_s/4$. At this frequency, as Fig. 9.53 shows, the STF is down by about 12 dB, further attenuating the replica, while the noise density is around −35 dB. Hence, at this frequency the remaining replica noise at the output of the delta-

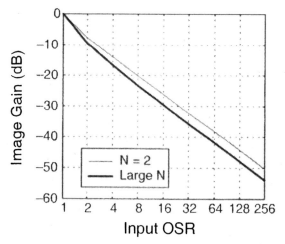

**Figure 9.54:** Image gain for $\left(\frac{1}{N}\right)\left(\frac{1-z^{-N}}{1-z^{-1}}\right)$.

---

[†] As shown in Fig. 2.25, for $k = 1$ the impulse response consists of $N$ replicas of the input impulse. Hence a $\text{sinc}^1$ interpolator acts simply as a sample-and-hold (zero-order hold) stage, and so requires no arithmetic operations in its implementation.

sigma loop will be around $43 + 12 - 35 = 20$ dB below the quantization noise. This is more than enough, and justifies the use of the simplest sinc filter in this stage.

As shown above, a zero-order hold suffices for taking the data rate from $f_s/4$ to $f_s$. However, a zero-order hold is not sufficient for taking the data rate from $f_s/8$ to $f_s/4$, since the separation between the STF and the shaped quantization noise is about 40 dB at $f_s/8$ (Fig. 9.53), whereas a zero-order hold only provides about 35 dB of attenuation. Thus a first-order hold, or *linear interpolator*, with transfer function $\text{sinc}_2^2$, should be used to increase the data rate from $f_s/8$ to $f_s/4$. (See Section 2.9.2 for a discussion of this filter.)

To increase the data rate from $f_s/16$ to $f_s/8$, Fig. 9.53 indicates that approximately 60 dB of image attenuation is required. Fig. 9.54 shows that at $OSR_{in} = 16$, a $\text{sinc}_2^1$ transfer function provides about 25 dB of attenuation. Thus, in order to meet the 60-dB requirement, a $\text{sinc}_2^3$ function is required for this interpolator stage. Similarly, a $\text{sinc}_2^5$ function is needed to raise the sampling rate from $f_s/32$ to $f_s/16$ ($8f_{s,in}$ to $16f_{s,in}$), while $\text{sinc}_2^7$ is needed to go from $4f_{s,in}$ to $8f_{s,in}$.

Fig. 9.55 plots the individual and composite frequency responses for the above interpolation stages, and compares the final result with the scaled NTF. In the diagram, the top curve shows the response of the $\text{sinc}_4^1$ filter; the next curve down shows the output response of the modulator, which includes also the STF of the modulator. In the next diagram, the top curve is the response of the preceding ($\text{sinc}_2^2$) stage, the bottom one is the same curve combined with the responses of all stages which follow it, including the STF of the modulator. The curves in the rest of the diagrams are similarly constructed. The last diagram shows also the PSD of the quantization noise; even without the image-narrowing effect of the yet-to-be discussed interpolation filters which precede the sinc filters, many of the images are already below the quantization noise curve, as desired.

The reader may be concerned that the use of high-order sinc filters implies high arithmetic complexity and a loss of passband flatness, but this is not the case. Considering the arithmetic complexity, Table 9.10 lists the computational requirements

## 9 Example Modulator Systems

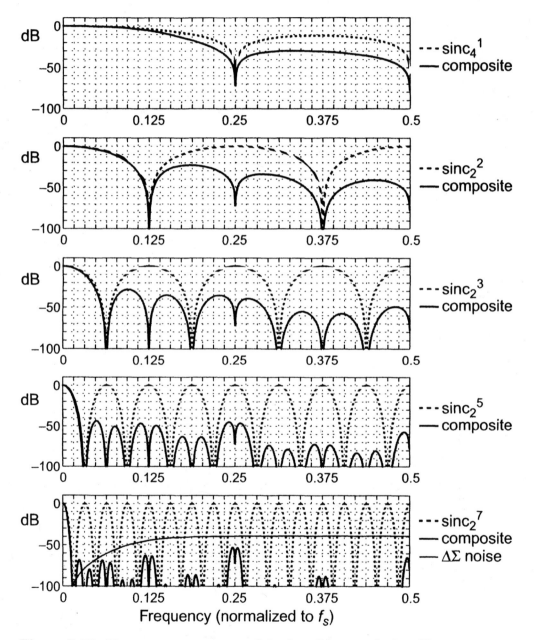

**Figure 9.55:** Frequency responses of the last 5 interpolation filter stages. The composite responses include the modulator's STF.

of direct implementations[†] of these filters. Since the total number of additions is less than 256 per input sample, the computational requirements of the modulator (8 additions per output sample, i.e. 2048 additions per input sample) are considerably greater. Furthermore, since the sinc filters only require a total of 13 registers, the storage requirements of the sinc filters are also reasonable. Implementations of the sinc cascade range from a direct implementation with dedicated hardware for each operation (the lowest-power implementation) to a micro-controlled accumulator-plus-register-bank (which requires the smallest chip area).

| Parameter | $sinc_2^7$ | $sinc_2^5$ | $sinc_2^3$ | $sinc_2^2$ | $sinc_4^1$ |
|---|---|---|---|---|---|
| Output Sample Rate | $8 f_{s,in}$ | $16 f_{s,in}$ | $32 f_{s,in}$ | $64 f_{s,in}$ | $256 f_{s,in}$ |
| Sinc order | 7 | 5 | 3 | 2 | 1 |
| Number of Additions at the output sample rate (= Number of registers) | 6 | 4 | 2 | 1 | 0 |
| Number of Additions at $f_{s,in}$ | 48 | 64 | 64 | 64 | 0 |

Table 9.10. Sinc filter complexity (direct implementation).

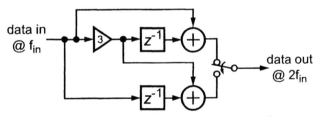

Figure 9.56: A polyphase implementation of a $sinc_2^3$ interpolator.

---

†. A direct implementation of a $sinc_2^N$ interpolator requires $N-1$ additions at the output rate. A polyphase implementation is sometimes more efficient. For example, Fig. 9.56 shows a polyphase implementation of a $sinc_2^3$ filter. This implementation requires only 3 additions at the input rate, or 1.5 additions at the output rate, which represents a 25% savings.

Considering next the passband flatness, Fig. 9.57 plots the passband response of the sinc filter cascade. The 1.2-dB passband droop can readily be corrected by one of the filter stages which precedes the sinc cascade.

The two remaining interpolation stages transform the data rate from $f_{s,in}$ to $2f_{s,in}$ and from $2f_{s,in}$ to $4f_{s,in}$. The latter operation can be accomplished by a Saramäki halfband filter [16], designed as in Fig. 9.58 using the toolbox function designHBF. Note that the stopband attenuation (the second argument to design-HBF) is specified as 70 dB since the sinc filters chosen above provide an additional 20 dB of attenuation at the edge of this filter's stopband. Table 9.11 lists the resulting coefficients, Fig. 9.59 plots the associated frequency response and Fig. 9.60 plots the associated impulse response. (Refer to the designHBF documentation on page 405 for a description of this function and for an illustration of the filter structure.) As Fig. 9.59 shows, this filter achieves 82 dB of image attenuation; the arithmetic complexity is 44 additions at its input rate, or 88 additions at the $f_{s,in}$ rate, and the filter requires 50 registers.

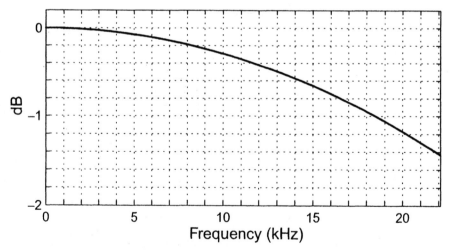

Figure 9.57: Passband response of the sinc filters.

# Audio DAC

We now turn our attention to the first interpolation filter. The situation is somewhat special for this stage. According to our design specifications, this filter must have a passband from dc to $f_{B0} = 20$ kHz, and must suppress the images above $f_{s,\,in}/2 = 22.05$ kHz by at least 90 dB[†]. A halfband filter would have a response with an odd symmetry around $f_{s,\,in}/2$, and hence would only provide 6 dB attenuation at that frequency. Since we also need to correct for the droop of the sinc fil-

```
[f1,f2,info] = designHBF(0.125,undbv(-70));
figure(1); clf
f = linspace(0,0.5,256);
plot(f*4, dbv(frespHBF(f,f1,f2)),'b','Linewidth',1.5)
figureMagic([0 2],0.25,2, [-140 10],10,2);
printmif('HBF_freq', [5 2.5], 'Helvetica10')

N = (2*length(f1)-1)*2*(2*length(f2)-1)+1;
y = simulateHBF([1 zeros(1,N-1)],f1,f2);
stem([0:N-1],y);
figureMagic([0 N-1],5,2, [-0.2 0.5],0.1,1)
printmif('HBF_imp', [5 2], 'Helvetica10')
```

**Figure 9.58:** Halfband filter design.

| $F_1$ Filter | $F_2$ Filter |
|---|---|
| $2^0 - 2^{-4} + 2^{-6}$ | $2^{-1} + 2^{-4} + 2^{-8}$ |
| $-2^{-1} - 2^{-3} - 2^{-5}$ | $-2^{-3} + 2^{-5} - 2^{-8}$ |
| $2^{-2} - 2^{-4} + 2^{-6}$ | |

**Table 9.11.** Coefficients for the Saramäki halfband filter.

---

†. The need for all image energy to be suppressed is debatable. If the input data is already bandlimited to $f_{B0}$, then the stopband edge can be placed at $f_{s,\,in} - f_{B0}$, which doubes the width of the transition band. For an audio application, it can also be argued that images above 20 kHz are inaudible, and thus unimportant. Nonetheless, the convention in high-performance commercial audio DACs is to suppress all image energy.

## 9 Example Modulator Systems

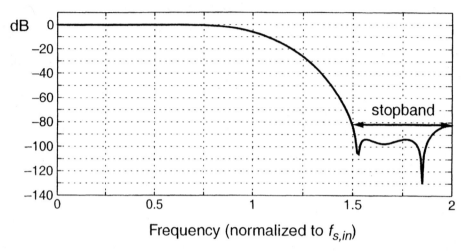

Figure 9.59: Frequency response of the Saramäki halfband filter.

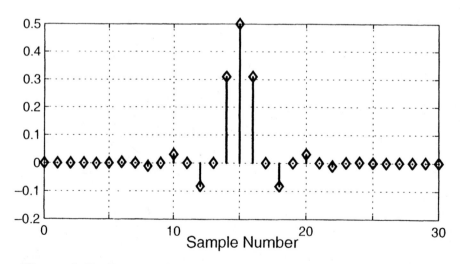

Figure 9.60: Impulse response of the Saramäki halfband filter.

ter cascade, the first filter needs to be a general FIR filter, rather than a halfband filter.

MATLAB's `remez` function can be used to design this FIR filter. The filter's order needs to be very high in order to provide the steep 20-kHz-to-22-kHz transition band called for in our design specifications. Fig. 9.61 shows the composite passband response with a $186^{th}$-order FIR first-stage. Since a realization of this filter requires 445 additions at its input rate and 186 registers, the first interpolation stage will occupy the bulk of the decimation filter's real estate. Fortunately, this complex filter operates at the lowest clock rate (namely 88.2 kHz) of all the stages. Even so, it will be responsible for approximately 2/3 of the interpolation filter's power consumption.

Fig. 9.62 shows the structure of the complete filter. As shown in Fig. 9.61, the passband variation of this filter is 0.08 dB, meeting the specifications. Fig. 9.63 demonstrates that the close-in images are attenuated by 90 dB, again in accordance with our design targets. All that remains is to attenuate the far-out images and quantization noise using the reconstruction filter.

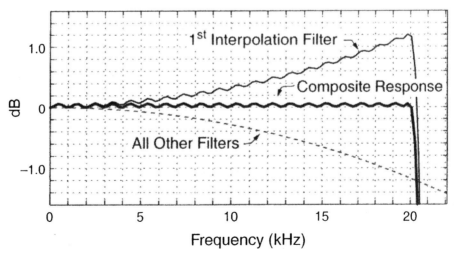

**Figure 9.61: Passband response of the complete interpolation filter.**

# 9 Example Modulator Systems

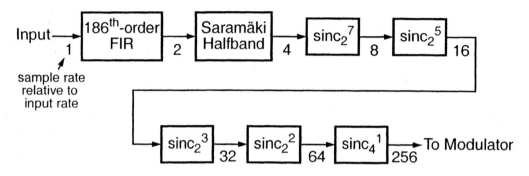

Figure 9.62: Structure of the interpolation filter.

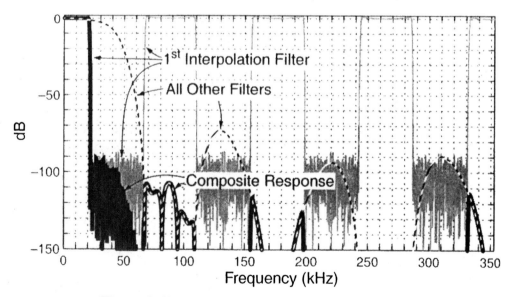

Figure 9.63: Frequency response up to $8f_{s,in}$.

### 9.6.3 DAC and Reconstruction Filter Design

Fig. 9.64 shows the structure of the DAC and reconstruction filter. The DAC has a current-mode output, while the reconstruction filter is a continuous-time single-amplifier biquad.

The transfer function of the filter circuit in Fig. 9.64 is

$$\frac{V_{out}(s)}{I_{in}(s)} = \frac{2R}{\left(\dfrac{s}{\omega_0}\right)^2 + \left(\dfrac{s}{\omega_0 Q}\right) + 1} \tag{9.36}$$

where

$$Q = \sqrt{\frac{C_1}{2C_2}} \tag{9.37}$$

and

$$\omega_0 = \frac{1}{\sqrt{2C_1 C_2} R}. \tag{9.38}$$

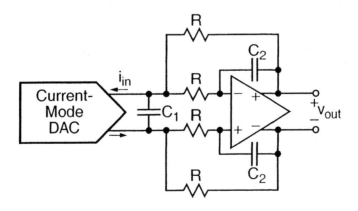

**Figure 9.64: DAC and reconstruction filter.**

For a second-order Butterworth response, $Q = 1/\sqrt{2}$, which implies $C_1 = C_2$. Setting the cut-off frequency to 80 kHz yields the quantization noise density shown in Fig. 9.65. Since the peak out-of-band quantization noise density is an acceptable −90 dBFS/22 kHz, the cut-off frequency does not need to be lowered. (The reconstruction filter would need a third-order transfer function to prevent a rise in the out-of-band noise with frequency.) Also, since the attenuation at 20 kHz is a mere 0.017 dB, while the group delay variation is less than 1% of a 44-kHz sample period, the passband droop and the group delay error associated with the reconstruction filter are sufficiently small for most audio applications.

The last step we will perform is component-value selection. If we assume that the output swing is 0.7 $V_p$, then a full-scale sine wave has an rms output of 0.5 V and so a 110-dB SNR requires a noise voltage of no more than $0.5 \cdot 10^{-110/20} = 1.6$ μV. Dividing this value by $\sqrt{20 \text{ kHz}}$ yields a required noise density of 11 nV/$\sqrt{\text{Hz}}$. Allocating half of our noise budget to the reconstruction filter (so that the DAC may take the other half) gives a specification of 8 nV/$\sqrt{\text{Hz}}$ for the output-referred noise of the reconstruction filter.

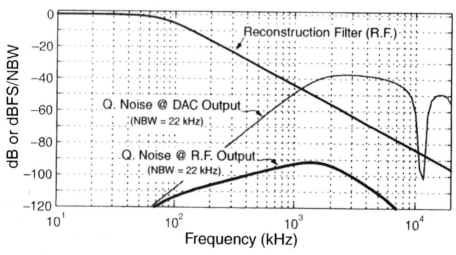

**Figure 9.65: Quantization noise density at the output of the reconstruction filter.**

Analysis of the circuit in Fig. 9.64 shows that, with an ideal op amp, the output-referred noise density of the reconstruction filter is $4\sqrt{kTR}$ at low frequencies. Equating this to the desired noise density leads to $R = 1$ k$\Omega$ and $C_1 = C_2 = 1.4$ nF. For a high-dynamic-range DAC such as that considered in this example, the need for large capacitors in the reconstruction filter typically precludes a monolithic implementation.

Since the low-frequency gain of the filter is $2R = 2$ V/mA, the peak DAC current for a 0.7-V peak output is $0.7/2 = 0.35$ mA. Hence, the differential DAC may be constructed from 8 pairs of current sources, each providing 44 µA.

## 9.7 Conclusions

This chapter examined five ADC systems and one DAC system. The ADCs ranged from a simple second-order, single-bit low-frequency switched-capacitor modulator (MOD2) to a sixth-order, multi-bit, mixed discrete-time/continuous-time bandpass modulator. Example circuits were given for a number of important building blocks, and sample calculations of block specifications were illustrated in the context of each system.

### 9.7.1 The ADC State-of-the-Art

We conclude our journey through the land of delta-sigma with a birds-eye view of the ADC landscape, using the following figure-of-merit (FOM)

$$FOM = DR_{dB} + 10\log\left(\frac{BW}{P}\right) \qquad (9.39)$$

to compare the performance of various ADCs. This figure-of-merit is essentially the same as that proposed in [17], namely $FOM = (4kT \times DR \times 2BW)/P$, in that a factor of 2 in power (P), a factor of 2 in bandwidth (BW) and a factor of 2 (3 dB) in dynamic range (DR) are all considered to be equivalent. Our FOM omits an $8kT$ factor and uses a dB scale in order to facilitate calculations. Section 9.7.2 offers several theoretical arguments to justify the exponents associated with these three performance metrics.

The example systems considered in this chapter span a 1000:1 range in DR, P and BW. Table 9.12 summarizes the estimated specifications for these example systems. Despite the wide range in ADC bandwidth, performance and power consumption, we see that the values of FOM achieved by these distinctly different architectures are remarkably close, hovering around $FOM = 160$.

Similarly, Table 9.13 lists the performance of ADCs which achieve the largest FOM for a given bandwidth, while Fig. 9.66 plots the FOM versus bandwidth for these and other ADCs reported in the open literature. Note that the first three entries in Table 9.13 have about the same bandwidth and achieve almost the same FOM value despite having dynamic ranges which span a 23-dB (200:1) range. This fact lends credence to our claim that dynamic range and power can be traded at the rate of 1 dB/dB. Table 9.13 shows that $\Delta\Sigma$ converters currently achieve the best FOM value for bandwidths up to approximately 10 MHz. Beyond this value, it becomes increasingly difficult to achieve a sufficiently high oversampling ratio and so pipeline and folding ADCs dominate.

The latter observation is also apparent in Fig. 9.66. As this figure indicates, a FOM "ceiling" of about 170 dB exists for bandwidths up to a few MHz; beyond this frequency, the achievable FOM drops due to technological limitations and the need to use architectures which are less efficient than $\Delta\Sigma$ modulation. As CMOS line-

| Design | BW (kHz) | DR (dB) | P (mW) | FOM |
|---|---|---|---|---|
| MOD2 (SC) | 1 | 100 | 0.05 | 173 |
| MOD5 (SC) | 20 | 110 | 200 | 160 |
| 2-0 Cascade (SC) [8] | 1250 | 89 | 550 | 157 |
| μpower (CT Active-RC) | 10 | 80 | 0.1 | 160 |
| Bandpass (CT LC) | 15 | 105 | 50 | 160 |
| | 333 | 90 | | 158 |

Table 9.12. Performance estimates for our example ADCs.

Conclusions

| Architecture | BW | DR (dB) | P (mW) | FOM |
|---|---|---|---|---|
| ΔΣ: 5(4b) SC [18] | 20 kHz | 111 | 27.5 | 170 |
| ΔΣ: 3(1b) SC [19] | 20 kHz | 88 | 0.14 | 170 |
| ΔΣ: 2(1b)-1(1b) SC [17] | 25 kHz | 99 | 2.5 | 169 |
| ΔΣ: 2(5b)-2(3b)-1(3b) [10] | 2 MHz | 95 | 150 | 166 |
| ΔΣ: 5(1b) gm-C [20] | 3.8 MHz | 78 | 9 | 164 |
| ΔΣ: 3(6b) gm-C [21] | 12 MHz | 79 | 75 | 161 |
| Pipeline [22] | 80 MHz | 66 | 290 | 150 |
| Pipeline [23] | 100 MHz | 51 | 135 | 140 |
| Folding [24] | 300 MHz | 45 | 200 | 137 |
| Folding [25] | 800 MHz | 46 | 1270 | 134 |

Table 9.13. Published ADCs achieving the highest FOM at a given BW.

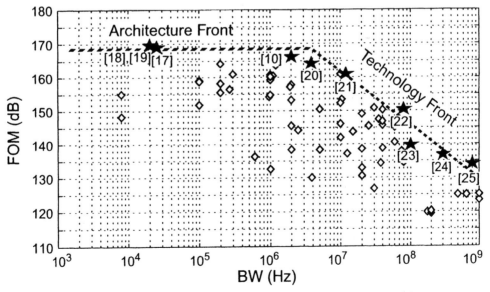

Figure 9.66: FOM vs. bandwidth for published ADCs.

359

widths continue to shrink, the "technology front" should therefore shift to the right. Achieving FOM values above 170 will require more efficient building blocks and/or architectural innovations.

## 9.7.2 FOM Justification

To see why 3 dB in DR should be equivalent to a factor of 2 in power, consider what happens when two identical ADCs are used to sample the same signal. Averaging the ADCs' outputs reduces the system noise by 3 dB, but to achieve this improvement we had to double the number of ADCs and hence double the power consumption. Similarly, if we take an ADC and "cut it in half" (so that the original ADC can be considered to be composed of two half-power ADCs operating in parallel), we can cut the power consumption by a factor of 2 and lose 3 dB in DR.[†]

The fact that this trade-off is (in principle) bidirectional implies that no better trade-off can exist for power-optimal ADCs. To see this, assume that a certain ADC (ADC1) is purported to be optimal in the sense that no other ADC consumes the same power $P$, achieves the same bandwidth BW and has higher dynamic range DR. Now assume that the structure of ADC1 is such that it offers a more favorable trade-off of (say) 6 dB in SNR for each doubling in $P$. Apply this trade-off to yield ADC2 which achieves DR + 6 dB with power $2P$. If we "cut ADC2 in half" to yield ADC3, then ADC3 will, according to the preceding discussion, achieve DR + 3 dB with power $P$, which is superior to ADC1. Consequently, ADC1 cannot be power-optimal.

Given any ADC, we can (in principle) construct other ADCs possessing the same BW and the same DR/P ratio.[‡] Alternatively, if two ADCs have the same BW, the one which has the larger DR/P ratio is superior since it can be used to construct an ADC which consumes the same power as the other ADC but at the same time achieves greater DR.

---

[†]. Admittedly, halving an ADC cannot be repeated many times before second-order effects such as peripheral capacitance and lithographic limits become significant. When such second-order effects are significant, a 3-dB reduction in SNR will not be accompanied by a full factor-of-2 reduction in power consumption.

[‡]. Here DR is a power ratio, on a linear scale.

As further justification for why 3 dB in DR is equivalent to a factor of 2 in power, consider cooling an ADC from a temperature $T$ to a temperature $T/2$. Since the power of all thermal noise sources is proportional $T$, cooling the ADC in this way will reduce all noise sources by a factor of 2, or 3 dB. However, the second law of thermodynamics tells us that in order to transport a power $P$ from a heat sink at temperature $T/2$ to a heat sink at a temperature $T$ requires the expenditure of at least a power $P$, i.e. we would need to supply $P$ to the circuit and $P$ to the refrigeration unit, for a total power consumption of $2P$.

Having established that 3 dB in DR is equivalent to a factor of 2 in power, let's consider BW. We can reduce the bandwidth of an ADC simply by applying linear filtering. Assuming that the ADC's noise is white, reducing the bandwidth by a factor of 2 increases the dynamic range by 3 dB. If we assume that such filtering can be done with negligible power, we can therefore trade bandwidth for dynamic range at constant power at the rate of +3 dB per octave reduction. The reverse trade-off is somewhat more tenuous, but in principle we can time-interleave two ADCs and thereby achieve twice the bandwidth at double the power. (Similarly, complex signal processing yields a 2x increase in bandwidth while requiring a 2x increase in power.) Applying the previously-established link between power and dynamic range leads to a bidirectional bandwidth/dynamic range trade-off of 3 dB per octave. Once again, since this trade-off is bidirectional, no better trade-off can exist.

The above arguments support our choice for an ADC figure-of-merit, but this figure-of-merit does not capture such important practical considerations as silicon area, manufacturability, or ease of use. For example, switched-capacitor modulators (especially those with large sampling capacitors) are notoriously hard to drive, and as much power may be needed for the driver as for the ADC itself. Also, since switched-capacitor ADCs require anti-alias filtering, whereas continuous-time modulators do not, a continuous-time ADC may simplify the architecture of the system in which it is embedded. Despite these shortcomings, the FOM described here can be used to classify instantly a given set of ADC specifications as easy ($FOM < 155$), hard ($FOM \approx 160$), very hard ($FOM \approx 165$) or beyond the state-of-the-art ($FOM > 170$).

## References

[1] Y. Yang, A. Chokhawala, M. Alexander, J. Melanson, and D. Hester, "A 114 dB 68 mW chopper-stabilized stereo multi-bit audio A/D converter," *ISSCC Digest of Technical Papers*, pp. 64-65, Feb. 2003.

[2] B. E. Boser and B. A. Wooley. "The design of sigma-delta modulation analog-to-digital converters," *IEEE Journal of Solid-State Circuits,* vol. 23, pp. 1298-1308, December 1988.

[3] B. P. Brandt, D. E. Wingard and B. A. Wooley, "Second-order sigma-delta modulation for digital-audio signal acquisition," *IEEE Journal of Solid-State Circuits,* vol. 26, pp. 618-627, April 1991.

[4] D. R. Welland, B. P. Del Signore, E J. Swanson, T. Tanaka, K. Hamashita, S. Hara, and K. Takasuka, "Stereo 16-bit delta-sigma A/D converter for digital audio," *Journal of the Audio Engineering Society*, vol. 37, pp. 476-486, June 1989.

[5] B. P. Del Signore, D. A. Kerth, N S. Sooch, and E. J. Swanson, "A monolithic 20b delta-sigma converter," *IEEE Journal of Solid-State Circuits,* vol. 25, pp. 1311-1317, December 1990.

[6] P. F. Ferguson Jr., A. Ganesan, R. W. Adams, S. Vincelette, R. Libert, A. Volpe, D. Andreas, A. Charpentier and J. Dattorro. "An 18b 20kHz dual sigma-delta A/D converter," *ISSCC Digest of Technical Papers,* vol. 34, February 1991.

[7] R. W. Adams, P. F. Ferguson, A. Ganesan, S. Vincelette, A. Volpe and R. Libert, "Theory and practical implementation of a fifth-order sigma-delta A/D converter," *Journal of the Audio Engineering Society*, vol. 39, pp. 515-528, July 1991.

[8] T. L. Brooks, D. H. Robertson, D. F. Kelly, A Del Muro and S. W. Harston, "A cascaded sigma-delta pipeline A/D converter with 1.25 MHz signal bandwidth and 89 dB SNR," *IEEE Journal of Solid-State Circuits*, vol. 32, no 12, pp. 1896-1906, December 1997.

[9] I. Fujimori, L. Longo, A. Hairapetian, K. Seiyama, S. Kosic, C. Jun and S. L. Chan, "A 90-dB SNR 2.5 MHz output-rate ADC using cascaded multibit delta-sigma modulation at 8x oversampling ratio," *IEEE Journal of Solid-State Circuits,* vol. 35, no. 12, pp. 1820-1828, December 2000.

[10] K. Vleugels, S. Rabii, and B. A. Wooley, "A 2.5-V sigma–delta modulator for broadband communications applications," *IEEE Journal of Solid-State Circuits,* vol. 36, no 12, pp. 1887-1899, Dec. 2001.

[11] Y. Geerts, M. Steyaert and W. Sansen, *Design of multi-bit delta-sigma A/D converters*, Kluwer academic publishers, Boston, Massachusetts, 2002.

[12] Y. Geerts, M. Steyaert and W. Sansen, "A high-performance multi-bit CMOS $\Delta\Sigma$ converter," *IEEE Journal of Solid-State Circuits*. vol. 35, no. 12, pp. 1829-1840, Dec. 2000.

[13] E. van der Zwan, E. Dijkmans and J. Huijsing, "A 0.2 mW $\Sigma\Delta$ modulator for speech coding with 80 dB dynamic range," *IEEE Journal of Solid-State Circuits,* vol. 31, no. 12, pp. 1873-1880, Dec. 1996.

[14] R. H. M van Veldhoven, B. J. Minnis, H. A. Hegt and A. H. M. van Roermund, "A 3.3-mW $\Sigma\Delta$ modulator for UMTS in 0.1µm CMOS with 70-dB dynamic range in 2-MHz bandwidth," *IEEE Journal of Solid-State Circuits*, vol. 37, no. 12, pp. 1645-1652, Dec. 2002.

[15] R. Schreier, J. Lloyd, L. Singer, D. Paterson, M. Timko, M. Hensley, G. Patterson, K. Behel. J. Zhou and W. J, Martin, "A 10-300 MHz IF-digitizing IC with 90-105 dB dynamic range and 15-333 kHz bandwidth," *IEEE Journal of Solid-State Circuits*, vol. 37, no. 12, pp. 1636-1644, Dec. 2002.

[16] T. Saramäki, "Design of FIR filters as a tapped cascaded interconnection of identical subfilters," *IEEE Transactions on Circuits and Systems*, vol. 34, no. 9, pp. 1011-1029, Sept. 1987.

[17] S. Rabii, and B. A. Wooley, "A 1.8-V digital-audio sigma–delta modulator in 0.8μm CMOS," *IEEE Journal of Solid-State Circuits*, vol. 32, no. 6, pp. 783-796, June 1997.

[18] Y. Yang, A. Chokhawala, M. Alexander, J. Melanson, and D. Hester, "A 114 dB 68 mW chopper-stabilized stereo multi-bit audio A/D converter," *ISSCC Digest of Technical Papers*, pp. 64-65, Feb. 2003.

[19] L. Yao, M. Steyaert and W. Sansen, "1V 88dB 20kHz ΣΔ modulator in 90nm CMOS," *ISSCC Digest of Technical Papers*, pp. 80-81, February 2004.

[20] R. H. M van Veldhoven, "A tri-mode continuous-time ΣΔ modulator with switched-capacitor feedback DAC for a GSMEDGE/CDMA2000/UMTS receiver," *ISSCC Digest of Technical Papers*, pp. 60-61, Feb. 2003.

[21] M. Moyal, M. Groepl, H. Werker, G. Mitteregger and J. Schambacher, "A 700/900mW/channel CMOS dual analog front-end IC for VDSL with integrated 11.5/14.5dBm line drivers," *ISSCC Digest of Technical Papers*, pp. 416-417, Feb. 2003.

[22] B. Murmann and B. Boser, "A 12b 75MS/s pipelined ADC using open-loop residue amplification," *ISSCC Digest of Technical Papers*, pp. 328-329, Feb. 2003.

[23] B. Hernes, A. Briskemyr, T. N. Andersen, F. Telstø, T. E. Bonnerud and Ø. Moldsvor, "A 1.2V 220MS/s 10b pipeline ADC implemented in 0.13μm Digital CMOS," *ISSCC Digest of Technical Papers*, pp. 256-257, Feb. 2004.

[24] G. Geelen and E. Paulus, "An 8b 600MS/s 200mW CMOS folding A/D converter using an amplifier preset technique," *ISSCC Digest of Technical Papers*, pp. 254-255, Feb. 2004.

[25] R. Taft, C. Menkus, M. R. Tursi, O. Hidri, V. Pons, "A 1.8V 1.6GS/s 8b self-calibrating folding ADC with 7.26 ENOB at Nyquist frequency," *ISSCC Digest of Technical Papers*, pp 252-253, Feb. 2004.

# 9 Example Modulator Systems

# APPENDIX A — *Spectral Estimation*

The purpose of this appendix is to demystify the procedure of analyzing $\Delta\Sigma$ data using the Fast Fourier Transform (FFT) [1]. The FFT is widely used to estimate the power spectral density of $\Delta\Sigma$ data, but is also sometimes abused in the process. When using the FFT to analyze $\Delta\Sigma$ data, the $\Delta\Sigma$ designer needs to be familiar with several important concepts, namely windowing, scaling, noise bandwidth and averaging. This appendix deals with each of these subjects in turn, applies them to an example, and concludes with a brief discussion of the mathematical background.

The FFT is a fast algorithm for computing the Fourier Transform

$$X(f) = \sum_{n=0}^{N-1} x(n) \cdot e^{-j2\pi f n} \qquad (A.1)$$

of a length-$N$ discrete-time sequence $x(n)$ at the $N$ frequency points, the *FFT bins*, $f = 0, 1/N, 2/N, \ldots, (N-1)/N$.[†]

---

[†] Here, as earlier, the sampling rate is assumed to be 1 Hz.

# APPENDIX A  Spectral Estimation

A discrete-time signal with period $N$ consists of a dc term and harmonics of the fundamental frequency $f_1 = 1/N$. Since the amplitude of the $i^{th}$ harmonic $f_i = i/N$ for $i \neq 0$ is given by $2|X(f_i)|/N$, an FFT can easily be used to compute the power spectrum of a periodic signal. Unfortunately, since $\Delta\Sigma$ data is typically not periodic, a direct application of the FFT to $\Delta\Sigma$ data is unwise at best.

We shall consider the "noise" associated with $\Delta\Sigma$ data to be like a random signal, for which a more technical term is *stochastic process* [2]. If the data comes from measurements, then it is bound to contain components which are true noise and our viewpoint is justified. For data obtained from simulation, the noise is the result of a *deterministic* process, and so it is not strictly proper to describe this process as random. However, since the process is complex, non-linear and often chaotic, the fact that the process is actually deterministic has little practical impact.

## A.1 Windowing

Windowing is the act of multiplying the signal to be analyzed by a *window function* $w(n)$ before subjecting it to an FFT. At first glance, it would appear that this operation would alter the spectral content of the signal and therefore be undesirable. Although it *is* true that windowing alters a signal's spectrum, some windowing is inevitable because we can never obtain an infinite-length record of modulator data. The best we can do is operate on a finite record of length $N$. Since a finite-length record can be thought of as the product of the infinite-duration modulator output and a rectangular window

$$w_{rect}(n) = \begin{cases} 1, & 0 \leq n \leq N-1 \\ 0, & \text{otherwise} \end{cases} \quad \text{(A.1)}$$

the damage caused by windowing the data has already been done. Thus the question is not "Should I window my data?" but rather "*How* should I window my data?"

# Windowing

The answer to this question lies in the relationship between the spectrum of the original data and that of the windowed data. Since multiplication in the time domain corresponds to convolution in the frequency domain, the spectrum of a windowed signal is, loosely speaking, the spectrum of the unwindowed signal convolved with the window's spectrum. In order to obtain an accurate spectrum, the designer must choose a window which introduces sufficiently low errors through spectral convolution.

Consider the three windowing functions illustrated in Fig. A.1. Table 1 lists their definitions, and summarizes various parameters which will be discussed in this Appendix. Since a rectangular window has discontinuities at its endpoints, whereas the Hann and Hann² windows are continuous (up to the 2$^{nd}$ and 4$^{th}$ derivatives, respectively), we would expect the rectangular window to have a great deal more high-frequency content than the other two windows. This suspicion in confirmed in Fig. A.2, which plots the magnitudes of the Fourier transform

$$W(f) = \sum_{n=0}^{N-1} w(n) \cdot e^{-j2\pi fn} \tag{A.2}$$

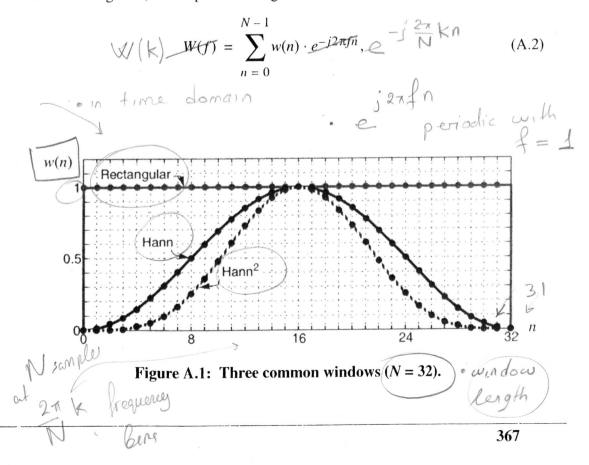

Figure A.1: Three common windows ($N = 32$).

# APPENDIX A  Spectral Estimation

normalized by the dc gain $W(0)$, for each of these windows. $N = 32$ points were used. As Fig. A.2 shows, the peaks of the high-frequency lobes in the spectrum of the rectangular window approach a constant value (which is proportional to $1/N$), whereas the peaks of the high-frequency lobes of the Hann and Hann$^2$ windows go to zero with –60 dB/decade and –100 dB/decade slopes, respectively. The high-

| Window | Rectangular | Hann | Hann$^2$ |
|---|---|---|---|
| $w(n)$, $n = 0, 1, ..., N-1$  ($w(n) = 0$ otherwise) | 1 | $\dfrac{1 - \cos\dfrac{2\pi n}{N}}{2}$ | $\left(\dfrac{1 - \cos\dfrac{2\pi n}{N}}{2}\right)^2$ |
| sum $w^2(n) = \|w\|_2^2$ | $N$ | $3N/8$ | $35N/128$ |
| No. of non-zero FFT bins | 1 | 3 | 5 |
| sum $w(n) = W(0)$ | $N$ | $N/2$ | $3N/8$ |
| NBW | $1/N$ | $1.5/N$ | $35/(18N)$ |

Table A.1: Properties of the three windows illustrated in Fig. A.1.

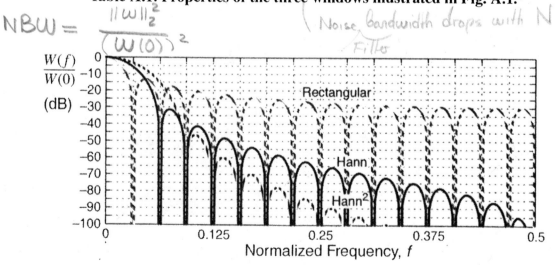

Figure A.2: Frequency responses of the windows shown in Fig. A.1.

frequency behavior of a window's spectrum is of critical importance in determining the magnitude of the error resulting from convolution.

As a demonstration of the convolution problem, Fig. A.3 shows the spectrum of some noise-shaped data, the Fourier transform of a 256-point rectangular window and the Fourier transform of the windowed data. As indicated in Fig. A.3, the skirts of the window convolve with the out-of-band noise, thereby filling in the noise notch and dramatically reducing the apparent SNR.

The $\Delta\Sigma$ designer must ensure that this *noise leakage* is small compared to the actual in-band quantization noise density. In the context of a high-accuracy $\Delta\Sigma$ modulator, the difference between the out-of-band noise density and the in-band noise density can be 80 dB or more. Fig. A.3 indicates that with $N = 256$, the observable difference between the out-of-band and in-band noise densities is only about 23 dB. Increasing $N$ improves the situation, but only at the rate of 3 dB per octave. According to this trend, a rectangular window would need to use more than $10^8$ points to reliably observe an 80 dB difference in noise densities!

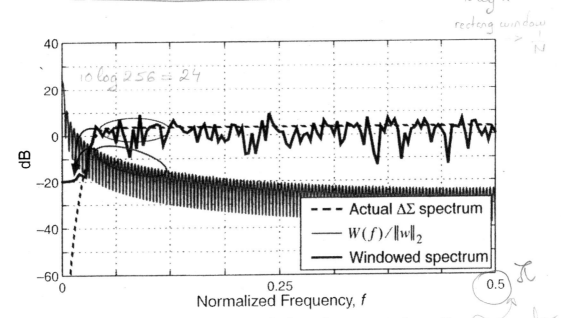

**Figure A.3: A rectangular window obscures a noise null.**

# APPENDIX A  Spectral Estimation

The $\Delta\Sigma$ designer is therefore compelled to use something other than a simple rectangular window. Many different windows exist (for example, see [3] or [4]) but the feature of greatest importance to the $\Delta\Sigma$ designer is the amount of high-frequency attenuation provided by the window. Windows which have finite high-frequency attenuation, such as the Hamming window, are less desirable than windows whose high-frequency attenuation increases without bound. In particular, a Hann window with $N = 512$ is able to resolve an 80 dB difference in noise density, while a Hann$^2$ window does similarly well with $N = 256$. Since the number of data points needed to provide sufficient frequency resolution is usually on the order of several thousands, a simple Hann window usually provides sufficient protection against noise leakage.

Another important consideration in the analysis of $\Delta\Sigma$ data is *signal leakage*. It is convenient, both in the lab and in simulation, to use sine-wave excitation. However, the frequency of that sine wave must be located precisely in an FFT bin, otherwise signal power will bleed into all bins. Fig. A.4 illustrates this phenomenon with several length-512 FFTs of two sine waves. The first has a frequency which is

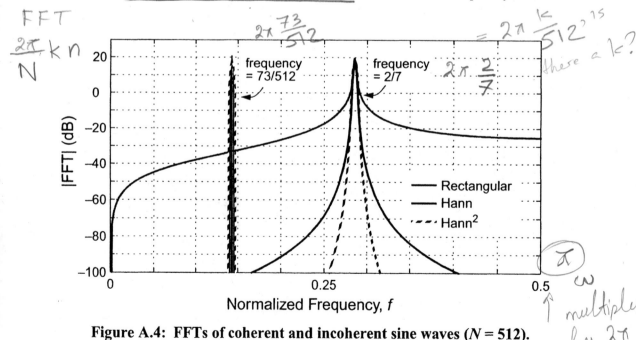

Figure A.4:  FFTs of coherent and incoherent sine waves ($N = 512$).

located precisely in an FFT bin (specifically, bin 73, which is close to a frequency of $1/7$), while the second has a frequency of exactly $2/7$ and so is not in an FFT bin. In the first case (the *coherent* case), the sine wave's power is concentrated in a small number of FFT bins (1 for a rectangular windowing, 3 for Hann, and 5 for Hann$^2$). In the *incoherent* case, the sine wave's power is smeared over all FFT bins. The severity of the spreading is determined by how far away the sine wave is from the nearest bin frequency, and by the shape of window's skirts. As was the case with noise leakage, the rectangular window exhibits the greatest signal leakage because its skirts are the most broad.

In simulation, it is a simple matter to place the signal frequency in an FFT bin and thereby eliminate signal leakage entirely. In the lab, signal leakage can be minimized by phase-locking the generators, and by setting the signal frequency accurately. If it is not possible to place the signal frequency in an FFT bin, windowing can be used to reduce spectral pollution. Alternatively, the signal's frequency and amplitude can be estimated, and the estimated signal can be subtracted from the data record to leave only the noise.

The final question regarding windowing which we will address is the length of the window required to obtain an accurate estimate of a modulator's SNR. The simplest way to estimate SNR is to compute the ratio of the power in the signal bins to the power in the in-band noise bins. Since the signal is usually placed in-band, it therefore occupies a few of the in-band bins. The number of bins occupied by the signal should be a relatively small fraction (less than 20%) of the in-band bins in order to have a small effect (less than 1 dB) on the SNR estimate. If we are using a Hann window, the signal will occupy 3 bins and thus we should have at least 15 in-band bins, which in turn requires $N \geq (30 \cdot OSR)$ bins in all, where OSR is the oversampling ratio.

If the noise is assumed to be flat in-band, the missing noise power can be accounted for by multiplying the power in the noise bins by $1/(1-a)$, where $a$ is the fraction of in-band bins occupied by the signal. Unfortunately, at least in simulation, the in-band quantization noise follows the NTF and so tends to be non-flat. Alternatively, the noise in the signal bins can be estimated by subtracting out the estimated signal component before or after performing the FFT.

# APPENDIX A  Spectral Estimation

Another consideration regarding the required length of the data record relates to the repeatability of the SNR measurement. Since the FFT of a random signal is itself a random quantity, the in-band noise power computed from the FFT is also a random quantity. Numerical experiments indicate that using $N = 30 \cdot OSR$ results in SNR estimates which have a standard deviation of about 1.4 dB. Using $N = 64 \cdot OSR$ results in a standard deviation of about 1.0 dB, while $N = 256 \cdot OSR$ is needed to reduce the standard deviation to 0.5 dB. This degree of repeatability is usually not required on an individual measurement, since it is common for many measurements to be carried out, as would be the case during an input amplitude sweep. The authors recommend using $N = 64 \cdot OSR$.

A tertiary consideration regarding the required length of the data record relates to the observable spurious-free dynamic range (SFDR). Using $N = 64 \cdot OSR$ is usually sufficient to reliably observe an SFDR which is about 10 dB greater than the SNR. In order to detect tones which are more than 10 dB below the total in-band noise, $N$ needs to be increased. Specifically, doubling $N$ increases the observable SFDR by 3 dB.

## A.2 Scaling and Noise Bandwidth

The astute reader may have already observed that the spectral spike associated with a sine-wave component in the data record has a height (as in Fig. A.4) that is dependent on both the window type and the window length. Most windows have $\max |W(f)| = W(0)$ (i.e. the peak in the window's spectrum occurs at dc), and so the height of the peak corresponding to a sine wave is $A \cdot W(0)/2$, where $A$ is the amplitude of the sine wave.

When displaying a spectral plot, it is customary to scale the FFT such that a full-scale sine wave yields a 0-dB spectral peak. If we denote the full-scale range by $FS$, the amplitude of a full-scale sine wave is $A = FS/2$ and thus the scaled version of the FFT which we would present is

## Scaling and Noise Bandwidth

$$\hat{S}_x'(f) = \frac{\left|\sum_{n=0}^{N-1} w(n) \cdot x(n) \cdot e^{-j2\pi fn}\right|^2}{(FS/4)W(0)} \quad (A.3)$$

Note that we have squared the magnitude in order to allow us to interpret $\hat{S}_x'(f)$ as a *power* spectral density (PSD). The power of a sine-wave signal relative to the power of a full-scale sine-wave signal is therefore given by $10\log \hat{S}_x'(f)$, where $f$ is the frequency of the signal. The units of this quantity are often given as dBFS (dB relative to full-scale) in order to emphasize that the reference power is the power of a full-scale sine wave, but we shall see shortly that these units omit an important detail. The caret on the symbol $\hat{S}_x'(f)$ indicates that the above expression is an estimate of the PSD, while the prime indicates that we have scaled the estimate so that sine-wave signals yield calibrated spike heights.

Although the above scaling is convenient for analyzing a signal which consists of sine waves, it is less convenient for analyzing a signal which contains noise. Fig. A.5 illustrates the problem by plotting $\hat{S}_x'(f)$ for a signal consisting of a 0 dBFS sine wave plus white noise having the same power as the sine wave, when different-length rectangular windows are used. (A rectangular window is safe here, since we are not interested in observing notches in $\hat{S}_x'(f)$.) Although the signal spike has the same height in each case, the average "noise floor" of $\hat{S}_x'(f)$ drops by 3 dB every time $N$ doubles. Since the noise power is actually 0 dBFS in each case, the location of the "noise floor" is not the only piece of information that the reader needs.

The problem with sine-wave scaling is that the noise power is, on average, evenly distributed over all FFT bins, whereas the sine-wave power is concentrated in only a few bins. With sine-wave scaling, the power of individual sine-wave components can be read directly from the spectral plot, but in order to determine the noise power, the powers of all the noise bins must be added together.

# APPENDIX A  Spectral Estimation

An alternative method of scaling, which is common in signal-processing texts, is to scale the FFT such that it provides a calibrated noise density. The appropriate scale factor in this case is $1/\|w\|_2^2$, where

$$\|w\|_2^2 = \sum_{n=0}^{N-1} |w(n)|^2 \qquad (A.4)$$

is the *energy* of the window.[†] When this scaling is used, the PSD estimate is

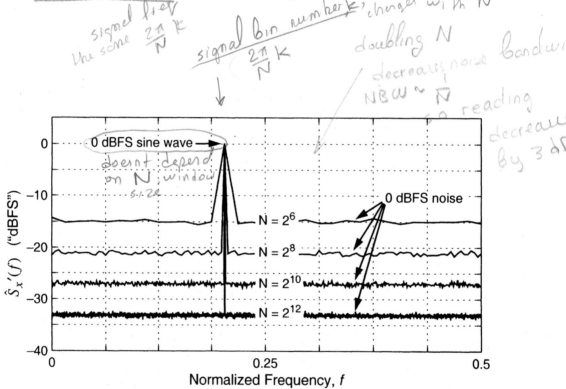

Figure A.5: FFTs of a sine wave plus white noise; sine-wave scaling.

---

† The notation $\|w\|_2$ means *2-norm*, a special case of the *p*-norm $\|w\|_p = \left( \sum_{n=0}^{N-1} |w(n)|^p \right)^{1/p}$.

## Scaling and Noise Bandwidth

$$\hat{S}_x(f) = \frac{\left| \sum_{n=0}^{N-1} w(n) \cdot x(n) \cdot e^{-j2\pi fn} \right|^2}{\|w\|_2^2} \qquad (A.5)$$

and as a result unit-power white noise yields a unit (0-dB) density, regardless of the window type or length. Unfortunately, scaling for noise in this way makes the height of a sine-wave spike dependent on both the window type and length.

Our resolution of the scaling dilemma follows the solution adopted in laboratory instruments, specifically in spectrum analyzers. A spectrum analyzer must contend with the problem of representing the *spectrum* of a periodic signal such as a sine wave on the same display as is used to represent the *spectral density* of a broadband signal such as noise. In a spectrum analyzer, the signal can be thought of as being processed by a bank of filters possessing identical filtering characteristics (gain, bandwidth, etc.), albeit with different center frequencies. The instrument measures the powers at the outputs of the filters and constructs a plot of power versus center frequency.

For a sine-wave input, the display shows a peak at the input frequency, the height of which equals the input power. Thus, the spectrum displayed by a spectrum analyzer is like an FFT with sine-wave scaling. For noise, the display indicates the power of the noise that lies in each filter's bandwidth. In other words, for a noise-like signal, the display gives the product of the noise density and the *noise bandwidth* of the filters.

For a filter with infinitely steep roll-off, the noise bandwidth (NBW) is equal to the filter's bandwidth, while for a filter with a single-pole roll-off, NBW is $\pi/2$ times the 3-dB bandwidth. In general, a filter's *NBW* is the bandwidth of an ideal brick-wall filter which has the same output power given a white noise input and which has the same mid-band gain as the filter under consideration. A spectrum analyzer solves the scaling problem by providing *NBW* along with each spectrum, thereby providing the designer with the information needed to convert the power shown on the display into a power density. NBW depends on various analyzer settings,

375

# APPENDIX A  Spectral Estimation

including those that determine the frequency range and resolution for the spectral plot.

The solution to the scaling problem in the case of a PSD obtained from a sine-wave scaled FFT is similarly simple. All we need do is provide the value of NBW.

The value of NBW is listed in Table 1 for the three windows considered in this appendix. For each window, *NBW* is inversely proportional to $N$, and thus doubling $N$ decreases the apparent level of the noise in a sine-wave scaled PSD by 3 dB, as indicated in Fig. A.5. Let us now illustrate the use of *NBW* in the calculation of the total noise power. From the top curve in Fig. A.5, we observe a "noise floor" of approximately $-15$ dBFS, which is the amount of power in bandwidth $NBW = 1/N = 2^{-6}$. The total power in the entire Nyquist band $[0, 0.5]$ is therefore $0.5/NBW = 2^5$ times (15 dB more than) the observed $-15$-dBFS value, or 0 dBFS, which is exactly the correct value.

The NBW, or at least sufficient information for calculating it (namely $N$ and the window type), should always be given for a sine-wave-scaled FFT. Furthermore, in order to emphasize that such an FFT represents a power spectral *density*, the units on the vertical axis should be shown as power per unit bandwidth. Since we report power in dBFS and since the unit of bandwidth is NBW, the vertical axis is usually labelled "dBFS/NBW."[†]

## A.3 Averaging

Our final point of discussion regarding the use of FFTs for spectral analysis is averaging. We noted earlier that the FFT of a random waveform is itself a random quantity. This quantity is random in both magnitude and phase, but since we are

---

[†] Although in common usage, units such as dBFS/NBW or dBm/Hz are deceptive: a density of 1 dBm/Hz does yield 1 dBm of power in a 1-Hz bandwidth but not 2 dBm of power in a 2-Hz bandwidth! Doubling the bandwidth doubles the power (a 3-dB increase), yielding a 4 dBm power in a 2-Hz bandwidth. The way out of this notational morass is to interpret "dBm/Hz" to mean "dB with respect to a 1 mW per Hz density." Similarly, "dBFS/NBW" means "dB with respect to a density equivalent to the power of a full-scale sine wave spread over a bandwidth NBW."

concerned with power we will only consider the magnitude. The magnitude of a particular frequency bin in an FFT of a random signal is a random value that has both a mean and a standard deviation. It turns out that the expected value[†] of the FFT magnitude equals the actual PSD (convolved with the window), as we would hope, but that the standard deviation of the magnitude is large. In fact the standard deviation is equal to the expected value!

The upshot of this property is that a single FFT results in a "noisy" spectral estimate, as demonstrated in Fig. A.6, where three length-64 FFTs of 0-dBFS white noise are shown. Each curve is expected to be a flat line at $-15$ dBFS, but this is not apparent in the Figure due to the large degree of variability in the individual bin magnitudes.

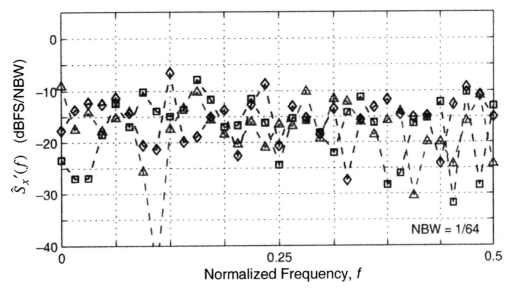

Figure A.6: Three length-64 FFTs of 0-dBFS white noise.

---

[†]. *Expected value* is another term for mean, or average, value.

# APPENDIX A  Spectral Estimation

If one is computing a noise power (by summing the powers over a range of FFT bins), the variability of an individual bin magnitude is not problematic, provided a sufficient number of bins are used. However, if one is trying to construct a clear graph of a spectral density, the erratic nature of the bin magnitudes results in a fuzzy plot such as that shown in Fig. A.7. Here, instead of a smooth curve, we see a broad black band spanning nearly 20 dB which obscures the true noise density. There are two solutions to this problem, namely averaging and integration. Averaging can be performed by either averaging many FFTs, or by averaging nearby bins in a single FFT.

Averaging multiple FFTs requires the use of multiple data records. When a single large data record is available, it can be partitioned into many (possibly overlapping) records which are individually windowed and transformed with an FFT. Averaging the squared magnitudes of these FFTs reduces the standard deviation and so improves the legibility of the spectral plot. (At this point, we owe the reader a confession: the spectra shown in Figs. A.3 and A.5 were averaged in this way for the sake of clarity.) Similarly, averaging the powers in adjacent bins will also

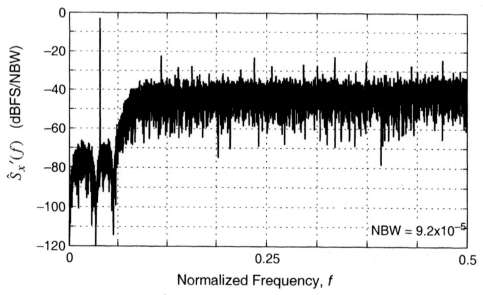

Figure A.7: A length-$2^{14}$ FFT without averaging.

smooth the FFT, effectively by "filtering" it. We present an example of this form of averaging in the next section.

The second method for increasing the legibility of a spectrum is to plot its accumulated value, scaled by $1/(N \cdot NBW)$. The resulting plot shows the amount of power contained in the band from dc up to the current frequency, and so saves the reader the effort of performing an integration to obtain a noise power. Of course, the designer is obliged to empty the signal bins before performing the integration, and the method must be changed when dealing with bandpass noise-shaping. Fig. A.8 illustrates the effectiveness of the technique in smoothing out the FFT of Fig. A.7. (Some might argue that the technique is too effective, since it hides high-frequency tones that are readily apparent in the original FFT.)

## A.4 An Example

We have now covered the essential topics regarding the use of the FFT to calculate a sine-wave scaled PSD, and so are ready to demonstrate them with an example. Fig. A.9 shows some example MATLAB code for producing a record of $\Delta\Sigma$ data and analyzing it according to the above procedures.

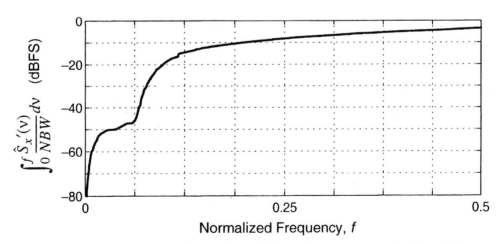

**Figure A.8:** Integrated version of the noise spectrum in Fig. A.7.

APPENDIX A  Spectral Estimation

The first block of code synthesizes a $5^{\text{th}}$-order NTF, creates the binary $\Delta\Sigma$ data, estimates the quantizer gain and computes the actual NTF (see Eq. (4.17) on page 100). The second block of code computes the scaled and windowed FFT, as well as NBW. The last two code blocks compute the SNR and create the plots shown in Fig. A.10.

The NTF is $5^{\text{th}}$-order, with zeros optimized for an oversampling ratio $OSR = 32$. The number of FFT points is set to $N = 64 \cdot OSR = 2048$, as advocated in Section A.1. Note that the half-scale input signal is placed precisely in an FFT bin (specifically, bin 11). Many practitioners use an odd FFT bin to ensure that the input data contains no repeated segments, but this is not strictly necessary. In a

```
% Compute modulator output and actual NTF
OSR = 32;
ntf0 = synthesizeNTF(5,OSR,1);
N = 64*OSR;
fbin = 11;
u = 1/2*sin(2*pi*fbin/N*[0:N-1]);
[v tmp1 tmp2 y] = simulateDSM(u,ntf0);
k = mean(abs(y)/mean(y.^2))
ntf = ntf0 / (k + (1-k)*ntf0);
% Compute windowed FFT and NBW
w = hann(N);    % or ones(1,N) or hann(N).^2
nb = 3;         % 1 for Rect; 5 for Hann^2
w1 = norm(w,1);
w2 = norm(w,2);
NBW = (w2/w1)^2
V = fft(w.*v)/(w1/2);
% Compute SNR
signal_bins = fbin + [-(nb-1)/2:(nb-1)/2];
inband_bins = 0:N/(2*OSR);
noise_bins = setdiff(inband_bins,signal_bins);
snr = dbp( sum(abs(V(signal_bins+1)).^2) / sum(abs(V(noise_bins+1)).^2) )
% Make plots
figure(1); clf;
semilogx([1:N/2]/N,dbv(V(2:N/2+1)),'b','Linewidth',1);
hold on;
[f p] = logsmooth(V,fbin,2,nb);
plot(f,p,'m','Linewidth',1.5)
Sq = 4/3 * evalTF(ntf,exp(2i*pi*f)).^2;
plot(f,dbp(Sq*NBW),'k--','Linewidth',1)
figureMagic([1/N 0.5],[],[], [-140 0],10,2);
```

**Figure A.9:** Example MATLAB code for performing spectral analysis.

Nyquist converter, it is wise to use an input which contains no repeated segments in order to exercise as many codes as possible, but in a $\Delta\Sigma$ converter the internal state of the converter typically ensures that the output data is not periodic.

As indicated in the second code block, different windows can be tried. The Hann window is used here. The number of non-zero signal bins for the Hann window is nb = 3. NBW is calculated directly for the window, using an expression which will be justified in the Section A.5. As indicated in Table 1 the result for the Hann window is $NBW = 1.5/N = 7.3 \times 10^{-4}$. The last line of the second code block calculates the FFT, and scales it by half the dc gain of the window to perform sine-wave scaling.

In the third code block, the SNR is calculated as the ratio of the total power in the signal bins to the total power in the noise bins, and the result for this simulation is $SNR = 81$ dB. If a rectangular window were used instead, the poor high-frequency roll-off of the window would corrupt the in-band portion of the spectrum

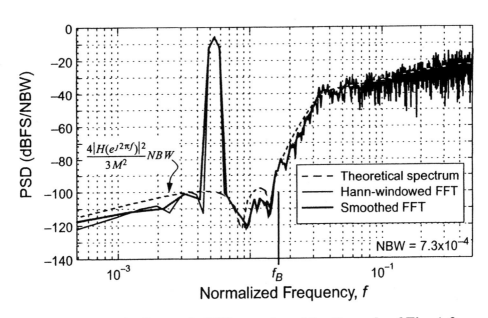

Figure A.10: **Example PSDs produced by the code of Fig. A.9.**

and yield a lower SNR. For this particular example, the shortfall was a disastrous 23 dB.

In the fourth and final code block, the raw and smoothed PSD estimates are compared graphically with the theoretical PSD. As expected, the raw FFT produces an erratic graph from which it would be difficult to tell if the observed PSD follows the expected PSD or not. Smoothing is performed with the $\Delta\Sigma$ Toolbox function `logsmooth`. This function averages the power across a number of bins in order to reduce the variance of the PSD, and also subsamples the result to yield points which are spaced fairly evenly on a logarithmic axis. This nearly-even spacing is achieved by using a number of bins which increases geometrically beyond a user-specified frequency (which defaults to the third harmonic of the input). See the "help" information associated with `logsmooth` for further details. With this function, it is feasible to present the results of a multi-million point simulation without having to plot millions of points in the spectrum. As Fig. A.10 shows, the agreement between the expected and observed PSDs is quite good. The justification of the formula used to calculate the expected PSD is also relegated to Section A.5.

As a final demonstration, we will use the PSD plot of Fig. A.10 to manually estimate the SNR. The signal power is read directly from the graph: –6 dBFS. The noise power is computed by multiplying the noise density by the bandwidth, or in logarithmic terms, by adding $10\log(BW/NBW)$ to the noise density in dB. For $OSR = 32$, $BW = 0.5/OSR = 1.6\times10^{-2}$, and since $NBW = 7.3\times10^{-4}$ the conversion factor from average noise density to total noise power is $10\log(BW/NBW) = 13$ dB. Since the average noise density in the passband is –100 dBFS/NBW, the estimated SNR is $-6 - (-100 + 13) = 81$ dB.

## A.5 Mathematical Background

Until now, the emphasis in this Appendix has been on using the FFT to perform spectral estimation for the signals in a $\Delta\Sigma$ modulator, while keeping the mathematics to a minimum. The mathematical theory associated with spectral estimation involves a number of concepts from stochastic processes which are worthy of chapters in themselves. We cannot do justice to such concepts in a subsection of an

Appendix, so we content ourselves to list the key results and to use these as justification for a number of formulae which appeared earlier in the Appendix without justification. See [2] for background.

The autocorrelation function of a discrete-time stationary random process[†] $x$ is defined as

$$r_x(k) = E[x(n)x(n+k)], \qquad (A.6)$$

where $E[\ ]$ denotes expectation ("average"). The Z-transform of $r_x$ is

$$R_x(z) = \sum_{n=-\infty}^{\infty} r_x(n) z^{-n} \qquad (A.7)$$

and the PSD $S_x(f)$ of $x$ is defined as

$$S_x(f) = R_x(e^{j2\pi f}). \qquad (A.8)$$

In other words, *the PSD is defined as the Fourier transform of the autocorrelation function*. The link between this definition and the more intuitive definition that $S_x(f)$ is amount of power between frequencies $f$ and $f+df$ divided by $df$ is established by the following two properties of $S_x$:

1. $P_x = \int_0^1 S_x(f) df$, where $P_x = E[|x(n)|^2]$ is the power of $x$.
2. If the output of a linear system having a transfer function $H(z)$ is $y$ when the input $x$, then $S_y(f) = |H(e^{j2\pi f})|^2 S_x(f)$.

The first property says that integrating the power spectrum over all frequencies yields the power in the signal. The second says that filtering the signal multiplies its power spectrum by the squared magnitude of the filter's transfer function. The

---

† A *stationary* random process is a random process whose statistical properties (mean, variance, etc.) do not vary with time.

# APPENDIX A  Spectral Estimation

intuitive definition of $S_x(f)$ results from considering an ideal filter $H$ having a bandwidth $df$ about a frequency $f$.

Note that the first property required $S_x$ to be integrated over the range $[0, 1]$. Since real signals have symmetric spectra, it is common in practice to use the range $[0, 0.5]$ and double the density

$$P_x = \int_0^{0.5} 2S_x(f) df. \tag{A.9}$$

Since the range $[0.5, 1]$ is equivalent to the range $[-0.5, 0]$, this convention is similar to the conventional use of single-sided spectral densities when dealing with continuous-time signals.

Next, we consider the estimate of $S_x$ (repeated from (A.5))

$$\hat{S}_x(f) = \frac{\left|\sum_{n=0}^{N-1} w(n) \cdot x(n) \cdot e^{-j2\pi fn}\right|^2}{\|w\|_2^2} \tag{A.10}$$

(We can obtain $\hat{S}_x(f_i)$ for $f_i = i/N$ from a length-$N$ FFT of $x$ windowed by $w$.)

The properties of this estimate are

1. $E[\hat{S}_x(f)] = S_x(f) * \dfrac{S_w(f)}{\|w\|_2^2},$

  where $S_w(f) = |W(f)|^2$ and $*$ denotes circular convolution.

2. $E\left[\displaystyle\sum_{i=0}^{N-1} \dfrac{\hat{S}_x(i/N)}{N}\right] = P_x.$

384

3. $\text{Var}[\hat{S}_x(f)] \approx [S_x(f)]^2$, where $\text{Var}[y]$ denotes the variance of a statistical quantity $y$.

The first property states that $\hat{S}_x(f)$ is a *biased estimator* of $S_x(f)$, i.e. that the expected value of $\hat{S}_x(f)$ is not equal to the true value of $S_x(f)$. For example, if a rectangular window is used, then the expected value of the estimated PSD is equal to the actual PSD convolved with

$$\frac{S_w(f)}{\|w\|_2^2} = \frac{1}{N}\left(\frac{\sin(N\pi f)}{\sin(\pi f)}\right)^2. \tag{A.11}$$

The second property effectively states that summing the values of $\hat{S}_x(f_i)$ contained in the FFT, and dividing by $N$, (effectively, integrating $\hat{S}_x(f_i)$ over $[0, 1]$) yields an unbiased estimate of the power in $x$. This property justifies summing the powers in the in-band bins of an FFT to produce an estimate of the in-band noise power.

The third property states that the PSD estimate is very "noisy," in that the standard deviation of the estimate is as large as the quantity being estimated. This property necessitates the use of averaging, as discussed in Section A.3.

We are now in the position to calculate the NBW of a sine-wave-scaled FFT employing a window $w$. We assume that $|W(f)|$ has a peak at $f = 0$ and that the full-scale range is $[-1, 1]$. The sine-wave-scaled PSD estimate is, from (A.3),

$$\hat{S}_x'(f) = \left|\frac{\sum_{n=0}^{N-1} w(n) \cdot x(n) \cdot e^{-j2\pi fn}}{W(0)/2}\right|^2. \tag{A.12}$$

Thus, $\hat{S}_x'$ is related to $\hat{S}_x$ (as given in Eq. (A.10)) by

## APPENDIX A  Spectral Estimation

$$\hat{S}_x'(f) = \frac{\hat{S}_x(f)\|w\|_2^2}{|W(0)/2|^2}. \tag{A.13}$$

Since we want the integral of $\hat{S}_x'(f)$ to yield the power in $x$ relative to the power in a full-scale sine wave, which is 0.5, we need to find the value of NBW which makes the following equation true

$$E\left[\int_0^{0.5} \frac{\hat{S}_x'(f)}{NBW} df\right] = \frac{P_x}{0.5} \cdot NBW \tag{A.14}$$

Since $E\left[\int_0^{0.5} 2\hat{S}_x(f) df\right] = P_x$,

$$NBW = \frac{\|w\|_2^2}{|W(0)|^2}. \tag{A.15}$$

This expression was used to tabulate NBW for the three windows in Table A.1. If we assume that $w(n) \geq 0$, then $|W(0)| = \|w\|_1$ and we arrive at the compact result

$$NBW = \left(\frac{\|w\|_2}{\|w\|_1}\right)^2 \tag{A.16}$$

which was used in the code of Fig. A.9. Since the full-scale range affects both the definition of $\hat{S}_x'(f)$ and the reference power of 0.5 proportionally, NBW is independent of the full-scale range.

The final formula which requires explanation is the formula for the expected PSD of the shaped quantization noise of a $\Delta\Sigma$ modulator, which was also used in the code of Fig. A.9. Since the power of quantization noise is $\Delta^2/12$, where $\Delta$ is the step size, the power of the quantization noise for an $M$-step quantizer with a step size $\Delta = 2$, relative to the power $M^2/2$ of a full-scale sine wave is $2/(3M^2)$. Assuming the quantization noise is white, its one-sided PSD is twice this amount, or $4/(3M^2)$. Thus the PSD of the shaped quantization noise is

$$S_q(f) = \frac{4|H(e^{j2\pi f})|^2}{3M^2}. \tag{A.17}$$

For consistency with a plot of $\hat{S}_x'(f)$, $S_q(f)$ must be multiplied by NBW.

## References

[1] O. Brigham, *The Fast Fourier Transform and Its Applications*, Prentice-Hall, Englewood Cliffs, New Jersey, 1988.

[2] A. Papoulis, *Probability, Random Variables and Stochastic Processes*, McGraw-Hill, New York, NY, 1984.

[3] A. V. Oppenheim and R. W. Schafer, *Digital Signal Processing*, Prentice-Hall, Englewood Cliffs, New Jersey, 1989.

[4] f. j. harris, "On the use of windows for harmonic analysis with the discrete Fourier transform," *Proceedings of the IEEE*, vol. 66, no. 1, pp. 51-83, Jan. 1978.

# APPENDIX B  *The Delta-Sigma Toolbox*

### Getting Started

To obtain your free copy of the Delta-Sigma Toolbox, go to the MathWorks web site (http://www.mathworks.com/matlabcentral/fileexchange), find the Controls category and select delsig. To improve simulation speed, compile the simulateDSM.c file by typing mex simulateDSM.c at the MATLAB prompt. Do the same for simulateESL.c and ai2mif.c. The Delta-Sigma Toolbox requires the Signal Processing Toolbox and the Control Systems Toolbox; the clans function also requires the Optimization Toolbox. These and other toolboxes may be purchased from the MathWorks.

### Toolbox Conventions

Frequencies are normalized; $f = 1$ corresponds to the sampling frequency ($f_s$).

Default values for function arguments are shown following an equals sign (=) in the parameter list. To use the default value for an argument, omit the argument if it is at the end of the list, otherwise use NaN (not-a-number) or [] (the empty matrix) as a placeholder.

A matrix is used to describe the loop filter of a general single-quantizer delta-sigma modulator. See "MODULATOR MODEL DETAILS" on page 410, for a description of this *ABCD* matrix.

## DEMONSTRATIONS AND EXAMPLES

| | |
|---|---|
| `dsdemo1` | Demonstration of the `synthesizeNTF` function. Noise transfer function synthesis for a $5^{th}$-order lowpass modulator, both with and without optimized zeros, plus an $8^{th}$-order bandpass modulator with optimized zeros. |
| `dsdemo2` | Demonstration of the `simulateDSM`, `predictSNR` and `simulateSNR` functions: time-domain simulation, SNR prediction using the describing function method of Ardalan and Paulos, spectral analysis and signal-to-noise ratio. Lowpass, bandpass, multi-bit lowpass examples are given. |
| `dsdemo3` | Demonstration of the `realizeNTF`, `stuffABCD`, `scaleABCD` and `mapABCD` functions: coefficient calculation and dynamic range scaling. |
| `dsdemo4` | Audio demonstration of MOD1 and MOD2 with $\text{sinc}^n$ decimation. |
| `dsdemo5` | Demonstration of the `simulateESL` function: simulation of the element selection logic of a mismatch-shaping DAC. |
| `dsdemo6` | Demonstration of the `designHBF` function. Hardware-efficient halfband filter design and simulation. |
| `dsdemo7` | Demonstration of the `findPIS` function: positively-invariant set computation. |
| `dsexample1` | Discrete-time lowpass modulator design example. |
| `dsexample2` | Discrete-time bandpass modulator design example. |

# SUMMARY of KEY FUNCTIONS

`ntf = synthesizeNTF(order=3,OSR=64,opt=0,H_inf=1.5,f0=0)`     page 393
`ntf = clans(order=4,OSR=64,Q=5,rmax=0.95,opt=0)`     page 394
    Synthesize a noise transfer function.

`[snr,amp,k0,k1,sigma_e2] = predictSNR(ntf,OSR=64,amp=...,f0=0)`     page 395
    Predict the SNR vs. input power curve using the describing function method of Ardalan and Paulos.

`[v,xn,xmax,y] = simulateDSM(u,ABCD,nlev=2,x0=0)`     page 396
`[v,xn,xmax,y] = simulateDSM(u,ntf,nlev=2,x0=0)`
    Simulate a delta-sigma modulator with a given input.

`[snr,amp] = simulateSNR(ntf,OSR,amp=...,f0=0,nlev=2,f=1/(4*R),k=13)`     page 398
    Determine the SNR vs. input power curve by simulation.

`[a,g,b,c] = realizeNTF(ntf,form='CRFB',stf=1)`     page 400
    Convert a noise transfer function into coefficients for a specific structure.

`ABCD = stuffABCD(a,g,b,c,form='CRFB')`     page 401
    Calculate the ABCD matrix given the parameters of a specified modulator topology.

`[a,g,b,c] = mapABCD(ABCD,form='CRFB')`     page 401
    Calculate the parameters of a specified modulator topology given the ABCD matrix.

`[ABCDs, umax] = scaleABCD(ABCD,nlev=2,f=0,xlim=1,ymax=nlev+2))`     page 402
    Perform dynamic range scaling on a delta-sigma modulator described by ABCD.

`[ntf,stf] = calculateTF(ABCD,k=1)`     page 403
    Calculate the NTF and STF of a delta-sigma modulator described by the ABCD matrix, assuming a quantizer gain of $k$.

`[sv,sx,sigma_se,max_sx,max_sy]`
`= simulateESL(v,mtf,M=16,dw=[1...],sx0=[0...])`     page 404
    Simulate the element-selection logic in a mismatch-shaping DAC.

`[f1,f2,info] = designHBF(fp=0.2,delta=1e-5,debug=0)`     page 405
    Design a hardware-efficient half-band filter for use in a decimation or interpolation filter.

`y = simulateHBF(x,f1,f2,mode=0)`     page 408
    Simulate a Saramäki half-band filter in the time domain.

`[s,e,n,o,Sc] = findPIS(u,ABCD,nlev=2,options)`     page 409
    Find a convex positively-invariant set for a delta-sigma modulator.

# SUMMARY OF OTHER SELECTED FUNCTIONS

## Delta-Sigma Utility
`mod1, mod2`
> Scripts for setting up the ABCD matrix, NTF and STF of the $1^{st}/2^{nd}$-order modulator.

`snr = calculateSNR(hwfft,f)`
> Estimate the SNR given the in-band bins of a Hann-windowed FFT and the location of the input signal.

`[sys, Gp] = mapCtoD(sys_c,t=[0 1],f0=0)`
> Map a continuous-time system to a discrete-time system whose impulse response matches the sampled pulse response of the original continuous-time system.

`[A B C D] = partitionABCD(ABCD, m)`
> Partition ABCD into A, B, C, D for an $m$-input state-space system.

`H_inf = infnorm(H)`
> Compute the infinity norm (maximum absolute value) of a z-domain transfer function. See `evalTF`.

`sigma_H = rmsGain(H,f1,f2)`
> Compute the root mean-square gain of the discrete-time transfer function H in the frequency band (`f1,f2`).

## General Utility
`dbv(), dbp(), undbv(), undbp(), dbm()`
> The dB equivalent of voltage/power quantities, and their inverse functions.

`window = hann(N)`
> A Hann window of length N. Unlike MATLAB's original `hanning` function, `hann` does not smear a tone which is located exactly in an FFT bin (i.e. the tone has an integer number of cycles in the given block of data). MATLAB 6's `hanning(N,'periodic')` function is the same as `hann(N)`.

## Graphing
`plotPZ(H,color='b',markersize=5,list=0)`
> Plot the poles and zeros of a transfer function.

`figureMagic(xRange,dx,xLab, yRange,dy,yLab, size)`
> Performs a number of formatting operations for the current figure, including axis limits, ticks and labelling.

`printmif(file,size,font,fig)`
> Print graph to an Adobe Illustrator file and then use `ai2mif` to convert it to FrameMaker MIF format. `ai2mif` is an improved version of the function of the same name originally written by Deron Jackson <djackson@mit.edu>.

`[f,p] = logsmooth(X,inBin,nbin)`
> Smooth the FFT, X, and convert it to dB. See also `bplogsmooth` and `bilogplot` for bandpass modulators.

# synthesizeNTF

**Synopsis:** `ntf = synthesizeNTF(order=3,OSR=64,opt=0,H_inf=1.5,f0=0)`
Synthesize a noise transfer function (NTF) for a delta-sigma modulator.

## Arguments

| | |
|---|---|
| *order* | The order of the NTF. *order* must be even for bandpass modulators. |
| *OSR* | The oversampling ratio. *OSR* is only needed when optimized NTF zeros are requested. |
| *opt* | A flag used to request optimized NTF zeros. *opt*=0 puts all NTF zeros at band-center (DC for lowpass modulators). *opt*=1 optimizes the NTF zeros. For even-order modulators, *opt*=2 puts two zeros at band-center, but optimizes the rest. |
| *H_inf* | The maximum out-of-band gain of the NTF. Lee's rule states that *H_inf*<2 should yield a stable modulator with a binary quantizer. Reducing *H_inf* increases the likelihood of success, but reduces the magnitude of the attenuation provided by the NTF and thus the theoretical resolution of the modulator. |
| *f0* | The center frequency of the modulator. *f0*≠0 yields a bandpass modulator; *f0*=0.25 puts the center frequency at $f_s/4$. |

## Output

| | |
|---|---|
| *ntf* | The modulator NTF, given as an LTI object in zero-pole form. |

## Example
Fifth-order lowpass modulator; zeros optimized for an oversampling ratio of 32.
`H = synthesizeNTF(5,32,1)`

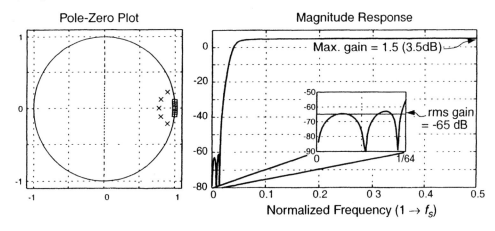

# clans

**Synopsis:** `ntf = clans(order=4,OSR=64,Q=5,rmax=0.95,opt=0)`
Synthesize a noise transfer function (NTF) for a lowpass delta-sigma modulator using the CLANS (Closed-loop analysis of noise-shaper) methodology [1]. This function requires the optimization toolbox.

[1] J. G. Kenney and L. R. Carley, "Design of multibit noise-shaping data converters," *Analog Integrated Circuits Signal Processing Journal*, vol. 3, pp. 259-272, 1993.

## Arguments

| | |
|---|---|
| *order* | The order of the NTF. |
| *OSR* | The oversampling ratio. |
| *Q* | The maximum number of quantization levels used by the fed-back quantization noise. (Mathematically, $Q = \|h\|_1 - 1$, i.e. the sum of the absolute values of the impulse response samples minus 1, is the maximum instantaneous noise gain.) |
| *rmax* | The maximum radius for the NTF poles. |
| *opt* | A flag used to request optimized NTF zeros. *opt*=0 puts all NTF zeros at band-center (DC for lowpass modulators). *opt*=1 optimizes the NTF zeros. For even-order modulators, *opt*=2 puts two zeros at band-center, but optimizes the rest. |

## Output

*ntf*   The modulator NTF, given as an LTI object in zero-pole form.

## Example

$5^{th}$-order lowpass modulator; time-domain noise gain of 5, zeros optimized for $OSR = 32$.
`H= clans(5,32,5,.95,1)`

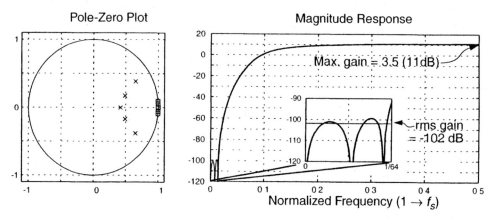

# predictSNR

**Synopsis:** `[snr,amp,k0,k1,sigma_e2] = predictSNR(ntf,OSR=64,amp=...,f0=0)`
Use the describing function method of Ardalan and Paulos [1] to predict the signal-to-noise ratio (SNR) in dB for various input amplitudes. This method is only applicable to binary modulators.

[1]  S. H. Ardalan and J. J. Paulos, "Analysis of nonlinear behavior in delta-sigma modulators," *IEEE Transactions on Circuits and Systems*, vol. 34, pp. 593-603, June 1987.

**Arguments**

| | |
|---|---|
| *ntf* | The modulator NTF, given in zero-pole form. |
| *OSR* | The oversampling ratio. *OSR* is used to define the "band of interest." |
| *amp* | A row vector listing the amplitudes to use. Defaults to [-120 -110...-20 -15 -10 -9 -8 ... 0] dB, where 0 dB means a full-scale sine wave. |
| *f0* | The center frequency of the modulator. *f0*≠0 corresponds to a bandpass modulator. |

**Output**

| | |
|---|---|
| *snr* | A row vector containing the predicted SNRs. |
| *amp* | A row vector listing the amplitudes used. |
| *k0* | The signal gain of the quantizer model; one value per input level. |
| *k1* | The noise gain of the quantizer model; one value per input level. |
| *sigma_e2* | The mean square value of the noise in the model of the quantizer. |

**Example**
See the example on page 399.

**The Quantizer Model:**
The binary quantizer is modeled as a pair of linear gains and a noise source, as shown in the figure below. The input to the quantizer is divided into signal and noise components which are processed by signal-dependent gains $k_0$ and $k_1$. These signals are added to a noise source, which is assumed to be white and to have a Gaussian distribution (the variance $\sigma_e^2$ is also signal-dependent), to produce the quantizer output.

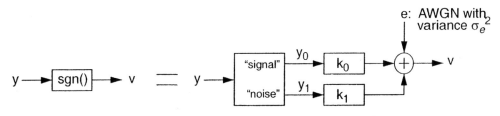

# simulateDSM

**Synopsis:** `[v,xn,xmax,y] = simulateDSM(u,ABCD,nlev=2,x0=0)` or
`[v,xn,xmax,y] = simulateDSM(u,ntf,nlev=2,x0=0)`

Simulate a delta-sigma modulator with a given input. For maximum speed, make sure that the mex file is on your search path (At the MATLAB prompt, type `which simulateDSM`).

## Arguments

| | |
|---|---|
| *u* | The input sequence to the modulator, given as a $m \times N$ row vector. $m$ is the number of inputs (usually 1). Full-scale corresponds to an input of magnitude $nlev-1$. |
| *ABCD* | A state-space description of the modulator loop filter. |
| *ntf* | The modulator NTF, given in zero-pole form. The modulator STF is assumed to be unity. |
| *nlev* | The number of levels in the quantizer. Multiple quantizers are indicated by making *nlev* an array. |
| *x0* | The initial state of the modulator. |

## Output

| | |
|---|---|
| *v* | The samples of the output of the modulator, one for each input sample. |
| *xn* | The internal states of the modulator, one for each input sample, given as an $n \times N$ matrix. |
| *xmax* | The maximum absolute values of each state variable. |
| *y* | The samples of the quantizer input, one per input sample. |

## Example

Simulate a 5$^{th}$-order binary modulator with a half-scale sine-wave input and plot its output in the time and frequency domains.

```
OSR = 32;
H = synthesizeNTF(5,OSR,1);
N = 8192; fB = ceil(N/(2*OSR)); f=85;
u = 0.5*sin(2*pi*f/N*[0:N-1]);
v = simulateDSM(u,H);
```

Time-domain plot:

```
t = 0:85;
stairs(t, u(t+1));
hold on;
stairs(t,v(t+1));
axis([0 85 -1.2 1.2]);
ylabel('u, v');
```

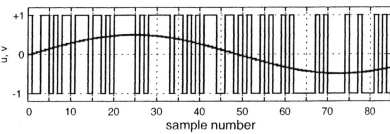

Frequency-domain plot:

```
spec=fft(v.*hann(N))/(N/4);
plot(linspace(0,1,N/2), dbv(spec(1:N/2)));
figureMagic([0 0.5],0.05,2, [-120 0],10,4);
ylabel('dB');
snr=calculateSNR(spec(1:fB),f);
s=sprintf('SNR = %4.1fdB\n',snr);
text(0.25,-90,s);
s=sprintf('NBW=%7.5f',1.5/N);
text(0.25, -110, s);
```

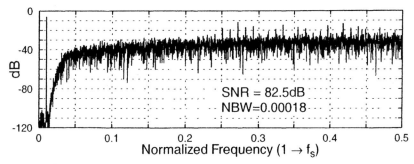

# simulateSNR

**Synopsis:** `[snr,amp] = simulateSNR(ntf,OSR,amp,f0=0,nlev=2,f=1/(4*OSR),k=13)`

Simulate a delta-sigma modulator with sine wave inputs of various amplitudes and calculate the signal-to-noise ratio (SNR) in dB for each input.

**Arguments**

| | |
|---|---|
| *ntf* | The modulator NTF, given in zero-pole form. |
| *OSR* | The oversampling ratio. *OSR* is used to define the "band of interest." |
| *amp* | A row vector listing the amplitudes to use. Defaults to [-120 -110...-20 -15 -10 -9 -8 ... 0] dB, where 0 dB means a full-scale sine wave, i.e. a sine wave whose peak value is *nlev*-1. |
| *f0* | The center frequency of the modulator. $f0 \neq 0$ corresponds to a bandpass modulator. |
| *nlev* | The number of levels in the quantizer. |
| *f* | The normalized frequency of the test sinusoid; a check is made that the test frequency is in the band of interest. The frequency is also adjusted so that it lies precisely in an FFT bin. |
| *k* | The number of time points used for the FFT is $2^k$. |

**Output**

| | |
|---|---|
| *snr* | A row vector containing the SNR values calculated from the simulations. |
| *amp* | A row vector listing the amplitudes used. |

**Example**

Compare the SNR vs. input amplitude curve for a fifth-order modulator determined by the describing function method with that determined by simulation.

```
OSR = 32;
H = synthesizeNTF(5,OSR,1);
[snr_pred,amp] = predictSNR(H,OSR);
[snr,amp] = simulateSNR(H,OSR);

plot(amp,snr_pred,'b',amp,snr,'gs');
figureMagic([-100 0], 10, 1, [0 100], 10, 1);
xlabel('Input Level, dB');
ylabel('SNR dB');
title('SNR curve');
s=sprintf('peak SNR = %4.1fdB\n', max(snr));
text(-49,15,s);
```

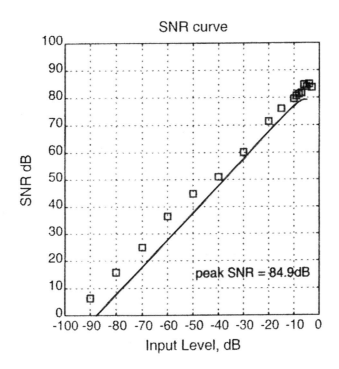

# realizeNTF

**Synopsis:** `[a,g,b,c] = realizeNTF(ntf,form='CRFB',stf=1)`
Convert a noise transfer function (NTF) into a set of coefficients for a particular modulator topology.

## Arguments

*ntf*  The modulator NTF, given in zero-pole form (i.e. a zpk object).

*form*  A string specifying the modulator topology.
    CRFB    Cascade-of-resonators, feedback form.
    CRFF    Cascade-of-resonators, feedforward form.
    CIFB    Cascade-of-integrators, feedback form.
    CIFF    Cascade-of-integrators, feedforward form.
Structures are described in detail in "MODULATOR MODEL DETAILS" on page 410.

*stf*  The modulator STF, specified as a zpk object. Note that the poles of the STF must match those of the NTF in order to guarantee that the STF can be realized without the addition of extra state variables.

## Output

*a*  Feedback/feedforward coefficients from/to the quantizer ($1 \times n$).

*g*  Resonator coefficients ($1 \times \lfloor n/2 \rfloor$).

*b*  Feed-in coefficients from the modulator input to each integrator ($1 \times n + 1$).

*c*  Integrator inter-stage coefficients. ($1 \times n$. In unscaled modulators, $c$ is all ones.)

## Example

Determine the coefficients for a $5^{th}$-order modulator with the cascade-of-resonators structure, feedback (CRFB) form.

```
>> H = synthesizeNTF(5,32,1);
>> [a,g,b,c] = realizeNTF(H,'CRFB')
a = 0.0007    0.0084    0.0550    0.2443    0.5579
g = 0.0028    0.0079
b = 0.0007    0.0084    0.0550    0.2443    0.5579    1.0000
c = 1    1    1    1    1
```

## stuffABCD

**Synopsis:** `ABCD = stuffABCD(a,g,b,c,form='CRFB')`
Calculate the ABCD matrix given the parameters of a specified modulator topology.

### Arguments
| | |
|---|---|
| *a* | Feedback/feedforward coefficients from/to the quantizer. $1 \times n$ |
| *g* | Resonator coefficients. $1 \times \lfloor n/2 \rfloor$ |
| *b* | Feed-in coefficients from the modulator input to the input of each integrator and to the input of the quantizer. $1 \times n+1$ |
| *c* | Integrator inter-stage coefficients. $1 \times n$ |
| *form* | see `realizeNTF` on page 400 for a list of supported structures. |

### Output
| | |
|---|---|
| *ABCD* | A state-space description of the modulator loop filter. |

## mapABCD

`[a,g,b,c] = mapABCD(ABCD,form='CRFB')`
Calculate the parameters for a specified modulator topology, assuming ABCD fits that topology.

### Arguments
| | |
|---|---|
| *ABCD* | A state-space description of the modulator loop filter. |
| *form* | see `realizeNTF` on page 400 for a list of supported structures. |

### Output
| | |
|---|---|
| *a* | Feedback/feedforward coefficients from/to the quantizer. $1 \times n$ |
| *g* | Resonator coefficients. $1 \times \lfloor n/2 \rfloor$ |
| *b* | Feed-in coefficients from the modulator input to each integrator. $1 \times n+1$ |
| *c* | Integrator inter-stage coefficients. $1 \times n$ |

# scaleABCD

**Synopsis:**  [ABCDs,umax]=scaleABCD(ABCD,nlev=2,f=0,xlim=1,ymax=nlev+5,umax,N=1e5)

Scale the ABCD matrix so that the state maxima are less than a specified limit. The maximum stable input is determined as a side-effect of this process.

## Arguments

| | |
|---|---|
| *ABCD* | A state-space description of the modulator loop filter. |
| *nlev* | The number of levels in the quantizer. |
| *f* | The normalized frequency of the test sinusoid. |
| *xlim* | The limit on the states. May be given as a vector. |
| *ymax* | The threshold for judging modulator stability. If the quantizer input exceeds *ymax*, the modulator is considered to be unstable. |
| *umax* | The maximum input which will be applied to the modulator during simulation. If *umax* is not supplied, it will be estimated by simulation. |
| *N* | The number of time steps used for determination of the state maxima. |

## Output

| | |
|---|---|
| *ABCDs* | The scaled state-space description of the modulator loop filter. |
| *umax* | The maximum stable input. Input sinusoids with amplitudes below this value should not cause the modulator states to exceed their specified limits. |

# calculateTF

**Synopsis:** `[ntf,stf] = calculateTF(ABCD,k=1)`
Calculate the NTF and STF of a delta-sigma modulator.

## Arguments
*ABCD*          A state-space description of the modulator loop filter.
*k*             The value to use for the quantizer gain.

## Output
*ntf*           The modulator NTF, given as an LTI system in zero-pole form.
*stf*           The modulator STF, given as an LTI system in zero-pole form.

## Example
Realize a fifth-order modulator with the cascade-of-resonators structure, feedback form. Calculate the ABCD matrix of the loop filter and verify that the NTF and STF are correct.

```
>> H = synthesizeNTF(5,32,1)
Zero/pole/gain:
       (z-1) (z^2 - 1.997z + 1) (z^2 - 1.992z + 1)
-----------------------------------------------------
(z-0.7778) (z^2 - 1.613z + 0.6649) (z^2 - 1.796z + 0.8549)
Sampling time: 1
>> [a,g,b,c] = realizeNTF(H)
a =
    0.0007    0.0084    0.0550    0.2443    0.5579
g =
    0.0028    0.0079
b =
    0.0007    0.0084    0.0550    0.2443    0.5579    1.0000
c =
    1    1    1    1    1
>> ABCD = stuffABCD(a,g,b,c)
ABCD =
    1.0000         0         0         0         0    0.0007   -0.0007
    1.0000    1.0000   -0.0028         0         0    0.0084   -0.0084
    1.0000    1.0000    0.9972         0         0    0.0633   -0.0633
         0         0    1.0000    1.0000   -0.0079    0.2443   -0.2443
         0         0    1.0000    1.0000    0.9921    0.8023   -0.8023
         0         0         0         0    1.0000    1.0000         0
>> [ntf,stf] = calculateTF(ABCD)
Zero/pole/gain:
       (z-1) (z^2 - 1.997z + 1) (z^2 - 1.992z + 1)
-----------------------------------------------------
(z-0.7778) (z^2 - 1.613z + 0.6649) (z^2 - 1.796z + 0.8549)
Sampling time: 1

Zero/pole/gain:
1
Static gain.
```

# simulateESL

**Synopsis:** `[sv,sx,sigma_se,max_sx,max_sy]`
`= simulateESL(v,mtf,M=16,dw=[1…], sx0=[0…])`

Simulate the element selection logic (ESL) of a multi-element DAC using a particular mismatch-shaping transfer function (`mtf`).

[1]   R. Schreier and B. Zhang "Noise-shaped multibit D/A convertor employing unit elements," *Electronics Letters*, vol. 31, no. 20, pp. 1712-1713, Sept. 28 1995.

## Arguments

| | |
|---|---|
| *v* | A vector containing the number of elements to enable. Note that the output of `simulateDSM` must be offset and scaled in order to be used here as *v* must be in the range $[0, \sum_{i}^{M} dw(i)]$. |
| *mtf* | The mismatch-shaping transfer function, given in zero-pole form. |
| *M* | The number of elements. |
| *dw* | A vector containing the weight associated with each element. |
| *sx0* | An $n \times M$ matrix containing the initial state of the element selection logic. |

## Output

| | |
|---|---|
| *sv* | The selection vector: a vector of zeros and ones indicating which elements to enable. |
| *sx* | An $n \times M$ matrix containing the final state of the element selection logic. |
| *sigma_se* | The rms value of the selection error, $se = sv - sy$. *sigma_se* may be used to analytically estimate the power of in-band noise caused by element mismatch. |
| *max_sx* | The maximum value attained by any state in the ESL. |
| *max_sy* | The maximum value attained by any component of the (un-normalized) "desired usage" vector. |

## Example

Run `dsdemo6.m`.

# designHBF

**Synopsis:** `[f1,f2,info]=designHBF(fp=0.2,delta=1e-5,debug=0)`
Design a hardware-efficient linear-phase half-band filter for use in the decimation or interpolation filter associated with a delta-sigma modulator. This function is based on the procedure described by Saramäki [1]. Note that since the algorithm uses a non-deterministic search procedure, successive calls may yield different designs.

[1] T. Saramäki, "Design of FIR filters as a tapped cascaded interconnection of identical subfilters," *IEEE Transactions on Circuits and Systems,* vol. 34, pp. 1011-1029, 1987.

## Arguments

| | |
|---|---|
| *fp* | Normalized passband cutoff frequency. |
| *delta* | Passband and stopband ripple in absolute value. |

## Output

| | |
|---|---|
| *f1,f2* | Prototype filter and subfilter coefficients and their canonical-signed digit (csd) representation. |
| *info* | A vector containing the following information data (only set when `debug=1`): |
| complexity | The number of additions per output sample. |
| n1,n2 | The length of the f1 and f2 vectors. |
| sbr | The achieved stop-band attenuation (dB). |
| phi | The scaling factor for the F2 filter. |

## Example

Design of a lowpass half-band filter with a cut-off frequency of $0.2f_s$, a passband ripple of less than $10^{-5}$ and a stopband rejection of at least $10^{-5}$ (-100 dB).

```
[f1,f2] = designHBF(0.2,1e-5);
f = linspace(0,0.5,1024);
plot(f, dbv(frespHBF(f,f1,f2)))
```

A plot of the filter response and the structure of this filter as a decimation or as an interpolation filter are shown on the next page. The filter achieves 109 dB of attenuation in the stopband and uses only 124 additions (no true multiplications) to produce each output sample.

# APPENDIX B  The Delta-Sigma Toolbox

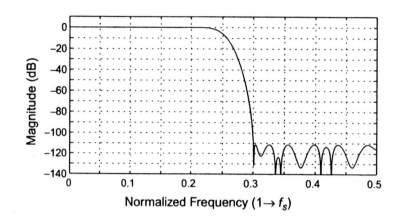

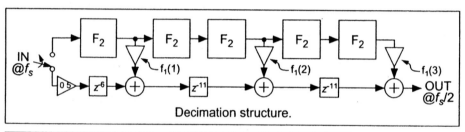

Decimation structure.

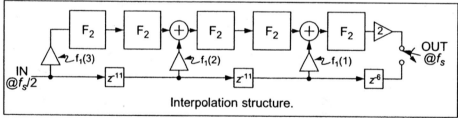

Interpolation structure.

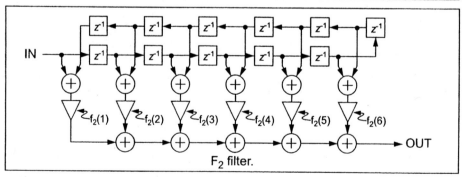

$F_2$ filter.

The coefficients and their signed-digit decompositions are

```
[f1.val]' =    [f2.val]' =    >> f1.csd              >> f2.csd
    0.9453         0.6211     ans =                  ans =
   -0.6406        -0.1895          0   -4   -7           -1   -3   -8
    0.1953         0.0957          1   -1    1            1    1   -1
                  -0.0508     ans =                  ans =
                   0.0269         -1   -3   -6           -2   -4   -9
                  -0.0142         -1   -1   -1           -1    1   -1
                               ans =                  ans =
                                   -2   -4   -7           -3   -5   -9
                                    1   -1    1            1   -1    1
                                                      ans =
                                                          -4   -7   -8
                                                          -1    1    1
                                                      ans =
                                                          -5    8  -11
                                                           1   -1   -1
                                                      ans =
                                                          -6   -9  -11
                                                          -1    1   -1
```

In the above signed-digit expansions, the first row contain the powers of two while the second row gives their signs. For example, $f_1(1) = 0.9453 = 2^0 - 2^{-4} + 2^{-7}$ and $f_2(1) = 0.6211 = 2^{-1} + 2^{-3} - 2^{-8}$. Since the filter coefficients for this example use only 3 signed digits, each multiply-accumulate operation shown in the diagram below needs only 3 binary additions. Thus, an implementation of this $110^{th}$-order FIR filter needs to perform only $3 \times 3 + 5 \times (3 \times 6 + 6 - 1) = 124$ additions at the low ($f_s/2$) rate.

# simulateHBF

**Synopsis:** `y = simulateHBF(x,f1,f2,mode=0)`
Simulate a Saramäki half-band filter (see `designHBF` on page 405) in the time domain.

## Arguments

| | |
|---|---|
| *x* | The input data. |
| *f1,f2* | Filter coefficients. f1 and f2 can be vectors of values or struct arrays like those returned from `designHBF`. |
| *mode* | The mode flag determines whether the input is filtered, interpolated, or decimated according to the following: |

    0   Plain filtering, no interpolation or decimation.
    1   The input is interpolated
    2   The output is decimated, even samples are taken.
    3   The output is decimated, odd samples are taken.

## Output

| | |
|---|---|
| *y* | The output data. |

## Example

Plot the impulse response of the HBF designed on the previous page.

```
N = (2*length(f1)-1)*2*(2*length(f2)-1)+1;
y = simulateHBF([1 zeros(1,N-1)],f1,f2);
stem([0:N-1],y);
figureMagic([0 N-1],5,2, [-0.2 0.5],0.1,1)
printmif('HBFimp', [6 3], 'Helvetica8')
```

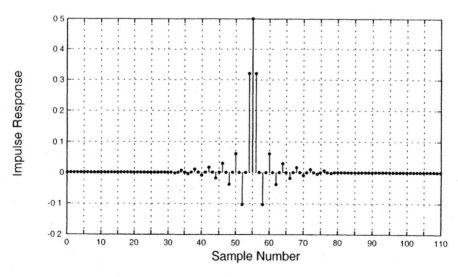

## findPIS, find2dPIS (in the `PosInvSet` subdirectory)

**Synopsis:**  `[s,e,n,o,Sc] = findPIS(u,ABCD,nlev=2,options)`
  `options = [dbg=0 itnLimit=2000 expFactor=0.005 N=1000 skip=100`
  `           qhullArgA=0.999 qhullArgC=.001]`
  `s = find2dPIS(u,ABCD,options)`
  `options = [dbg=0 itnLimit=100 expFactor=0.01 N=1000 skip=100]`

Find a convex positively-invariant set for a delta-sigma modulator. `findPIS` requires compilation of the `qhull` mex file; `find2dPIS` does not, but is limited to second-order systems.

### Arguments

| | |
|---|---|
| *u* | The input to the modulator. If $u$ is a scalar, the input to the modulator is constant. If $u$ is a $2 \times 1$ vector, the input to the modulator may be any sequence whose samples lie in the range $[u(1), u(2)]$. |
| *ABCD* | A state-space description of the modulator loop filter. |
| *nlev* | The number of quantizer levels. |
| *dbg* | Set dbg=1 to get a graphical display of the iterations. |
| *itnLimit* | The maximum number of iterations. |
| *expFactor* | The expansion factor applied to the hull before every mapping operation. Increasing `expFactor` decreases the number of iterations but results in sets which are larger than they need to be. |
| *N* | The number of points to use when constructing the initial guess. |
| *skip* | The number of time steps to run the modulator before observing the state. This handles the possibility of "transients" in the modulator. |
| *qhullArgA* | The 'A' argument to the `qhull` program. Adjacent facets are merged if the cosine of the angle between their normals is greater than the absolute value of this parameter. A negative value implies that the merge is performed during hull construction, rather than after. |
| *qhullArgC* | The 'C' argument to the `qhull` program. A facet is merged into its neighbor if the distance between the facet's centrum (the average of the facet's vertices) and the neighboring hyperplane is less than the absolute value of this parameter. As with *qhullArgA*, a negative value implies pre-merging whereas a positive value implies post-merging. |

### Output

| | |
|---|---|
| *s* | The vertices of the set ($dim \times n_v$). |
| *e* | The edges of the set, listed as pairs of vertex indices ($2 \times n_e$). |
| *n* | The normals for the facets of the set ($dim \times n_f$). |
| *o* | The offsets for the facets of the set ($1 \times n_f$). |
| *Sc* | The scaling matrix which was used internally to "round out" the set. |

## MODULATOR MODEL DETAILS

A delta-sigma modulator with a single quantizer is assumed to consist of quantizer connected to a loop filter as shown in the diagram below.

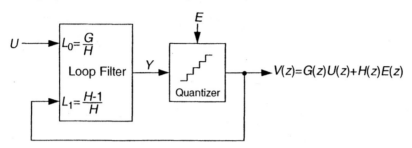

**The Loop Filter**

The loop filter is described by an ABCD matrix. For single-quantizer systems, the loop filter is a two-input, one-output linear system and ABCD is an $(n+1) \times (n+2)$ matrix, partitioned into $A$ $(n \times n)$, $B$ $(n \times 2)$, $C$ $(1 \times n)$ and $D$ $(1 \times 2)$ sub-matrices as shown below:

$$ABCD = \left[\begin{array}{c|c} A & B \\ \hline C & D \end{array}\right].  \quad (B.1)$$

The equations for updating the state and computing the output of the loop filter are

$$x(n+1) = Ax(n) + B\begin{bmatrix} u(n) \\ v(n) \end{bmatrix}$$

$$y(n) = Cx(n) + D\begin{bmatrix} u(n) \\ v(n) \end{bmatrix}. \quad (B.2)$$

This formulation is sufficiently general to encompass all single-quantizer modulators which employ linear loop filters. The toolbox currently supports translation from an ABCD description to coefficients and vice versa for the following topologies:

CIFB            Cascade-of-integrators, feedback form.
CIFF            Cascade-of-integrators, feedforward form.
CRFB           Cascade-of-resonators, feedback form.
CRFF           Cascade-of-resonators, feedforward form.

Multi-input and multi-quantizer systems are also described with an ABCD matrix and Eq. (2) still applies. For an $n^{th}$-order, $n_i$-input, $n_o$-output modulator, the dimensions of the sub-matrices are $A$: $n \times n$, $B$: $n \times (n_i + n_o)$, $C$: $n_o \times n$ and $D$: $n_o \times (n_i + n_o)$.

## The Quantizer

The quantizer is as described in Section 2.1, namely a symmetric binary quantizer with a step size of two. Quantizers with an even number of levels are of the mid-rise type and produce outputs which are odd integers. Quantizers with an odd number of levels are of the mid-tread type and produce outputs which are even integers.

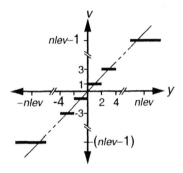

Transfer curve of a quantizer with an even number of levels.

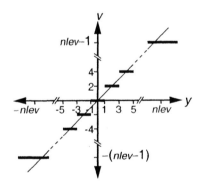

Transfer curve of a quantizer with an odd number of levels.

Illustrations of the four topologies for which the toolbox supports coefficient calculation are given on the following pages.

# APPENDIX B  The Delta-Sigma Toolbox

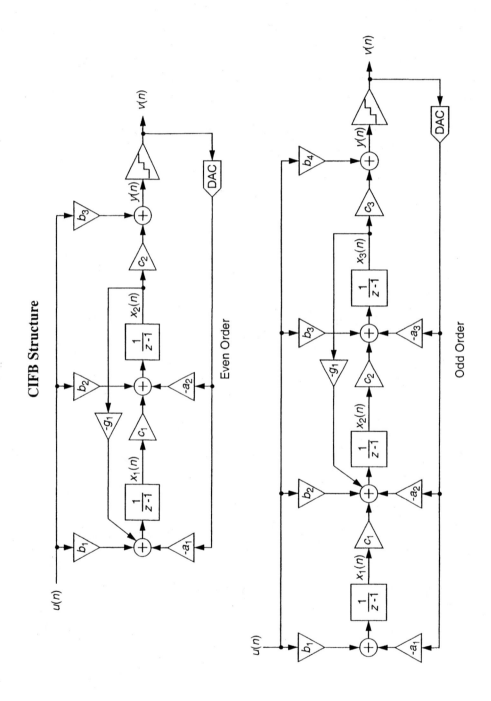

**CIFB Structure**

Even Order

Odd Order

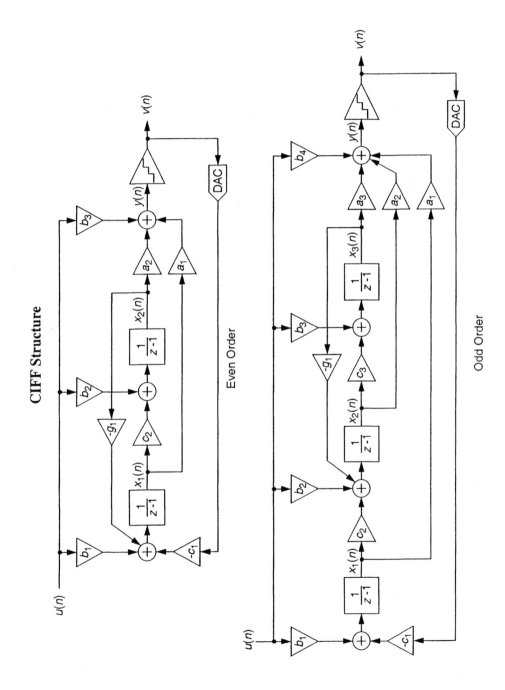

# APPENDIX B  The Delta-Sigma Toolbox

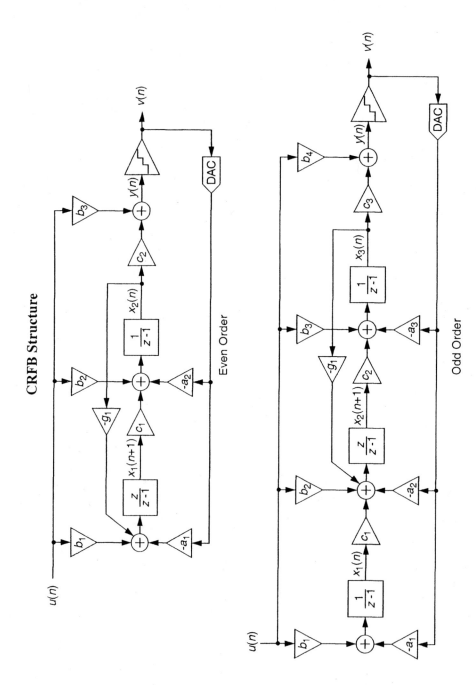

**CRFB Structure**

Even Order

Odd Order

Note that with NTFs designed using synthesizeNTF, omission of the $b_2...b_2...$ coefficients in the CRFB structure will yield a maximally-flat STF.

414

# CRFF Structure

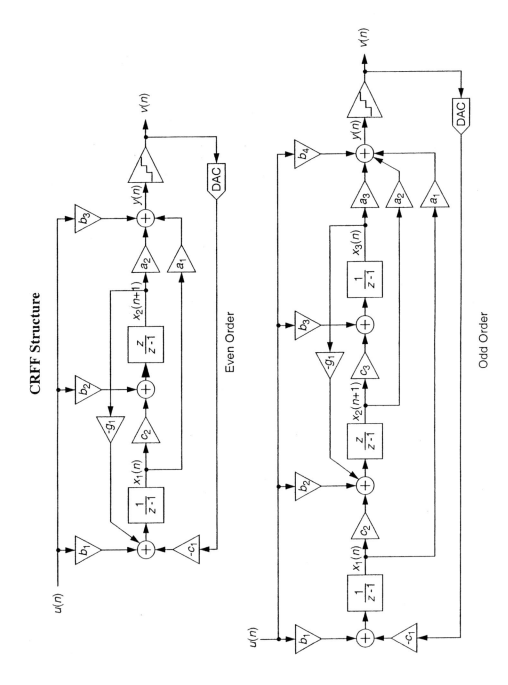

Even Order

Odd Order

# APPENDIX B   The Delta-Sigma Toolbox

# APPENDIX C
# *Noise in Switched-Capacitor Delta-Sigma Data Converters*

One of the main limitations of the performance of delta-sigma converters is noise. The sources of noise include the intrinsic noise generated in the MOS transistors, as well as the extrinsic (interference) noise originating, typically, from the on-chip digital circuitry, and coupled into the sensitive analog stages via the substrate and supply or ground lines.

In this Appendix, the analysis of the effects of intrinsic noise on delta-sigma converters using switched-capacitor (SC) circuits is discussed[†]. The reader is referred to [1] for information on the physics, modeling and reduction of extrinsic noise.

There are two important intrinsic noise effects in MOS transistors: thermal and flicker noise.[‡] Thermal noise is caused by the thermal motion of the charge carriers in the channel of the device. This causes a small amount of random fluctuation in

---

[†]. The authors are grateful for the assistance of Jesper Steensgaard and José Silva in the preparation of this Appendix. José Silva contributed Sections C.4 and C.5.

[‡]. Shot noise, which is due to the random flow of discrete charges under high electric field conditions, is significant mostly in forward-biased p-n junctions, which do not exist in common MOS devices. It may occur, however, in deep submicron devices if significant gate current flows.

## APPENDIX C  Noise in Switched-Capacitor Delta-Sigma Data Converters

*one sided*

*α = 1 for long channel*

the drain current. If the transistor operates in its triode region, as it does for a conducting switch, the noise can be represented by a voltage source in series with the device. The power spectral density (PSD) of its voltage is white; its estimated value is given by

$S_{id}(f) = 4kT\gamma g_{do}$

$S_{vg}(f) = 4kT\dfrac{\gamma}{\alpha g_m}$

$$S_{vt}(f) = 4kTR_{on} \quad (V^2/Hz). \qquad (C.1)$$

*channel resistance*   $R_{on}$   $\gamma = \dfrac{1}{2}$ linear (C.1)
$\dfrac{2}{3}$ saturation

Here, $k$ is the Boltzmann constant, $k = 1.38 \times 10^{-23}$ J/K, $T$ is the absolute temperature of the device in degrees Kelvin, and $R_{on}$ is its on-resistance in ohms. The mean value of the thermal noise is zero.

$\gamma = 2 - 3$ short channel ($V_{od} > E_{sat}$)

Note that here, and in the rest of this Appendix, all PSDs are regarded as one-sided distributions, so the noise power between $f_1$ and $f_2$ is obtained simply by integrating $S(f)$ between $f_1$ and $f_2$.

For a MOSFET operating in its active region, the thermal noise can be modeled by a current source in parallel with the channel. The PSD of the noise current is to a good approximation given by

$$S_{it}(f) = \dfrac{8}{3}kTg_m \quad (A^2/Hz). \qquad (C.2)$$

$\gamma = \dfrac{2}{3}$
$\alpha = 1$

Flicker noise or $1/f$ noise is caused by the charge carriers getting trapped and later released as they move in the channel. It is usually modeled by a series noise voltage source connected to the gate. The PSD of this voltage is approximately given by

$g_m = \mu_n C_{ox} \dfrac{W}{L}(V_{gs} - V_t)$  long channel

$$S_{vf}(f) = \dfrac{K}{WLC_{ox}f} \quad (V^2/Hz) \qquad (C.3)$$

$S_{vf, if} = \dfrac{K 2\mu_n I}{L^2 f}$

where $W$ and $L$ are the width and length of the channel, $C_{ox}$ the gate capacitance per unit area, $f$ is the frequency, and $K$ is a fabrication parameter. Note that $S_{vf}(f)$ is not white; most of its power is concentrated at low frequencies.

$S_{vf, if} = \dfrac{K v_{sat}}{L \cdot f}$  short channel ($g_m = W C_{ox} v_{sat}$)

The reader can find a more detailed discussion of intrinsic device noise, e.g., in Chapter 4 of [2].

In the following, the effects of thermal noise and $1/f$ noise on the components of switched-capacitor (SC) circuits will be discussed. Then, the noise analysis of a complete SC integrator will be performed. Finally, the noise analysis of delta-sigma converters using SC circuitry will be described.

## C.1 Noise Effects in CMOS Op Amps

CMOS op amps are sensitive to the effects of both thermal noise and $1/f$ noise. To reduce $1/f$ noise, several steps may be taken:

1. As can be seen from (C.3), a device with larger gate area $WL$ has less flicker noise. Since often the op-amp input devices dominate the noise, these should have large gate areas.
2. For some technologies, the PMOS devices have considerably smaller values for the constant $K$ than the NMOS ones. Hence, PMOS input devices are preferable for these technologies.
3. Using correlated double sampling (CDS) at the op-amp input, the input noise may be stored and then subtracted from the signal [2]. This introduces a highpass filtering of the noise, and suppresses it effectively at frequencies which are much lower than the sampling frequency $f_s$.
4. Chopper stabilization [3] can be used to modulate the $1/f$ noise out of the signal band.

These measures can usually reduce $1/f$ noise to the point where thermal noise effects dominate. Hence, the following discussions will concentrate on thermal noise.

Next, the estimation of thermal noise in op amps will be discussed. As an illustration, a differential pair (which may be the input stage of a multistage op amp) is shown in Fig. C.1, along with its thermal noise current sources. These represent the effect of the noise currents, given in Eq. (C.2). It can readily be seen that the noise current of the tail device Q5 does not contribute to the output current, since it

## APPENDIX C  Noise in Switched-Capacitor Delta-Sigma Data Converters

is present in the drain currents of both Q2 and Q4, and hence cancels in the output current under ideal matching conditions. (This is a good approximation only at frequencies where the gains and delays of the two signal paths match.†) The noise currents of the input devices (assumed to be perfectly matched) can be represented by equivalent noise voltage sources at their input terminals. From (C.2), the PSDs of these sources are given by $(8/3)kT/g_{m1}$. The noise current of Q3 can also be represented by a voltage source at the gate of Q1. The PSD of this source is $(8/3)kT g_{m3}/g_{m1}^2$. Similar considerations hold when the noise of Q4 is represented by an equivalent source at the input of Q2.

Combining these results, the total thermal noise of the stage may be represented by a single equivalent noise voltage source $v_{n,eq}$ at the gate of Q1 (or Q2). Its PSD is

$$S_{vt}(f) = \frac{16kT}{3g_{m1}}\left(1 + \frac{g_{m3}}{g_{m1}}\right) \approx \frac{16kT}{3g_{m1}} \quad (C.4)$$

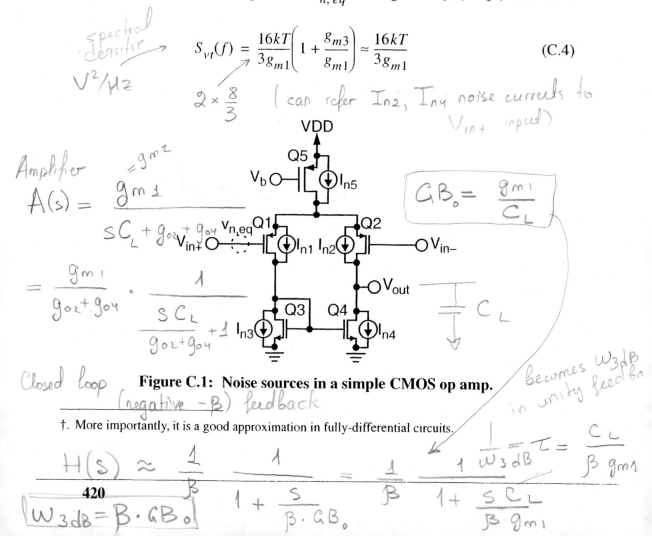

**Figure C.1:** Noise sources in a simple CMOS op amp.

† More importantly, it is a good approximation in fully-differential circuits.

420

# Noise Effects in CMOS Op Amps

The first equality of (C.4) suggests that for low noise $g_{m1} \gg g_{m3}$ should be used, which justifies the approximation made in the last part.

Consider next the op amp under negative feedback conditions (Fig. C.2). Note that the op-amp noise is represented by the equivalent source $v_{n,eq}$, and that no signal is present since only the noise amplification is analyzed. It will be realistically assumed that the op amp is properly compensated, so that in the frequency range of interest (where the gain is 1 or larger) the closed-loop transfer function can be approximated by the one-pole expression → second pole (and all other poles and zero is pushed higher than GB of amplifier)

$$H(s) = \frac{V_{out}(s)}{V_+(s)} = \frac{G_0}{1 + s\tau} \quad (C.5)$$

Here, $G_0$ is the dc gain of the stage, and $\tau$ is its settling time constant. $\tau = p_1$. $G_0$ is determined by the feedback factor $\beta = C_2/(C_1 + C_2)$ and by the dc gain $A_0$ of the op amp. Assuming $A_0 \gg 1/\beta$, $A_0 \beta \gg 1$

$$G_0 = \frac{1}{\beta + 1/A_0} \approx \frac{1}{\beta} = 1 + C_1/C_2 \quad (C.6)$$

$\omega_{3dB} = \omega_u \cdot 1/\beta$ makes it faster

Since the settling time is determined by how fast the op-amp output current can charge the load capacitances, $\tau$ will be proportional to $1/g_{m1}$. Also, since feedback reduces $\tau$, it may be written in the form $\tau = C_0/(\beta g_{m1})$, where $C_0$ depends

Amplifier (properly compensated)
$A(s) = \dfrac{A_o}{1 + \frac{s}{p_1}}$

Closed loop (negative feedback)
$H(s) = \dfrac{A_o}{1 + \frac{s}{p_1} - (-\beta)A_o}$

$= \dfrac{A_o}{1 + \beta A_o} \cdot \dfrac{1}{1 + \frac{s}{p_1(1+\beta A_o)}}$ $(\beta < 1)$

Feedback $= \dfrac{Z_1}{Z_1 + Z_2} = \dfrac{C_2}{C_1 + C_2}$ (large $C_2$ → shorts)

gain drops, bandwidth increases, $GB = \dfrac{g_{mI}}{C_c}$

$p_1 = \dfrac{1}{R_I(g_{mII} R_{II})C_c}$ For two-stage Miller compensated amp

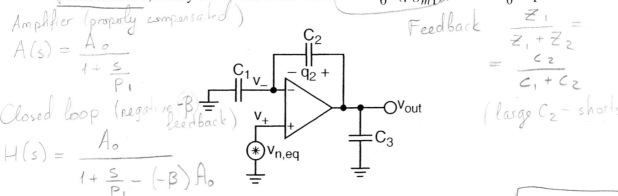

Figure C.2: An op-amp with capacitive feedback and capacitive loading.

# APPENDIX C  Noise in Switched-Capacitor Delta-Sigma Data Converters

on the structure of the op amp. For a two-stage op amp, $C_o = C_c$, where $C_c$ is the compensation capacitance connected between the stages. For a single-stage (folded- or telescopic-cascode) op amp, $C_o$ is the load capacitance at the output node of the op amp: $C_o = C_{load} = C_3 + C_1 C_2/(C_1 + C_2)$.

The PSD of the white input noise becomes shaped at the output by the first-order lowpass filter function given in (C.5). The mean-square (MS) value of the output noise may be calculated by integrating the shaped PSD from dc to infinite frequency:

$$\overline{v_{out}^2} = \int_0^\infty S_{v_i}(f)|H(j2\pi f)|^2 df = \left(\frac{16kT}{3g_{m1}}\right)\left(\frac{G_0^2}{4\tau}\right). \tag{C.7}$$

Substituting the values derived above for $G_0$ and $\tau$ gives

$$\overline{v_{out}^2} = \left(\frac{4kT}{3}\right)\left(\frac{1 + C_1/C_2}{C_0}\right) = \frac{4kT}{3\beta C_0}. \tag{C.8}$$

The MS value of the output noise voltage for unity-gain feedback can also be obtained, simply by setting $\beta = 1$ ($C_1 = 0$) in (C.8):

$$\overline{v_{out}^2} = \frac{4kT}{3C_0}. \tag{C.9}$$

Notice that even though the PSD of the op amp's input noise source is determined by $g_{m1}$, the expressions (C.8) and (C.9), which give the MS value of its output noise, do not contain $g_{m1}$. This is because the PSD is inversely proportional to $g_{m1}$, while the bandwidth of the stage is directly proportional to it. Hence, its effect cancels in the calculation of the MS output noise.

## C.2 Sampled Thermal Noise

The discussions of (C.5)-(C.7) can easily be generalized. As shown, when a thermal noise voltage $v_n$ with a white spectrum $S_v(f)$ is processed by a first-order filter with a transfer function

$$H(s) = \frac{G_0}{1 + s\tau} \quad \text{(Transfer function of feedback system)} \tag{C.10}$$

the result is a noise voltage $v_{no}$ with a lowpass spectrum. Its PSD has a 3-dB bandwidth at $f_{3dB} = 1/(2\pi\tau)$, and a MS value

$$\overline{v_{no}^2} = \frac{G_0^2 S_v}{4\tau} \tag{C.11}$$

The time constant $\tau$ determines the settling time of the system, since its transient follows the $e^{-t/\tau}$ response.

Assume next that the thermal noise source feeds a lowpass filter (active or passive) whose voltage output is sampled on a capacitor C (Fig. C.3), at a clock rate $f_s$. Let the continuous-time transfer function $H(s)$ from $v_n$ to $v_{no}$ when C is connected to $v_{no}$ be given by (C.10). If the system must have an N-bit performance, then the settling error of the LPF must be less than half an LSB. For a 2-phase system, this requires that the condition

$$e^{-\frac{1/(2f_s)}{\tau}} < 2^{-(N+1)} = \frac{1}{2}\left(\frac{1}{2^N}\right) \text{LSB} \tag{C.12}$$

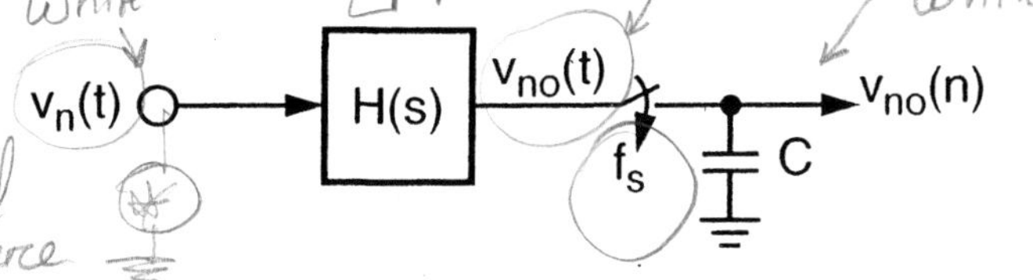

**Figure C.3:** Noise signals in a sampled system.

## APPENDIX C  Noise in Switched-Capacitor Delta-Sigma Data Converters

or equivalently

$$\frac{f_{3dB}}{f_s} > \frac{(N+1)\ln 2}{\pi} \quad (C.13)$$

be satisfied. For example, if $N = 18$, (C.13) requires that $f_{3dB}/f_s > 4.2$. Thus, the PSD of the $H(s)$ block output will be heavily aliased. Due to folding, the spectrum of the sampled noise signal $v_{nos}(n)$ will be very nearly white.[†]

The aliasing described increases the PSD of the noise significantly. The MS value of the sampled signal $v_{nos}(n)$ remains the same as that of $v_{no}(t)$, since its samples are taken from values of $v_{no}(t)$. Since $v_{nos}(n)$ is a (nearly) white noise, its PSD can be found from

$$S_{no}(f) = \frac{\overline{v_{no}^2}}{f_s/2} = \frac{G_0^2 S_v}{2\tau f_s} \quad (C.14)$$

Since the low-frequency PSD of $v_{no}(t)$ is $G_0^2 S_v$, the PSD has increased by a factor $1/(2\tau f_s)$, which by (C.12) is larger than $(N+1)\ln 2$. For $N = 18$ bits, the noise power is magnified by a factor of 13.2, or 11.2 dB. This represents an inherent disadvantage of sampled-data analog signal processors, such as switched-capacitor and switched-current filters, as compared to continuous-time ones.

Note that in oversampled SCFs followed by a unity-gain-passband digital LPF, the MS value of the oversampled noise is reduced by the OSR after digital filtering.

---

[†] Another possible explanation for this phenomenon is that under conditions (C.12)-(C.13) the memory span of the LPF is much shorter than $1/f_s$. Hence, the samples of $v_{no}(t)$ taken at times $n/f_s$, $n = 1, 2, \ldots$ are very nearly uncorrelated. This property results in a white spectrum for $v_{nos}(n)$.

## C.3 Noise Effects in an SC Integrator

Consider next the stray-insensitive SC integrator shown in Fig. C.4. As shown, the sampling instants $nT$, $(n+1)T$, ... are at the end of the $\phi_1$ clock phases. For ideal components, the input capacitor $C_1$ samples the input voltage $v_{in}$ when $\phi_1$ is high, and stores a charge $q_1(n) = C_1 v_{in}(n)$ at the end of the phase. During $\phi_2 = 1$, $C_1$ discharges $q_1$ into the virtual ground created by the op amp. This causes the charge stored in $C_2$ at the end of $\phi_2$ to become

$$q_2(n + 1/2) = q_2(n + 1) = q_2(n) + C_1 v_{in}(n) \tag{C.15}$$

Hence, the output voltage $v_{out}$ satisfies

$$v_{out}(n + 1) = v_{out}(n) + (C_1/C_2) v_{in}(n) \tag{C.16}$$

Next, we consider the noise contributed by the switches $S_1$ through $S_4$ (Fig. C.5). The only significant noise effect introduced by these switches is thermal noise.[†] For simplicity, $v_{in} = 0$ will be assumed. We shall first find the noise in $v_{C1}(n)$

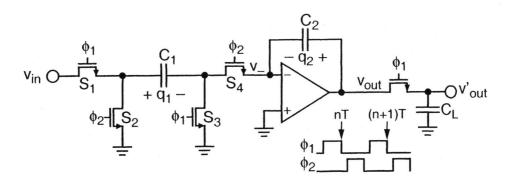

**Figure C.4:** A stray-insensitive switched-capacitor integrator.

---

† Since the current flow in these switches consists of short pulses occurring at the clock rate, the $1/f$ noise (which is caused by the trapping and release of charge carriers occurring at long intervals) has only negligible effect here.

## APPENDIX C  Noise in Switched-Capacitor Delta-Sigma Data Converters

due to the thermal noises generated in switches $S_1$ and $S_3$. While $\phi_1$ is high, the circuit containing $C_1$ can be represented by the branch shown in Fig. C.5a. Here, the conducting switches $S_1$ and $S_3$ have been replaced by their noise voltages and on-resistances. As illustrated in Fig. C.5b, the noise voltages and resistors can be pairwise combined. Assuming that all switches have the same $R_{on}$, the combined switch resistance is $2R_{on}$ and (since $v_{n1}$ and $v_{n3}$ are uncorrelated) the PSD of the associated noise voltage $v_n$ is $4kT(2R_{on}) = 8kTR_{on}$.

The PSD of the noise voltage $v_{c1}$ across $C_1$ can be expressed in terms of the PSD of $v_n$ as follows:

$$S_{c1}(f) = \frac{2S_{vt}(f)}{1+(\omega\tau_0)^2} = \frac{8kTR_{on}}{1+(\omega\tau_0)^2} \tag{C.17}$$

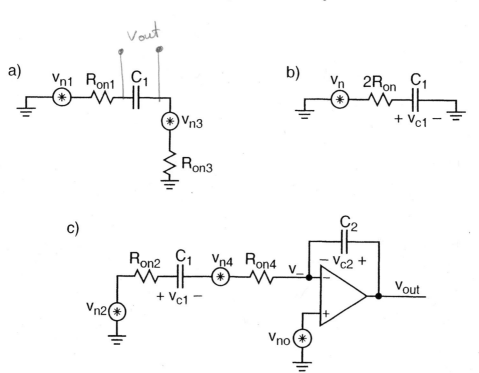

Figure C.5: Noise analysis for the integrator of Fig. C.4.

where $\tau_o = 2R_{on}C_1$ is the time constant of the $C_1$ branch during $\phi_1 = 1$. $S_{c1}(f)$ has a lowpass-filtered spectrum; it is no longer white. Its dc value is $8kTR_{on}$, while its 3-dB (half-power) frequency is $\omega_o = 1/\tau_0$.

The total power (mean-square value) of $v_{c1}(t)$ can be obtained by integrating $S_{c1}(f)$ for all frequencies from 0 to infinity. This gives

- noise bandwidth
$$\frac{\pi}{2} f_{3dB} = \frac{\pi}{2} \frac{\omega_{3dB}}{2\pi} = \frac{1}{4\tau}$$

$$\overline{v_{c1}^2} = \frac{kT}{C_1} = \frac{8kTR_{on}}{4 \cdot (2R_{on} \cdot C_1)} \tag{C.18}$$

which is independent of $R_{on}$. The explanation is that the dc value of the PSD is proportional to $R_{on}$, while its bandwidth $\omega_0$ is proportional to $1/R_{on}$.[†] The noise charge stored in $C_1$ is $q_1(t) = C_1 v_{c1}(t)$; its mean-square value is $C_1^2 kT/C_1 = kTC_1$.

At the end of the $n^{th}$ $\phi_1 = 1$ period (i.e., at $t = nT$), switches $S_1$ and $S_3$ will open, and the charge $q_1(n)$ is trapped in $C_1$. The sequence $q_1(n), n = 1, 2, ...$ has the same mean-square value $kTC_1$ as $q_1(t)$, since it is constructed from the samples of $q_1(t)$. However, due to the sampling, the PSD of $q_1(n)$ is nearly perfectly white. The reason for this is aliasing, as discussed in Section C.2.

Next, when $\phi_2$ rises, switches $S_2$ and $S_4$ close. The resulting noisy circuit is shown in Fig. C.5c. To make the analysis more specific, we shall assume a single-stage op amp represented by the model shown in Fig. C.6a, compensated by its capacitive load. The diagram of Fig. C.6b illustrates the resulting circuit during the time when $\phi_2$ is high. Note that the on-resistances and noise voltages of the two switches have again been combined.

In the analysis of the circuit, it is reasonable to assume that the loop gain of the stage satisfies the condition $\beta g_{m1} R_L \gg 1$, where $\beta = C_2/(C_1 + C_2)$; this is nec-

$\beta \cdot A_o \gg 1 \qquad \beta = \frac{Z_1}{Z_1 + Z_2}$

---

[†] An alternative derivation appeals to the physical principle known as Equipartition of Energy, which states that the average energy associated with any degree of freedom in a system at thermal equilibrium is $kT/2$. For a capacitor $C$ charged to a voltage $v$, the electrical energy is $E = Cv^2/2$ and thus $\overline{E} = kT/2$ implies $\overline{v^2} = kT/C$.

essary for suppressing nonlinear signal distortion caused by op-amp nonlinearity. Under this condition, $1/R_L = 0$ may be assumed, and the calculation simplified.

Using the Laplace transform, the noise voltage across $C_1$ may be found as

$$V_{c1}(s) = \frac{V_n(s) - V_{no}(s)}{1 + s\tau} \qquad (C.19)$$

where

$$\tau = (2R_{on} + 1/g_{m1})C_1. \qquad (C.20)$$

Utilizing the results of Section C.2, the MS value of the $C_1$ noise power caused by the switch noise $v_n$ can be calculated:

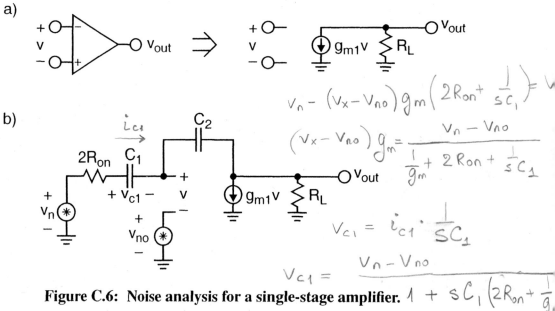

Figure C.6: Noise analysis for a single-stage amplifier.

$$\overline{v_{c1,sw}^2} = \frac{S_n}{4\tau} = \frac{8kTR_{on}}{4(2R_{on} + 1/g_{m1})C_1} = \frac{kT/C_1}{1 + 1/x}. \qquad (C.21)$$

Here, the parameter $x = 2R_{on}g_{m1}$ was introduced.

The noise power in $C_1$ due to op-amp noise can also be found:

$$\overline{v_{c1,op}^2} = \frac{S_{no}}{4\tau} = \frac{(16/3)kT/g_{m1}}{4(2R_{on} + 1/g_{m1})C_1} = \left(\frac{4}{3}\right)\frac{kT/C_1}{1 + x}. \qquad (C.22)$$

Consider next the total noise power stored in $C_1$. At the end of clock phase $\phi_1$, as illustrated in Fig. C.5b, $C_1$ had acquired a noise voltage $v_{c1,1}$ whose power was given by (C.18). During phase $\phi_2$, the noise voltage becomes $v_{c1,sw} + v_{c1,op}$, which have the noise powers given in (C.21) and (C.22). The change in $v_{c1}$ during $\phi_2$ is hence $v_{c1,sw} + v_{c1,op} - v_{c1,1}$. Since the three noise voltages are uncorrelated, their powers are added. Hence, the total noise power is

$$\overline{v_{c1}^2} = \frac{kT}{C_1}\left(1 + \frac{x}{1+x} + \frac{4/3}{1+x}\right) = \frac{kT}{C_1}\left(\frac{7/3 + 2x}{1+x}\right) = \boxed{\frac{2kT}{C_1}\left(1 + \frac{1/6}{1+x}\right)} \qquad (C.23)$$

The total noise power is minimized, and becomes $2kT/C_1$, if $x \gg 1$, i.e. if the condition $g_{m1} \gg 1/R_{on}$ holds. Under these conditions, all of the noise is contributed by the switches. The absence of noise from the op amp is due to the fact that (as (C.4) shows) the input-referred noise of the op amp is inversely proportional to $g_{m1}$.

For a given capacitor size (i.e. for a fixed $C_1$), the above condition yields the minimum achievable noise. For a circuit whose area is dominated by its capacitors, this condition corresponds to the minimum-area solution. However, since realizing a large value of $g_{m1}$ requires a large current, the most power-efficient solution is the one which minimizes $g_{m1}$ subject to constraints on the noise (C.23) and the settling time (C.20). Combining these relations leads to

## APPENDIX C   Noise in Switched-Capacitor Delta-Sigma Data Converters

$$\boxed{\overline{V_{c1}^2} = \frac{2kT}{C_1}\left(\frac{7}{6}+x\right)}, \text{ assume } x \ll 1$$

$x = 2R_{ON} \cdot g_{m1} = 0$

$R_{ON} = 0$

$\overline{V_{c1}^2} = \frac{2kT}{\tau g_{m1}}(1+g_m \overset{0}{R_{ON}}) \cdot \left(\frac{7}{6}+x\right)$

$g_{m1} = \frac{kT}{\tau v_{c1}^2}(7/3+2x), \qquad (C.24)$

$= \frac{2kT}{\tau g_{m1}}\left(\frac{7}{6}+x\right)$

which is clearly minimized for $x = 0$, i.e. for vanishingly-small switch resistance. Since, according to (C.23), the total noise power associated with this solution is $2.33 kT/C_1$, the size of $C_1$ in the minimum-power configuration is only about 17% larger than the size of $C_1$ in the minimum-area configuration.

From (C.23), the mean-square noise charge stored in $C_1$ is given by

$q = CV$

$$\overline{q_{c1}^2} = kTC_1\left(\frac{7/3+2x}{1+x}\right) = \begin{cases} 2kTC_1, & x \gg 1 \\ \frac{7}{3}kTC_1, & x \ll 1 \end{cases} \qquad (C.25)$$

Since $C_1$ and $C_2$ become series-connected as $\phi_2$ rises, $C_2$ acquires the same noise charge power as $C_1$. Hence, the added noise charge for $C_2$ is $q_{c2} = q_{c1}$, and the MS noise voltage of $C_2$ is increased by    ?

$$\overline{(\Delta v_{c2})^2} = \frac{kTC_1}{C_2^2}\left(\frac{7/3+2x}{1+x}\right) \qquad (C.26)$$

during each $\phi_2$ clock phase.

It is often useful to represent the effect of the noise source in a stage by one or more equivalent voltage sources at the input and/or output of an otherwise ideal and noiseless circuit. As discussed above, the effect of thermal noise in the transistors of the switches and the op amp in an SC integrator is an added noise charge $q_{c2}$, with an MS value given in (C.25). In a noiseless and ideal SC integrator, this charge can be delivered into $C_2$ by connecting an equivalent noise voltage source $v_{n,\,in}$ to the input of the integrator. The MS value of $v_{n,\,in}$ is the same as that of $v_{c1}$, given in (C.23); $v_{n,\,in}$ is a sampled-data signal with a white spectrum.

$$\overline{V_{c1}^2} = \overline{V_{n,in}^2} = \frac{2kT}{C_1}\left(1+\frac{1/6}{x}\right)$$

430

Frequently, there are several SC input branches in an integrator. Assuming for simplicity that $g_{m1} \gg 1/R_{on}$, the total noise charge contributed by the switched input capacitors $C_{1a}, C_{1b}, \ldots$ to $C_2$ is given by

$$\overline{q_{c2}^2} = 2kT(C_{1a} + C_{1b} + \ldots) \tag{C.27}$$

Thus, a large input capacitor contributes more noise than a small one.

In the input stage of an SC delta-sigma loop, it is possible to use either a single capacitor to enter both input and DAC signals, or to use two separate ones. As (C.27) demonstrates, the latter solution introduces twice as much thermal noise power as the first one, so the SNR is reduced by 3 dB. It may, however, reduce signal-dependent disturbances in the DAC reference voltage.

The input-referred noise source $v_{n,in}$ models the noise charge entering $C_2$ in every clock period, and hence the noise voltage $v_{c2}$. However, the output voltage is the sum of $v_{c2}$ and the voltage $v_-$ at the inverting input terminal of the op amp (Fig. C.5c). During $\phi_1 = 1$, when the output voltage is sampled, $v_- = v_{out} = v'_{out}$. Thus the circuit is in a unity-gain configuration, and the MS value of $v_-$ may be obtained from (C.9). Since $v_-$ is added only to the output voltage, it may be represented by an output-referred sampled-data voltage source $v_{n,out} = v_-$. The MS value of $v_{n,out}$ is given by (C.9); for the op-amp representation of Fig. C.6a, $C_o = C_{load}$ should be used. The overall model of thermal noise effects is shown in Fig. C.7, where the center box represents an ideal SC integrator.

As will be illustrated in the example given below, the effect of $v_{n,out}$ is usually negligible compared to that of $v_{n,in}$. Hence, the relative contributions of the

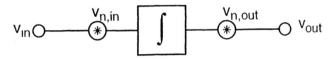

**Figure C.7: Equivalent noise sources for the integrator of Fig. C.4.**

switches and the op amp to the integrator noise can be predicted from (C.23), as functions of $x = 2R_{on}g_{m1}$. The resulting curves are shown in Fig. C.8. As the curves illustrate, the noise due to the op amp is comparable to the noise due to the switches when $x$ is small, but when $x$ is large the switch noise dominates.

The described noise calculations assume single-ended circuit configurations. They can, however, be easily adapted to fully-differential configurations by splitting all calculated capacitances in half, and assigning them to the positive and negative side of the circuits. Although this causes the thermal noise to increase by 3 dB for each side, or by 6 dB in the differential mode, this is balanced out by the fact that the signal range is also increased by 6 dB.

The total device size is nearly the same for both configurations. Differential circuits need a slightly larger area for common-mode feedback control and additional interconnections. However, the differential configuration is preferable due to its robustness to extrinsic (common-mode) noise and linearity enhancement properties (suppression of even harmonics).

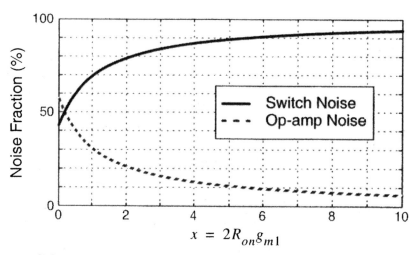

**Figure C.8:** Relative contributions from the switches and op amp to the total integrator thermal noise.

Finally, it should be noted that the noise estimation process discussed above remains a good approximation up to a frequency $\omega = \beta g_{m1}/C_o$ even if the op-amp model includes an output capacitance $C_o$ in parallel with $R_L$ (Fig. C.6).

## C.4 Integrator Noise Analysis Example

The following example illustrates how to calculate the output noise for the integrator shown in Fig. C.4, given some specific values. It is assumed that $C_1 = 1$ pF, $C_2 = 1$ pF and $C_L = 0.5$ pF. The switches have an on-resistance of 250 $\Omega$, so $2R_{on} = 500\ \Omega$. The amplifier is an OTA with transconductance $g_m = 4$ mA/V and output impedance $R_L = 250$ k$\Omega$. Hence, the dc gain is $A = 60$ dB. The clock frequency is $f_s = 100$ MHz. Note that both time constants $2R_{on}C_1$ and $C_o/(\beta g_m)$ are 0.5 ns, and thus 20 times smaller than the clock period $1/f_s = 10$ ns. This should guarantee adequate settling accuracy for the stage.

The PSD of the output referred noise is given by

$$S_{vo} = S_{vin} \cdot |H|^2 + S_{vout} \qquad (C.28)$$

In this equation, the power of the input referred noise source $v_{n,in}$ is given by (C.22), while that of the output referred noise source $v_{n,out}$ is given by (C.9). $S_{vin}$ ($S_{vout}$) is obtained dividing the power of $v_{n,in}$ ($v_{n,out}$) by $f_s/2$.

In (C.28), $H(z)$ is the integrator transfer function. It can be found using available SC analysis programs, such as SWITCAP, or analytically, as shown below. For the integrator with a finite-gain op amp, $H(z)$ is given by [5]:

$$H(z) = \frac{kz^{-1}}{1 + \mu(1 + k) - (1 + \mu)z^{-1}} \qquad (C.29)$$

The parameter $k$ is the integrator gain factor; here, $k = C_1/C_2 = 1$. Also, $\mu = 1/A = 10^{-3}$. The parameter $x$ in the input-referred noise source is $x = 2g_m R_{on} = 2$.

## APPENDIX C  Noise in Switched-Capacitor Delta-Sigma Data Converters

The output referred noise power can be calculated by integrating the output PSD in the signal band, from 0 to $f_B = f_s/(2OSR)$. The resulting equation is very complicated, and tedious to calculate manually. However, a symbolic analysis tool, such as Maple™ [6], can be used for this calculation. The integrated noise power is then found to be

$$N_{out}^2 = 2.78 \times 10^{-6} \tan^{-1}\left(2003 \tan\left(\frac{\pi}{2OSR}\right)\right) + \frac{5.52 \times 10^{-9}}{OSR}$$
$$\approx 2.78 \times 10^{-6} \tan^{-1}\left(2003 \tan\left(\frac{\pi}{2OSR}\right)\right) \quad \text{(C.30)}$$

Alternatively, the output noise power can be simulated by a dedicated CAD tool such as SpectreRF [7]. This circuit simulator has built-in analysis routines that can handle discrete-time circuits.

Fig. C.9 shows the calculated and simulated output noise powers, as functions of $f_B$. It also shows the noise contribution of $v_{n,\,out}$, which is negligible here for all values of $f_B$ shown, as predicted earlier. The good agreement between the calcu-

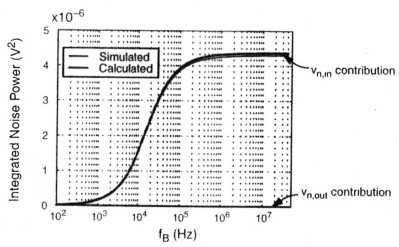

**Figure C.9:** Calculated and simulated integrated noise powers at the output of the integrator.

lated noise and the simulated one verifies the accuracy of the theory discussed earlier.

## C.5 Noise Effects in Delta-Sigma ADC Loops

Once all noise sources have been identified and their power spectral densities calculated, the next step is to find their contributions to the overall noise performance of the delta-sigma ADC. This depends on the location of the noise source in the loop (which determines how its PSD is shaped), and on the oversampling ratio (which specifies the signal bandwidth).

To determine how the PSD is shaped, the transfer function from its source to the input (determining the input-referred noise PSD) may be calculated. Alternatively, the output-referred noise transfer function can be calculated. The input-referred noise transfer function can be obtained by dividing the output-referred one by the signal transfer function (STF). Since the signal transfer function STF typically has a flat inband magnitude response, the input- and output-referred noise PSDs are usually close to each other. Often, the input-referred noise is easier to estimate [8] than the output-referred one.

The in-band noise power can then be found by integrating the shaped PSD over the desired bandwidth. Denoting the noise power of the $j^{th}$ noise source by $\overline{v_j^2}$, and assuming that all noise sources provide sampled white noise, the PSD of the source is

$$S_{vj} = \frac{\overline{v_j^2}}{f_s/2} \tag{C.31}$$

and hence the inband portion of the output noise power due to the $j^{th}$ noise source is given by

$$\overline{v_{oj}^2} = \frac{2}{f_s} \int_0^{f_s/(2OSR)} \overline{v_j^2} \cdot |NTF_j(f)|^2 df \tag{C.32}$$

where $NTF_j$ is the transfer function from the $j^{th}$ noise source to the output.

After the in-band noise power has been calculated for all noise sources, the total noise is obtained by adding all their contributions:

$$\overline{v_n^2} = \sum_{j=1}^{N} \overline{v_j^2} \qquad (C.33)$$

Finally, the maximum SNR of the system is given by:

$$SNR_{max} = \frac{v_{max}^2}{\overline{v_{nt}^2}} \qquad (C.34)$$

where $v_{max}^2$ is the maximum signal power, and $\overline{v_{nt}^2}$ is the total noise power due to all intrinsic and extrinsic sources. The total noise power is thus

$$\overline{v_{nt}^2} = \overline{v_{nq}^2} + \overline{v_{sw}^2} + \overline{v_{oa}^2} + \overline{v_{ex}^2} \qquad (C.35)$$

This expression includes the noise contributions of the quantization noise ($v_{nq}$), switch noise ($v_{sw}$), op-amp thermal noise ($v_{oa}$) and all external noise sources ($v_{ex}$). The latter term can include, for example, noise generated in the digital section of the chip.

When designing a delta-sigma ADC, it is important to find a good balance between the contributions of all noise sources. A good balance means that all noise sources are scaled in a way that makes the circuit implementation economical. For example, if the quantization noise power takes up 90% of the total noise budget, then the capacitor sizes that satisfy the remaining 10% $kT/C$ noise target could be excessively large, resulting in large chip area and power consumption. In addition, a large quantization noise power is not desirable. Its behavior is not truly random, and this may compromise the noise performance in certain applications, such as in high-fidelity audio systems. This may occur since, under some circumstances, the human ear can detect tones even with amplitudes as much as 20 dB below the total noise level. Hence, usually the thermal noise is allowed the largest share in the budget. An example of a reasonable noise budget is shown in Fig. C.10.

For most applications, $OSR \gg 1$, and hence the first integrator dominates the thermal noise contributions. The remaining noise sources have their PSDs attenuated by at least the first integrator gain, so their contributions are usually negligible. However, this is not true for very low OSR (wide bandwidth) applications, because then the gain of an integrator is low at the high end of the signal band, and so the noise is not significantly attenuated.

As an example, we shall find next the thermal noise in a delta-sigma ADC using the feedforward topology. Consider the second-order modulator loop shown in Fig. C.11. We shall calculate the capacitor sizes that satisfy a specified total noise target.

The first step is to identify the thermal noise sources in the topology. There are five switched-capacitor branches: one in the first integrator, another in the second integrator, and the remaining ones in the adder preceding the quantizer. Also, there are two op amps, one in each integrator.

An equivalent representation of the noise sources associated with these devices is shown in Fig. C.11. As the Figure shows, there are three distinct points in the

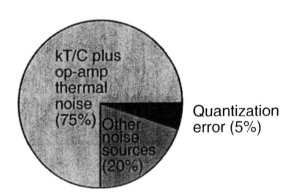

**Figure C.10: Noise budget.**

topology where noise is generated. The sources representing the output referred op-amp noise are not included since their contributions are negligible.

This leads to the second step, which is to calculate the PSDs of the noise generated at each of these points. The noise at the input of the first integrator contains a contribution from the sampling capacitor, and also from the op-amp input-referred thermal noise:

$$\overline{v_{ni1}^2} = \left(\frac{kT}{C_{s1}}\right) \cdot \frac{(7/3 + 2x)}{(1 + x)}. \tag{C.36}$$

The parameter $x$ was introduced previously as $x = 2R_{on}g_{m1}$. It will be assumed here that $x = 2$, which corresponds to equal contributions to the settling error by the switched-capacitor branches and the op amps.

The input of the second op amp contains the noise power

$$\overline{v_{ni2}^2} = \left(\frac{kT}{C_{s2}}\right) \cdot \frac{(7/3 + 2x)}{(1 + x)}. \tag{C.37}$$

The noise contributions of the three switched-capacitor branches in the adder can be combined into an equivalent noise power:

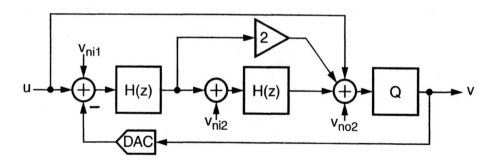

Figure C.11: Noise sources in the feedforward topology.

$$\overline{v_{no2}^2} = \frac{2kT}{C_{f1}} \cdot \left(1 + \frac{C_{f2}}{C_{f1}} + \frac{C_{f3}}{C_{f1}}\right) = \frac{2kT}{C_{f1}}(1 + 2 + 1) = \frac{8kT}{C_{f1}}. \quad (C.38)$$

Since all these sources generate sampled white noise, it follows that their PSDs are given by their MS values divided by $f_s/2$.

The third step is to calculate the gain responses from each of these sources to the output. Again, this can be done using dedicated SC analysis programs, or analytically. Here we shall do the latter. Note that the integrators blocks, denoted by $H(z)$, are delaying (forward-Euler) structures, with transfer functions:

$$H(z) = \frac{z^{-1}}{1 - z^{-1}}. \quad (C.39)$$

The noise transfer function from the input of the first integrator to the output of the modulator is

$$NTF_{i1}(z) = \frac{H^2 + 2H}{1 + 2H + H^2} = 2z^{-1} - z^{-2}. \quad (C.40)$$

The NTF from the input of the second integrator is

$$NTF_{i2}(z) = \frac{H}{1 + 2H + H^2} = z^{-1}(1 - z^{-1}). \quad (C.41)$$

Finally, the NTF from the output of the second integrator is

$$NTF_{o2}(z) = \frac{1}{1 + 2H + H^2} = (1 - z^{-1})^2. \quad (C.42)$$

The frequency responses can be found by substituting $z = e^{j2\pi fT}$. The results are illustrated in Fig. C.12, which indicates that (as expected) at low frequencies $NTF_{i1}$ dominates the other gain responses.

The fourth step is to integrate each noise PSD in the band of interest. Again, to save time, it is recommended that a symbolic analysis tool be used for these calculations. The resulting output noise powers are:

$$\overline{N_{i1}^2} = \frac{2\overline{v_{ni1}^2}}{f_s} \int_0^{f_s/(2OSR)} |NTF_{i1}|^2 df = \overline{v_{ni1}^2} \left[ \frac{5}{OSR} - \frac{4}{\pi}\sin\left(\frac{\pi}{OSR}\right) \right] \quad (C.43)$$

$$\overline{N_{i2}^2} = \overline{v_{ni2}^2} \left[ \frac{2}{OSR} - \frac{2}{\pi}\sin\left(\frac{\pi}{OSR}\right) \right] \quad (C.44)$$

$$\overline{N_{o2}^2} = \overline{v_{no3}^2} \left[ \frac{6}{OSR} + \frac{2}{\pi}\sin\left(\frac{\pi}{OSR}\right)\cos\left(\frac{\pi}{OSR}\right) - \frac{8}{\pi}\sin\left(\frac{\pi}{OSR}\right) \right] \quad (C.45)$$

Finally, the total $(kT)/C$ and op-amp thermal noise power can be found as the sum of all these contributions:

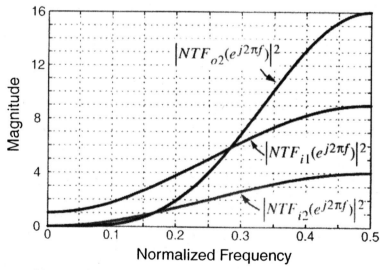

Figure C.12: **Frequency responses of the noise transfer functions.**

$$\overline{N_{total}^2} = \overline{N_{i1}^2} + \overline{N_{i2}^2} + \overline{N_{o2}^2} \tag{C.46}$$

As a numerical example, assume that this loop operates with an oversampling ratio of $OSR = 256$, its maximum input signal power is $1\ V^2$, and a 16-bit performance is required. Then the total permissible noise power is $N_{total} = 12.5\ V_{rms}$.

Replacing $OSR = 256$ in the previous equations, we get for the total thermal noise power:

$$\overline{v_{th}^2} = 8.25 \times 10^{-3} \frac{kT}{C_{s1}} + 5.88 \times 10^{-7} \frac{kT}{C_{s2}} + 1.42 \times 10^{-10} \frac{kT}{C_{f1}} \tag{C.47}$$

This expression shows that capacitors $C_{s2}$ and $C_{f1}$ have a negligible contribution to the total noise power, so they will be ignored. This makes sense, since the noise of these capacitors is shaped by first- and second-order high-pass transfer functions, and the oversampling ratio is large.[†]

By allotting to the thermal noise 75% of the total noise power ($N_{total}^2$), the value of $C_{s1}$ becomes

$$C_{s1} = 8.25 \times 10^{-3} \frac{kT}{0.75 N_{total}^2} \approx 0.29\ \text{pF} \tag{C.48}$$

As mentioned in Section C.1, it can be expected that the effects of flicker noise are much less than those due to thermal noise. If this is not the case, techniques such as correlated double sampling [3] can be used to suppress the $1/f$ noise in the first-stage output signal.

---

† . In the cases where more than one capacitor significantly contributes to the total noise, an optimization algorithm can be used to find the minimum capacitor sizes that satisfy the noise target, and result in minimum total capacitance.

# References

[1] N. K Verghese, T.J. Schmerbeck and D.J Allstot, *Simulation Techniques and Solutions for Mixed-Signal Coupling in Integrated Circuits*, Kluwer Academic Publishers, Boston, 1995.

[2] D. Johns and K.W. Martin, *Analog Integrated Circuits*, Wiley, New York, 1997.

[3] C. Enz and G.C. Temes, "Circuit techniques for reducing the effects of op-amp imperfections," *Proceedings of the IEEE*, vol. 84, no. 11, Nov. 1996, pp. 1584-1614.

[4] R. Gregorian and G.C. Temes, *Analog MOS Integrated Circuits for Signal Processing*, Wiley, New York, 1986, Sec.7.4.

[5] W. Ki and G. C. Temes, "Offset-compensated switched-capacitor integrators," *International Symposium on Circuits and Systems*, May 1990, vol. 4, pp. 2829-2832.

[6] A.Heck, *Introduction to Maple*, Springer Verlag, 2003.

[7] K. Kundert, "Simulating switched-capacitor filters with SpectreRF," `http://www.designers-guide.com/Analysis/sc-filters.pdf`.

[8] G. C. Temes and J. Silva, "Simple and efficient noise estimation algorithm," *Electronics Letters*, vol. 40, no. 11, May 27, 2004, pp. 640-641.

# Index

**A**
ABCD matrix 271, 401, 410
Accumulate-and-dump 57
Autocorrelation function 383

**B**
Bandpass ADC 141, 326
Bandpass modulator 13
    center frequency selection 145
    NTF 142, 145
    OSR definition 142
    spectrum 147
    topology 151

**C**
C model, modulator 343
Capacitive T 307
Cascade 10
Chaotic modulator 53
Chopper stabilization 419
Circuits
    clock generator 297
    comparator 297
    current-mode DAC 323
    DAC latch 323
    low-power comparator 326
    op amp 294
    two-stage amplifier 325
Clock generator 297
Comb filter 55
Comparator 297
Conditional stability 101
Continued fraction expansion 47
Continuous-time modulator 161, 205
    comparator delay in 214
    jitter 215
    STF 211
    feedback delay 318
Correlated double sampling (CDS) 419

**D**
DAC
    based on MOD1 34
    calibrated 237
    delta-sigma 16, 220
    digitally-corrected 237
    dual-truncation 230
    example 252, 253
    multi-bit 229
    segmented 234
    segmented and scrambled 235
    single-stage 223

# Index

Data-weighted averaging 189
Dead band 51
Dead zones, MOD1 51
Decimation filter 54
Decimation, MOD2 86
Delta modulation 4
Digital correction 199
Direct charge transfer stage 251
Discrete Fourier transform 365
Dither 45
Dynamic element matching 184, 404
Dynamic range scaling 266, 268, 300, 402

## E

Effective number of bits (ENOB) 3, 9, 10
Element rotation 189
Error-feedback structure 35, 81, 224
Excess loop delay 215

## F

Fast Fourier Transform, see FFT 365
Feedback delay 318
Feedforward topology 309
FFT
    averaging 377
    bins 365
    noise leakage 370
    required length 372
    scaling 373
    signal leakage 371
    smoothing 377
Figure of merit, ADC 357
Filter
    architectures 247
    Bessel 248, 249
    half-band 242
    interpolation 239
    reconstruction 243
Finite op-amp gain in MOD1 50
Flicker noise 418
Folded-cascode op amp 294
Fourier transform, definition 365
Full scale 24

## G

General modulator structure 92

## H

Halfband filter 87
    design 405

Halfband filter simulation 408
History of delta-sigma 17
Hogenauer structure 88

## I

Image gain 345
Image rejection ratio (IRR) 168
Image rejection, function of amplitude and phase error 168
Image transfer function 171
Impulse response check 277, 290, 310
Individual-level averaging 191
Infinity norm 99
Inherent anti-aliasing 206, 212, 319, 333
Inherent linearity 7
Instability detection 311
Integrator, noise 425
Integrator, SC 245, 274, 300, 316
Interpolation filter 220, 239, 344

## K

kT/C noise 280

## L

Least significant bit (LSB) 23
Lee criterion 98
Leslie-Singh structure 123
Linear interpolator 347
Loop filter
    arbitrary NTF and STF 92
    differentiating NTF 93
Lossless discrete integrator (LDI) 155

## M

MASH 10, 17, 127, 182, 226, 311
    2-stage 127
    3-stage 131
    noise leakage 132
Mismatch shaping 15, 186, 232
    second order 195, 198
    first order 189, 198
    individual-level averaging 191
    tree structure 196
    vector-based 192
Mixer 328
MOD1
    ADC 29
    DAC 34
    dc input 41
    dead zones 51

MOD1 (cont'd)
    finite op-amp gain 50
    geometric interpretation 45
    irrational input 44
    linear model 36
    NTF 37
    rational input 43
    robustness 32
    simulation 38
    SQNR 38
    SQNR vs. input amplitude 40
    stability 49
    STF 37
    tones 42, 48
MOD2 from MOD1 64
MOD2
    Boser-Wooley structure 79
    dead zones 77
    decimation 86
    distortion 71
    error-feedback structure 81
    harmonic distortion 71
    in-band noise power vs. dc input 73
    linear model 64
    NTF 64
    NTF zero optimization 84
    optimal 84
    output spectrum 68
    quantizer gain 68, 71
    SC realization 283
    Silva-Steensgaard structure 80
    SQNR 65
    SQNR vs input amplitude 70
    SQNR vs. OSR 66
    stability 74
MODn
    NTF pole optimization 111
    NTF zero optimization 107
    SQNR vs. OSR 112
    stability 97
Modulator, ADC or DAC 30
Modulator, dual-quantizer 182
Modulator, simulation 396
Multi-bit modulator 179
Multi-stage modulator, see MASH

## N
NBW, arbitrary window function 386
Noise bandwidth (NBW) 39, 376, 386
Noise budget 437

Noise leakage 132
Noise shaping 6
Noise transfer function (NTF) 8
Noise, kT/C 280, 292, 301, 427
Noise, op amp 419
N-path transformation 149
NTF zero optimization
NTF
    as a function of quantizer gain 69
    cookbook procedure 113
    function of quantizer gain 100
    pole optimization 111
    realization 266
    realizability constraint 97
    realization 400
    stability constraint 97
    synthesis 257, 393
    zero optimization 85, 107

## O
Op amp, finite gain 50, 77, 315
Op amp, folded cascode 294
OSR 8
OSR, bandpass modulator 142
OSR, quadrature modulator 144
Overload 10
Oversampling ratio 8

## P
Positively invariant set 409
Post-filter, DAC 243
Power spectral density (PSD) 7, 38, 266, 383

## Q
Quadrature ADC 144
Quadrature filter 169
Quadrature filter, mismatch 170
Quadrature mixing 166
Quadrature modulator 172
    definition of OSR 144
    NTF 172
    spectrum 174
Quadrature resonator 172
Quadrature signal 166
Quantization 21
Quantization error 23
Quantization noise 21
    FFT of 25
Quantizer gain 22
Quantizer gain, binary 29

# Index

Quantizer transfer curve (QTC) 71
Quantizer,
    binary 28
    bipolar 23
    mid-rise 22
    mid-tread 22
    overload 10
    unipolar 22

## R

Rational of best approximation 47
Realizability constraint 96
Realization 400
Receiver 140, 326
Reconstruction filter 243
Reset 98, 311
Resonator
    2-path 156
    active-RC 159
    delaying integrators 119
    Gm-C 159
    LC 160
    LDI 119, 155
    SC 275
Root locus 100

## S

Scaling 288
Sigma delta vs. delta sigma 7
Signal leakage 371
Signal to quantization noise ration (SQNR) 38
Sinc filter 55
    Hogenauer structure 88
    image attenuation 346, 350
    two-stage 58
Single-bit quantization 7
Slewing 294

SQNR limits
    1-bit modulators 112
    2-bit modulators 113
    3-bit modulators 113
SQNR simulation 299
SQNR prediction 395
Stability 409
    invariant set method 409
    Lee criterion 98
Stability, multi-bit modulator 104
Stable input range 97
STF
    continuous-time modulator 211
    example continuous-time bandpass 333
    maximally-flat, all-pole 261
Superheterodyne 326

## T

Thermal noise, MOSFET 418
Thermal noise, resistor 418
Timing 276, 286
Timing check 304
Topology
    feedback 115
    feedforward 121, 303
Two-path transformation 149

## V

Verification 289

## W

White noise approximation 8
Window
    Hann 368
    rectangular 366
Windowing 265, 366